AF594970

Complex Concentrated Alloys (CCAs)

Complex Concentrated Alloys (CCAs)—Current Understanding and Future Opportunities

Editor

Sundeep Mukherjee

MDPI • Basel • Beijing • Wuhan • Barcelona • Belgrade • Manchester • Tokyo • Cluj • Tianjin

Editor
Sundeep Mukherjee
University of North Texas
USA

Editorial Office
MDPI
St. Alban-Anlage 66
4052 Basel, Switzerland

This is a reprint of articles from the Special Issue published online in the open access journal *Metals* (ISSN 2075-4701) (available at: https://www.mdpi.com/journal/metals/special_issues/complex_concentrated_alloys).

For citation purposes, cite each article independently as indicated on the article page online and as indicated below:

LastName, A.A.; LastName, B.B.; LastName, C.C. Article Title. *Journal Name* **Year**, *Article Number*, Page Range.

ISBN 978-3-03943-474-9 (Hbk)
ISBN 978-3-03943-475-6 (PDF)

Contents

About the Editor

Sundeep Mukherjee is currently an Associate Professor at the Department of Materials Science and Engineering at University of North Texas. Prior to joining UNT, he worked as a Post-Doctoral Research Associate at Yale University (2011–2012) and Senior Engineer at Intel Corporation (2005–2011). Prof Mukherjee obtained his BS (1998) from Indian Institute of Technology, and his MS (2003) and PhD (2005) from California Institute of Technology. Prof Mukherjee has published more than 100 papers in reputed international journals and given many invited talks and keynote lectures at conferences and universities. Prof Mukherjee has organized symposiums in several international conferences and currently serves on the editorial board of three journals. His research interests include understanding the structure–property relationships in metallic glasses and concentrated alloys.

Editorial

Complex Concentrated Alloys (CCAs)—Current Understanding and Future Opportunities

Sundeep Mukherjee

Department of Materials Science and Engineering, University of North Texas, Denton, TX 76203, USA; sundeep.mukherjee@unt.edu; Tel.: +940-565-4170

Received: 11 September 2020; Accepted: 15 September 2020; Published: 17 September 2020

1. Introduction and Scope

Complex concentrated alloys with multiple principal elements represent a new paradigm in alloy design by focusing on the central region of a multi-component phase space and show a promising range of properties unachievable in conventional alloys. High configurational entropy leads to single-phase solid solutions in a certain subset of these multi-principal element alloys (MPEAs), which have been termed as high entropy alloys (HEAs). The core effects of high configurational entropy, lattice distortion, sluggish diffusion, and simultaneous activation of multiple deformation mechanisms lead to a gamut of attractive properties including high strength-ductility combination, resistance to oxidation, corrosion/wear resistance, and excellent high-temperature properties. The published papers in this special issue aim to report on the current state of research in complex concentrated alloys as well as compelling future opportunities in wide ranging applications.

2. Contributions

Eight research articles, four reviews, and one perspective have been published in this special issue of Metals. The subjects are multidisciplinary and divided into several topics including: (i) simulation and modeling for predicting structure and properties [1], (ii) unique deformation mechanisms in multi-principal alloys [2–4], (iii) microstructure and properties resulting from various processing routes [5–10], (iv) corrosion and surface degradation behavior [11,12], and (v) perspectives on ways to design mechanically and functionally advanced concentrated alloys [13]. Beyramali Kivy et al. [1] reviewed the computational tools for studying the structure and properties of multi-principal alloys and identified the advantages as well as limitations of simulations in accelerating design and development of new alloys. The unique deformation mechanisms of concentrated alloys are discussed in one review article and two research papers in this special issue [2–4]. Brechtl et al. [2] reviewed the serration behavior during deformation in HEAs and its dependence on composition, microstructure, and testing condition (i.e., strain rate and temperature). They highlighted the complex dynamic behavior of serrated flow and its occurrence due to dislocation pinning by solute atoms, nanoparticles, twinning, grain boundaries, and phase transformation during deformation. Mridha et al. [3] evaluated the activation volume and activation energy of multi-principal alloys using a first-order shear bias statistical model. Results indicated heterogeneous dislocation nucleation, higher activation volume, and higher activation energy for concentrated alloys in comparison to pure metals, which suggests complex cooperative motion of atoms for deformation in MPEAs. Sadeghilaridjani et al. [13] studied creep behavior of multi-principal alloys as a function of temperature and load. Dislocation glide was reported as the dominant mechanism for creep in MPEAs. The creep stress exponent decreased with increasing temperature due to thermally activated dislocations and increased with load due to higher density and entanglement of dislocations.

The processing-microstructure-properties relationship in complex concentrated alloys are discussed in one review article and five research papers in this special issue [5–10]. Dong et al. [5]

reviewed the structure, evolution in properties, and deformation behavior of HEAs after high-pressure treatment (HPT). A certain group of concentrated alloys showed phase transformation under high pressure while others remained stable, which indicates unique structural changes and phase transitions in complex concentrated alloys (CCAs) under high pressure. Kafexhiu et al. [6] investigated the microstructure and wear behavior of dual phase HEAs as a function of aging time. Maximum hardness and negligible wear were seen for intermediate aging time. Zhang et al. [7] evaluated the distribution in microstructure and mechanical properties of HEAs obtained after spark plasma sintering (SPS). Sha et al. [8] and Zhao et al. [9] researched the effect of chemical composition on the microstructure and properties of $Al_xCoCrFe_{2.7}MoNi$ and $Al_xCoCrFeNiTi_{0.5}$ HEAs, respectively. Increase in Al content led to an increase in hardness and wear resistance [8] and a change in erosion mode from ductile to brittle [9]. Lu et al. [10] studied the effect of SiC reinforcement on microstructure and mechanical properties of Nb solid solution. They demonstrated that flexural strength, compressive strength, and Vickers hardness of Nb matrix composites increased, while fracture toughness decreased with an increase in the fraction of SiC particles.

Surface degradation and corrosion behavior of complex concentrated alloys are discussed in one review article and one research paper in this special issue [11,12]. Ayyagari et al. [11] reviewed the corrosion, erosion, and wear behavior of complex concentrated alloys. The superior corrosion, erosion, and wear resistance of CCAs compared to conventional alloys based on an extensive literature survey was attributed to multiple elements participating in complex passive-layer formation and associated improvement in surface degradation characteristics. Ren et al. studied the corrosion behavior of additively manufactured HEA [12]. The additively manufactured HEA showed higher pitting and polarization resistance when compared to a conventional as-cast counterpart, which was attributed to the homogeneous elemental distribution and lower defect density. Lastly, the perspective article by Basu and De Hosson [13] explore the interplay of defect topologies and large composition gradients in CCAs essential for designing future alloys for potential use in diverse applications.

3. Conclusions and Outlook

A wide range of research topics have been collected in the current special issue of Metals, providing comprehensive insights into recent advances in diverse aspects of complex concentrated alloys. I would like to sincerely thank all the authors for submitting their high-quality work to this special issue and anonymous reviewers for their contributions. I would also like to give special thanks to all the staff in the Metals editorial office, especially to Ms. Sunny He, for managing and facilitating this Special Issue.

Conflicts of Interest: The author declares no conflict of interest.

References

1. Beyramali Kivy, M.; Hong, Y.; Asle Zaeem, M. A Review of Multi-Scale Computational Modeling Tools for Predicting Structures and Properties of Multi-Principal Element Alloys. *Metals* **2019**, *9*, 254. [CrossRef]
2. Brechtl, J.; Chen, S.; Lee, C.; Shi, Y.; Feng, R.; Xie, X.; Hamblin, D.; Coleman, A.M.; Straka, B.; Shortt, H.; et al. A Review of the Serrated-Flow Phenomenon and Its Role in the Deformation Behavior of High-Entropy Alloys. *Metals* **2020**, *10*, 1101. [CrossRef]
3. Mridha, S.; Sadeghilaridjani, M.; Mukherjee, S. Activation volume and energy for dislocation nucleation in multi-principal element alloys. *Metals* **2019**, *9*, 263. [CrossRef]
4. Sadeghilaridjani, M.; Mukherjee, S. High-Temperature Nano-Indentation Creep Behavior of Multi-Principal Element Alloys under Static and Dynamic Loads. *Metals* **2020**, *10*, 250. [CrossRef]
5. Dong, W.; Zhou, Z.; Zhang, M.; Ma, Y.; Yu, P.; Liaw, P.K.; Li, G. Applications of High-Pressure Technology for High-Entropy Alloys: A Review. *Metals* **2019**, *9*, 867. [CrossRef]
6. Kafexhiu, F.; Podgornik, B.; Feizpour, D. Tribological Behavior of As-Cast and Aged $AlCoCrFeNi_{2.1}$ CCA. *Metals* **2020**, *10*, 208. [CrossRef]

7. Zhang, M.; Peng, Y.; Zhang, W.; Liu, Y.; Wang, L.; Hu, S.; Hu, Y. Gradient distribution of microstructures and mechanical properties in a FeCoCrNiMo high-entropy alloy during spark plasma sintering. *Metals* **2019**, *9*, 351. [CrossRef]
8. Sha, M.; Jia, C.; Qiao, J.; Feng, W.; Ai, X.; Jing, Y.-A.; Shen, M.; Li, S. Microstructure and Properties of High-Entropy $Al_xCoCrFe_{2.7}MoNi$ Alloy Coatings Prepared by Laser Cladding. *Metals* **2019**, *9*, 1243.
9. Zhao, J.; Ma, A.; Ji, X.; Jiang, J.; Bao, Y. Slurry erosion behavior of $Al_xCoCrFeNiTi_{0.5}$ high-entropy alloy coatings fabricated by laser cladding. *Metals* **2018**, *8*, 126. [CrossRef]
10. Lu, Z.; Lan, C.; Jiang, S.; Huang, Z.; Zhang, K. Preparation and performance analysis of Nb matrix composites reinforced by reactants of Nb and SiC. *Metals* **2018**, *8*, 233. [CrossRef]
11. Ayyagari, A.; Hasannaeimi, V.; Grewal, H.S.; Arora, H.; Mukherjee, S. Corrosion, erosion and wear behavior of complex concentrated alloys: A review. *Metals* **2018**, *8*, 603. [CrossRef]
12. Ren, J.; Mahajan, C.; Liu, L.; Follette, D.; Chen, W.; Mukherjee, S. Corrosion behavior of selectively laser melted CoCrFeMnNi high entropy alloy. *Metals* **2019**, *9*, 1029. [CrossRef]
13. Basu, I.; Hosson, D. High Entropy Alloys: Ready to Set Sail? *Metals* **2020**, *10*, 194. [CrossRef]

Review

A Review of Multi-Scale Computational Modeling Tools for Predicting Structures and Properties of Multi-Principal Element Alloys

Mohsen Beyramali Kivy [1], Yu Hong [2] and Mohsen Asle Zaeem [2,*]

[1] Materials Engineering Department, California Polytechnic State University, 1 Grand Ave, San Luis Obispo, CA 93407, USA; mbeyrama@calpoly.edu

[2] Mechanical Engineering Department, 1610 Illinois St., Colorado School of Mines, Golden, CO 80401, USA; yuhong@mines.edu

* Correspondence: zaeem@mines.edu

Received: 14 January 2019; Accepted: 9 February 2019; Published: 20 February 2019

Abstract: Multi-principal element (MPE) alloys can be designed to have outstanding properties for a variety of applications. However, because of the compositional and phase complexity of these alloys, the experimental efforts in this area have often utilized trial and error tests. Consequently, computational modeling and simulations have emerged as power tools to accelerate the study and design of MPE alloys while decreasing the experimental costs. In this article, various computational modeling tools (such as density functional theory calculations and atomistic simulations) used to study the nano/microstructures and properties (such as mechanical and magnetic properties) of MPE alloys are reviewed. The advantages and limitations of these computational tools are also discussed. This study aims to assist the researchers to identify the capabilities of the state-of-the-art computational modeling and simulations for MPE alloy research.

Keywords: multi-principal element alloys; computational models; first-principles calculations; molecular dynamics; phases; properties

---

1. Introduction

Multi-principal element (MPE) alloys, are a specific class of multicomponent alloys consisted of five or more alloying elements usually with near equi-atomic compositions [1]. The development of MPE alloys can be retrospect to one and a half decades ago, opening the door to alloy design with a vast variety of compositions, microstructures and properties [2]. Since then, numerous research projects have been conducted regarding the design, development, and study of different aspects of these alloys. However, the majority of these studies have been done using experiments involving trials and errors [3]. Due to the different type of principal elements in MPE alloys, designing of these alloys to achieve desired microstructures and properties is a challenging task. In addition to the severe lattice distortions and very sluggish elemental diffusions due to different neighboring atomic sites in each lattice [4,5], cocktail effects and high order of elemental interactions make designing of these alloys difficult [1]. In traditional design of MPE alloys only substitutional elements were considered, and the latest designs of MPE alloys that consider interstitial elements or precipitants [6–10] have increased the design intricacy of these alloys. Fundamental studies on phase formations, microstructure evolutions, and structural transformations of these alloys are required.

Figure 1 shows how the number of publications on MPE alloys has dramatically increased over the years since the first publications in 2004. The data used in Figure 1 is extracted from the Scopus abstract and citation database in December 2018, using high entropy alloy and/or multi principal element alloy keywords. Based on this data, more than 2300 research articles have been published in various journals and a number of publications are growing remarkably every year.

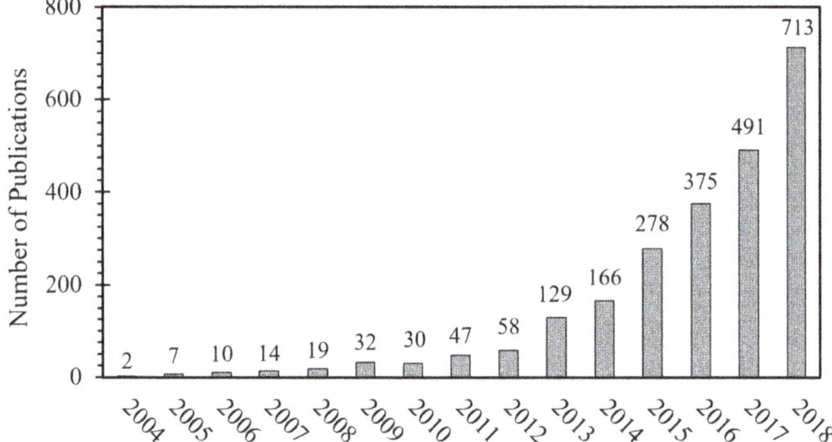

Figure 1. Number of publications on multi-principal element (MPE) alloys per year.

According to the comprehensive review article of Miracle and Senkov [1] in 2017, 37 different elements have been used so far to produce MPE alloys. These elements were 22 transition metals, six lanthanides, three metalloids, two basic metals, two alkaline earth metals, one alkali metal, and one non-metal [1]. Figure 2 shows the most common elements used in MPE alloys [1] with their value rates and price fluctuations (volatility) over 10 years period [11]. As it can be seen, most of these common elements are low-priced elements, however elements with intermediate prices (e.g., Co) or even high price (e.g., Ta) are used to produce MPE alloys as well.

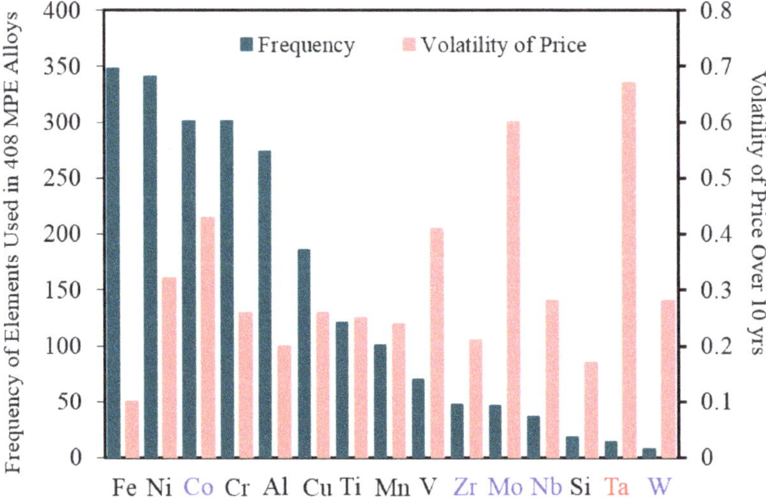

Figure 2. The most common elements used in MPE alloys [1] and their volatility of prices over a period of 10 years. Black elements have a low price (~$0.5/mole of average 50 years), blue elements have an intermediate price (~$5/mole of average 50 years), and the red element has a high price (~$100/mole of average 50 years) [11].

Traditionally, the term high entropy alloys (HEAs) have been employed broadly by researchers to describe the MPE alloys. This high entropy definition is based on the ideal configurational molar entropy, calculable by Boltzmann equation [12]. Based on this definition, metallic alloys with $\Delta S_{config}^{ideal} \geq 1.61R$ (where R is gas constant) are called HEAs. The primary problem with this description is that it denotes a single value of configurational entropy for each alloy system [1]. This assumption can be viable if the atomic positions on lattices are totally random. However, alloy configurational entropy can constantly change with temperature due to the change in atomic positions, formation of different phases, elemental segregations, and possible phase transformations [1,13]. Another approach to defining MPE alloys that have been used in the literature is compositional-based definition instead of the entropy-based concept. In this approach, these alloys are called complex concentrated alloys (CCAs). Based on this definition, alloys with non equi-molar compositions, and/or alloys with four alloying elements are also considered as CCAs [1].

Figure 3 illustrates the relationship between the properties/price ratio and the entropy of common alloys [14]. According to this data, MPE alloys (HEAs in this figure) can be expensive (medium to low cost effective), but some of them may belong to the most cost-effective zone which includes metallic glasses and super alloys, and further research is needed for their design and development [14]. It should be mentioned that the authors in Figure 3 have employed the ideal configurational entropy. Therefore, considering the complexity of configurational entropy of MPE alloys as a function of composition, atomic configuration, and temperature, the projected property-price relationship and its variation with alloy entropy can change.

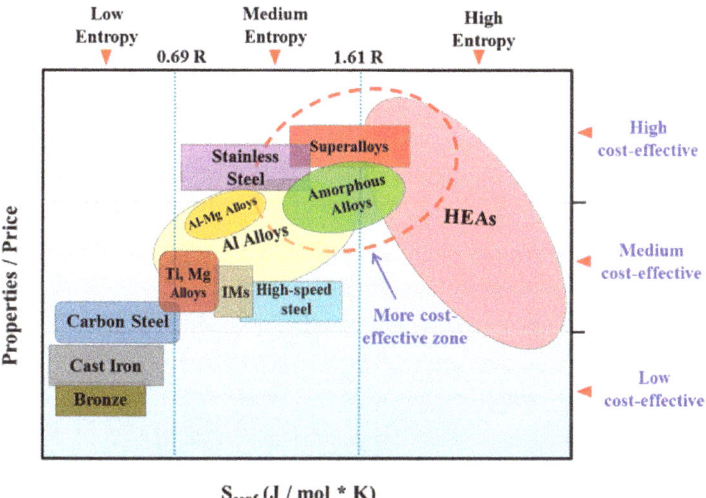

Figure 3. Relationship between property/price ratio and the configurational entropy of common alloys (this figure is from Reference [14]; IMs: Intermetallics or metallic compounds).

To overcome the complexities in experimental design of MPE alloys, reduce the inevitable expenses in empirical investigations, and decrease the number of time-consuming trials and errors, developing and utilizing computational modeling and simulations seem to be credible. Over the years various computational modeling tools have been developed or utilized in order to predict the microstructures and properties of different MPE alloys. In this study, we review the research articles published in the last decade studying different features of the MPE alloys by using computational modeling and simulations. This study is prepared in two major sections. First, the studies on phase equilibria and crystal structures of the MPE alloys are discussed, and then the research about predicting the properties of the MPE alloys are reviewed. Furthermore, in each section and shortcomings of each of the computational modeling tools are clarified. The purpose of this review article is to provide the current state of the art in computational modeling and simulations of MPE alloys, and assist the researchers to gain more knowledge about their capabilities and shortcomings in the study and design of these alloys.

2. Phase Equilibria and Crystal Structures of MPE Alloys

Similar to other alloy systems, nano- and microstructures of MPE alloys essentially affect their properties. In spite of being compositionally complex due to having multiple principal elements, the majority of MPE alloys are designed to form simple microstructures, mainly solid solutions (SS) [15]. Miracle and Senkov [1] reported that from 643 different studied MPE alloys, SS microstructures were the most common with 48% of all the cases. SS plus intermetallics (IM) microstructures are 42%, and IM microstructures are only 10%. Moreover, within the reported SS microstructures for various MPE alloys under different processing conditions (e.g., as-cast, heat treated, etc.), single-phase SS is the most common SS microstructure (25% of the cases) and double-phase SS ranks second (17%). Body-centered cubic (bcc) and face-centered cubic (fcc) phases are the most common phases in the experimentally studied MPE alloys (43% and 56%, respectively), while MPE alloys with an hcp phase are very rare (1%) [1]. Therefore, regardless of thermodynamic stabilities/instabilities of these reported microstructures discussed by E.J. Pickering and N.G. Jones [16], the strong tendency to form simple microstructures could make it easy to theoretically predict the microstructures and investigate the phase equilibria of MPE alloys. On the other hand, the presence of multiple principal elements increases the order of interactions between the elements which can result in the complexity of calculations and computational simulations.

Calculating the phase equilibria, predicting the stable phases and crystal structures, and investigating the phase formations/transformations are the initial steps in computational materials design. Over the last few years various computational techniques and theoretical approaches have been utilized to study and predict the phase equilibria and crystal structures of MPE alloys. These theoretical and computational studies provide valuable information about the various aspects of the crystal structures and microstructure evolutions of MPE alloys, however, they have their limitations and challenges.

Zhang et al. study on the relationship between SS stability, atomic size difference (δ), and enthalpy of mixing (ΔH_{mix}) was one of the earliest theoretical studies attempted to predict the crystal structures and phases of MPE alloys [17]. According to their study, SS in MPE alloys forms when $1 \leq \delta \leq 6$ satisfying Hume-Rothery rule [18], and $-15kJ.mol^{-1} \leq \Delta H_{mix} \leq 15kJ.mol^{-1}$ [17]. Figure 4 shows how SS and ordered SS microstructures in MPE alloys can be achieved when $1 \leq \delta \leq 6$ and $-15kJ.mol^{-1} \leq \Delta H_{mix} \leq 5kJ.mol^{-1}$ [17].

Later, Guo et al. [19] reported a successful study about the relationship between valance electron concentration (VEC) and crystal structures of the MPE alloys. Their study suggested that a single bcc phase forms for $VEC < 6.87$, while for $VEC \geq 8$, a single fcc phase is stabilized. Consequently at $6.87 \leq VEC < 8$, both fcc and bcc phases can co-exist in the microstructure of an MPE alloy [19]. The VEC criterion has been also applied to determine the presence of σ-phase in Cr and V containing MPE alloys [20]. According to Tsai et al. [20], σ-phase can be stabilized in Cr and V containing MPE alloys if $6.88 \leq VEC \leq 7.84$. These criteria have been applied to many different MPE alloys and the results seem to be mostly consistent with experiment [21].

Thermodynamics based approaches are some of the most popular techniques that can be applied to theoretically study the phase equilibria of materials without laborious and rather expensive experiments. Various thermodynamic methods and approaches are available to study different characteristics of alloys, including MPE alloys. For instance, Ye et al. [13] studied the trend of the elemental segregation in the number of SS MPE alloys by calculating the deviation of a designed composition from the optimum composition at different temperatures using their modified thermodynamic regular solution model [13]. In another example, Morral and Chen [22] studied the miscibility gaps and their stabilities in MPE alloys using thermodynamical criteria. Furthermore, the semi-empirical calculation of phase diagram (CALPHAD) method, which method is based on minimization of the free energy of the system, was introduced in 1970 by Kaufman and Bernstein [23,24]. CALPHAD method is known to be the most general and extensible method capable of predicting the phase equilibria of MPE alloy systems [25].

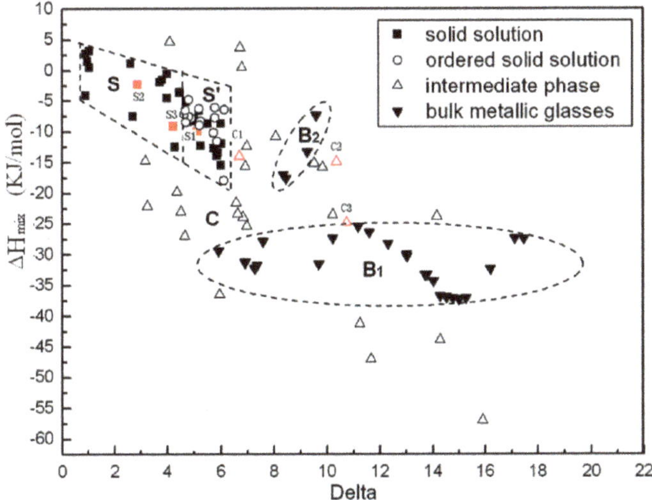

Figure 4. Microstructures of MPE alloys and Bulk Metallic Glasses as a function of enthalpy of mixing and atomic size difference (this figure is from Reference [17]).

It has been shown that in multicomponent systems, three separate Gibbs free energies can be considered to determine the total Gibbs free energy of a phase (Equation (1)) [25]:

$$G^{Phase} = G^0 + G^{ideal} + G^{excess}, \tag{1}$$

where G^0 is the reference Gibbs energy equivalent to the mechanical mixture of the constitute components of the phase, G^{ideal} is the ideal Gibbs energy corresponding to the ideal solution entropy of mixing, and G^{excess} is the excess Gibbs energy for a regular solution.

CALPHAD method is available in some thermodynamic software packages, such as Factsage [26], Thermo-Calc [27,28], and Pandat [29]. In addition to predicting structures and phases of MPE alloys, thermodynamic methods were also used to calculate the mixing enthalpy, configurational entropy, mismatch entropy, and other thermodynamic aspects of high entropy bulk metallic glasses (HE-BMGs) [30,31].

Thermo-Calc has recently become arguably the best commercial tool to study the phase equilibria of multicomponent systems, including MPE alloys. For example, Tang et al. [32] were able to determine the stabilities of different Al-transition metal binary phases in some MPE alloys by calculating their enthalpy of mixing using the Thermo-Calc SSOL5 database. Sonkusare et al. [33] calculated sudo-binary phase diagram of CoCuFeMnNi, Saal et al. [34] studied the phase fractions of CoCrFeMnAl, Yao et al. [35] computed the mixing Gibbs free energies and phase fraction of four refractory MPE alloys, Fang et al. [36] determined the phase molar fractions of selected light-weight MPE alloys, and Stepanov et al. [37] calculated the phase molar fractions of $AlCr_xNbTiV$, using TCHEA1, TCHEA2, TCNI7, TCNI8, and TTTI3 databases of Thermo-Calc, respectively. Moreover, B. Gwalani et al. [38] have comprehensively studied the effects of thermo-mechanical processes on microstructures of $Al_{0.3}CoCrFeNi$ MPE alloy and their discrepancies from thermodynamic predictions by TCHEA database of Thermo-Calc. According to this study, nanoscale ordered $L1_2$ phase was formed experimentally when the alloy was solutionized at 1150 °C before annealing. This meta-stable $L1_2$ phase was not captured in the equilibrium thermodynamic calculations due to different driving forces for nucleation, as well as different nucleation barriers in different phases [38]. According to the recent comprehensive study of Abu-Odeh et al. [39], the calculated phase equilibria of 71% of alloys (153 alloys out of 216 total studied alloys) using TCHEA1 (the database deigned for MPE alloys) matched the experiments. In another study, Tancret et al. [40] have reported the predictive capability of TCNI7, TTNI8, TCFE8, and SSOL5 databases of Thermo-Calc. According to this study, TCFE8 was the most successful database for MPE alloys, while SSOL5 was the most accurate for non-MPE alloys [40]. Figure 5 shows the accuracy of different CALPHAD databases in predicting the phase formations in MPE and non-MPE alloys [41]. It should be noted that current databases of Thermo-Calc are not able to capture the phase formations in most of the experimentally studied MPE alloys.

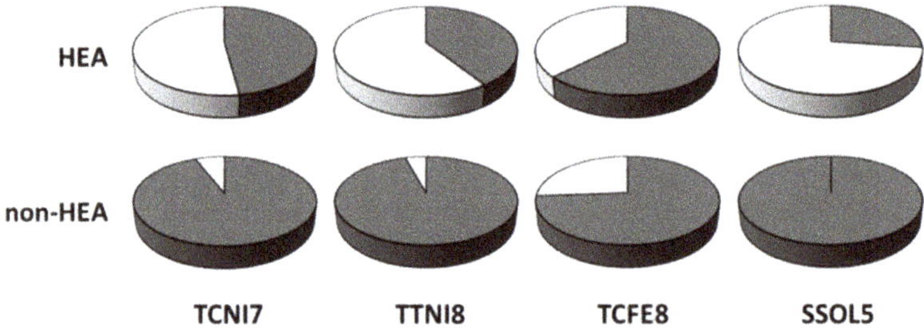

Figure 5. Capability of four different CALPHAD databases in predicting the formation of MPE alloys (HEA) and others (non-HEA). Dark sectors show success and light sectors represent the failure of the data bases (this figure is from Reference [41]).

The Pandat thermodynamic software package has also been utilized to study the phase equilibria of a few MPE alloys. For instance, Liu et al. [42] calculated the phase molar fractions and a phase diagram of $AlCrCuFeNi_2$ alloys using the PanHEA database of Pandat. Nevertheless, some researchers prefer to develop their own thermodynamic databases [43,44] or use the available CALPHAD-based databases (e.g., SGTE) directly without using commercial software [45,46]. This will allow these frameworks to study MPE alloys for which there are not enough experimental data available in the databases of commercial software. A statistical study was performed by Miracle and Senkov [1] illustrating the predictive capability of the CALPHAD calculations with respect to the experimental results. They compared two CALPHAD datasets (f_{AB} = 1 and f_{AB} = All) at two different temperatures (600 °C and melting temperature (T_m)). According to their findings, CALPHAD predicts more phases in the microstructures of the MPE alloys than the experimental observations. Also, some phases are over predicted while some phases are neglected. In addition to predicting structures and phases of MPE alloys, thermodynamic methods and approaches were also used to calculate the mixing enthalpy, configurational entropy, mismatch entropy, and other thermodynamic aspects of high entropy bulk metallic glasses (HE-BMGs) [30,31].

First-principles approaches (electronic scale simulations), including density functional theory (DFT) calculations and ab-initio molecular dynamics (AIMD) simulations, are other strong theoretical tools which can provide the thermodynamic information and phase stabilities of MPE alloys with good precisions. These approaches are the smallest scale simulations in the multi-scale computational materials framework and are known to be the most successful methods to study the electronic structures of materials [47]. Different commercial and open-access ab-initio based software packages, such as VASP [48], Quantum Espresso [49], ABINIT [50], and CASTEP [51], are available to study different aspects of the MPE alloys. First-principles approaches have been used to study the phase equilibria and crystal structures of different MPE alloys. For instance, Ma et al. [52] studied the thermodynamic properties of the CoCrFeMnNi using DFT and AIMD. They found that the ferromagnetic hcp structure is stable at ground state with respect to other structures (bcc or fcc) and other magnetic states (non-magnetic or disordered local moments). They also calculated the temperature dependent free energy values of different crystal structures and magnetic states, and temperature dependent vibrational entropy, magnetic entropy, and electronic entropy [52]. In other examples, Niu et al. [53] calculated the mixing enthalpy and lattice constant of NiFeCrCo, Yalamanchili et al. [54] studied the configurational entropy of (AlTiVNbCr)N, Jiang and Uberuaga [55] developed an ab-initio based small set of ordered structures (SSOS) and calculated the instability energies as energy difference between fcc and bcc structures for different bcc MPE alloys, Tian et al. [56] studied the similar energy difference for number of CoMoW-based MPE alloys, Li et al. [57] calculated the hcp-fcc energy difference, lattice vibration, and magnetic entropy of $Co_{20}Cr_{20}Fe_{40\text{-}x}Mn_{20}Ni_x$ MPE alloys, Heidelmann et al. [58] determined the stable structures for matrix and inter-grain phases of ZrNbTiTaHf, and Mu et al. [59] computed the lattice constants of five different refractory MPE alloys, all using first-principles calculations.

Moreover, first-principles approaches can provide information about phase evolutions and structural properties of the MPE alloys. For example, Zhang et al. [60] determined an unstable sluggish pressure-induced phase transformation from fcc to hcp in NiCoCrFe alloy by calculating the Gibbs free energies and lattice constants of fcc and hcp phases at different pressures and lattice volumes. In another study, Tian et al. [61] investigated the phase stabilities of paramagnetic $(NiCoFeCr)_{1-y}Al_y$ MPE alloys by calculating Gibbs free energies and structural energies of different phases as a function of Al content, Middleburgh et al. [62] studied the segregation and migration of species in CrCoFeNi alloys by calculating the vacancy energy and defect energy versus lattice binding energies, Yu et al. [63] investigated the nano-scale phase separation in some fcc MPE alloys using ground state formation energies, and Leong et al. [64] applied the rigid band approximation (RBA), a simplification of density functional theory approach, to investigate the phase formation behaviors in MPE alloys and several phases were successfully predicted.

In addition to crystal structures and phase formations, other nano/microstructure and phase equilibria related properties can be calculated using first-principles approaches. For example, Gutierrez et al. [65] calculated the melting temperature, chemical energy, and specific heat of CoCrFeNiMn MPE alloys. They also calculated the temperature dependent thermal expansion and lattice constants. Due to the complexity of the MPE alloys with high chemical disorder and composition fluctuation, energy dissipation and defect evolution have also been important topics to study, especially for designing radiation resistant materials used in extreme radiation environments. Based on first-principles approaches, Y. Zhang et al. [66] applied ab initio Korringa–Kohn–Rostoker coherent-potential-approximation (KKR-CPA) method for electron structure calculations of NiCoFeCr to study the controlling factors in radiation resistance mechanisms. Based on their first-principles and experiments, they suggested that exploiting single-phase concentrated solid solutions of MPEs could be an effective approach to modifying intrinsic material transport properties, which can affect equilibrium and non-equilibrium defect dynamics [66].

First-principles calculations can also be integrated with thermodynamic CALPHAD models to study the phase stabilities and phase diagrams of MPE alloys. For instance, enthalpy of mixing for different structures and compositions can be calculated using DFT methods and the results can be considered in the CALPHAD formulations [67,68]. The advantage of incorporating first principles with CALPHAD approaches is to improve the predictive capabilities of CALPHAD by adding more physical factors into the calculations [69].

Despite the capabilities of the first-principles approaches in studying the structures and phase equilibria of the MPE alloys [70], small scale (pm to nm), small number of atoms (a few hundred atoms), and relatively high computational costs are the main limitations of these methods. Therefore, using combinations of thermodynamics and first-principles approaches have shown better promising results in studying the MPE alloys [71–73].

In addition to the first-principles and thermodynamics approaches as the most popular methods to study the crystal structures and phase equilibria of MPE alloys, other multi-scale computational methods, such as molecular dynamics (MD), Monte-Carlo, and some analytical-numerical methods have also been applied to study different features of the microstructures of some MPE alloys. For instance, Choi et al. [74] studied the migration vacancy energies of the species in CoCrFeMnNi alloy, and Sharma et al. [75] calculated the crystallization temperatures and stable phases in Al_XCrCoFeNi alloy, using MD simulations with considering second nearest-neighbor modified embedded atom method (2NN-MEAM) and Lennard-Jones (LJ) potentials, respectively. The main problem in using the MD simulations is generally the unavailability of the interatomic potentials. If the interatomic potentials are not specifically developed for the target materials or the temperature of interest, the results may not be useful or accurate. Developing interatomic potentials can be a time-consuming and computationally expensive process.

Some researchers have utilized Monte Carlo simulations which are a broad range of computerized mathematical algorithms to study the probability of different objectives [76]. For example, Anzorena et al. [77] investigated the evolution probabilities with temperature for different coordination matrices in MoTaVWZr, and Feng et al. [78] calculated the pair correlation functions in AlCoCrFeNi; both work employed Monte Carlo simulations. In another work, hybrid Monte Carlo and MD (hybrid MC/MD) approach was used to simulate the structure and calculate the partial radial distribution functions in $Al_{1.33}$CoCrFeNi MPE alloy (Figure 6) [79]. The large amounts of the required data in Monte Carlo simulations usually increase the complexity of such simulations.

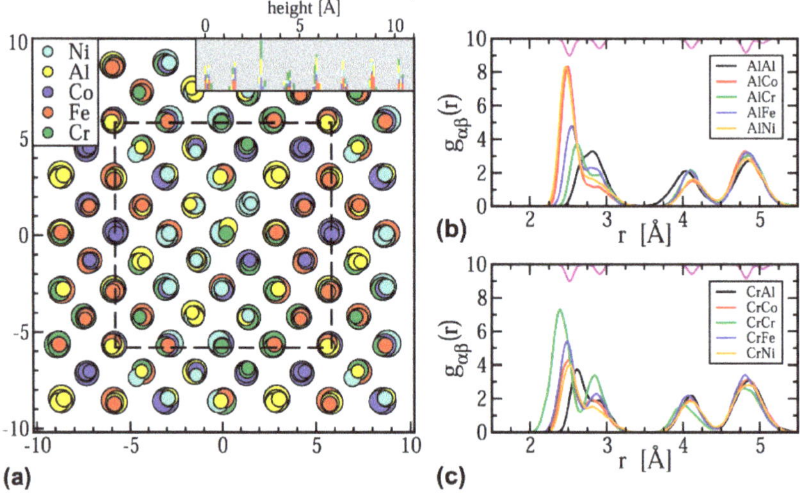

Figure 6. (**a**) Simulated structure of $Al_{1.33}$CoCrFeNi at T = 300 K viewed along the [001] direction. (**b**) and (**c**) Partial radial distribution functions $g_{\alpha\beta}(r)$ showing the atomic distributions around Al and Cr atoms, respectively (this figure is from Reference [79]).

It should be noted that some characteristics and properties of the MPE alloys have been studied by other numerical, analytical, and statistical analyses, such as lattice distortion by an analytical calculation method [80], lattice constants and short range order matrices by a theoretical atomistic model [81], densities and melting temperatures by a combination of physical and statistical models [82], and elemental diffusivities by combinations of different theoretical methods, including Darken and Miedema's scheme [83].

Although the reviewed multi-scale computational tools improve the research capabilities in study and design MPE alloys, due to the configurational disorders in MPE alloys, these approaches still have uncertainties and errors, either in modeling or calculations. Various models have been utilized in the literature to study the structures and properties of different systems. For instance, coherent potential approximation (CPA) to study the chemical and magnetic disorders, special quasi-random structure (SQS) as one of the most successful models known for binary and ternary alloys, and coarse-grain cluster expansion (CE) to investigate short-range order effects in Monte-Carlo simulations [55]. In two recent studies, Fernandez-Caballero et al. [84,85] investigated the configurational entropy as a functional temperature, and calculated the multi-body ordering probabilities and short range ordering (SRO) for Cr-Fe-Mn-Ni and MNbTaVW MPE systems, by developing a hybrid combinations of effective cluster interactions (ECIs) derived from DFT and Semi-Canonical Monte Carlo Simulations.

Due to the unique characteristics of MPE alloys, such as a large number of alloying elements, high interaction order, and large local atomic displacements (lattice distortions), there are some noticeable discrepancies between computational data and experiments [55]. The uncertainty of computational data must be reported, which is often ignored, and more efforts need to be dedicated to developing more accurate models.

3. Properties of MPE Alloys

3.1. Mechanical Properties

MPE alloys can be designed to have exceptional mechanical properties. As it can be seen in Figure 7 [86], MPE alloys can show excellent damage tolerance (strength combined with toughness) when compared to conventional alloys. Therefore, a more comprehensive understanding of the mechanical behaviors of MPE alloys can result in optimization the design of these alloys.

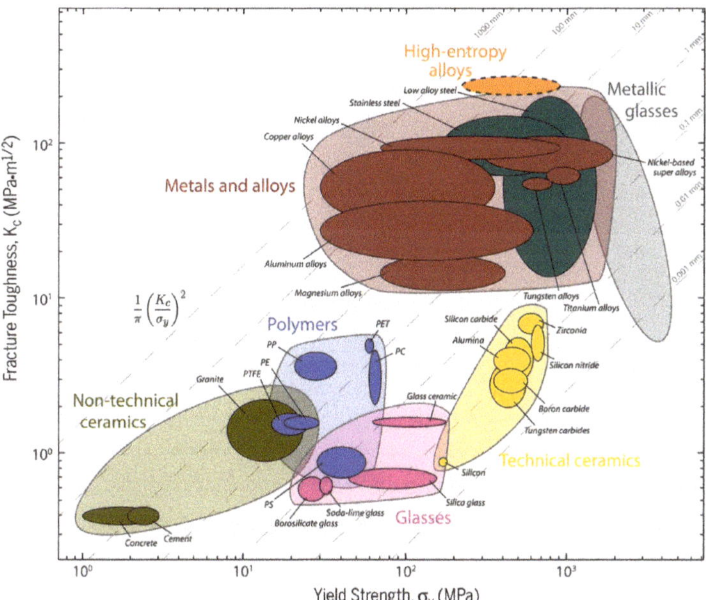

Figure 7. Fracture toughness as a function of yield strength for a wide variety of materials, including MPE alloys (high entropy alloys) (this figure is from Reference [86]).

Generally, elastic constants (C), bulk modulus (B), Young's modulus (E), shear modulus (G), and the Poisson ratio (υ) are some of the most important mechanical properties of alloys. To measure the elastic constants through experimental technique, large and homogeneous single crystal samples without any defects are required. As a result, limited experimental data on the elastic constants of MPE alloys have been reported in the literature. Due to the great challenges of experimental studies in this area, computational modeling tools have become essential alternatives to study the properties of MPE alloys in a time-efficient and low-cost manner. In the following subsections, different mechanical properties of MPE alloys predicted by computational approaches are reviewed.

3.1.1. Elastic Properties

Elastic properties are determined by applying external forces bellow the yield limit on computer models.

Starting with first-principles calculations at the electronic scale, exact muffin-tin orbital (EMTO) method [87] is an efficient approach to solve Kohn–Sham equations, while coherent potential approximation (CPA) [88] is a technique which can powerfully treat the substitutional disorder for degrees of freedom of chemical and magnetic. It was demonstrated that combined EMTO and CPA (EMTO-CPA) can successfully predict the elastic properties of materials. For example, Tian et al. [89] employed the EMTO-CPA method to determine the elastic properties and equilibrium volume of CoCrFeMnNi based MPE alloys. They also studied the effects of alloying elements on electronic structures and elastic properties of $TiZrNbMoV_x$ MPE alloys [90]. Moreover, they indicated that the VEC value of about 4.72 is critical for the elastic isotropy in these refractory MPE alloys. In another study, Li et al. [91] utilized first-principles calculations to determine the effects of crystallographic directions on the Young's modulus of some fcc and hcp MPE alloys (see Figure 8).

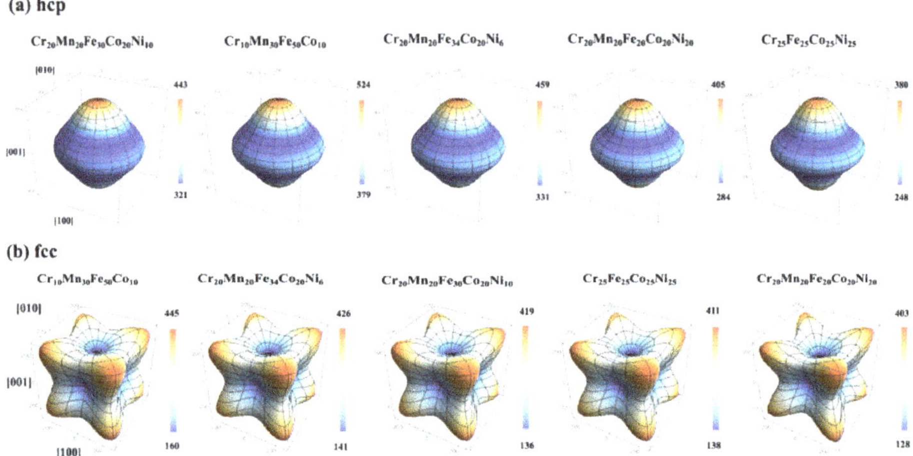

Figure 8. Directional dependence of Young's modulus E (in GPa) of some MPE alloys in hcp and fcc structures (this figure is from Reference [91]).

In another first-principles study, Cao et al. [92] investigated the equilibrium bulk properties of $Al_xMoNbTiV$ alloys by employing EMTO-CPA, in which the fraction of Al was controlled between x = 0 and x = 1.5. They have shown that the υ, B/G ratio and ab-initio Cauchy pressure of these alloys decreased with the increase of Al content. Moreover, these alloys were predicted to become isotropic when VEC $\approx$ 4.82 or x $\approx$ 0.4. Employing the EMTO-CPA method, Li et al. [93] studied the equilibrium volume, the ideal tensile strength, and elastic properties of some bcc-phase MPE alloys, ZrVTiNb, ZrNbHf, ZrVTiNbHf and ZrTiNbHf. The obtained results were expected to provide a guideline to design refractory MPE alloys with controlled strength level [93]. Furthermore, Li et al. [94] investigated the variation behaviors of the ideal tensile strength (ITS) and the ideal shear strength (ISS) in terms of the composition of elements. The ab-initio EMTO-CPA calculations were combined with the quasi-harmonic Debye-Grüneisen model by Ge et al. [95] to investigate the equilibrium bulk properties, thermo-elastic properties, and the Curie temperature of ferromagnetic and paramagnetic CoCrFeMnNi alloys. In their study, the elastic moduli were found to linearly decrease and the ductility increase with the temperature increase [95]. All of these studies demonstrated that integrated the EMTO-CPA method is capable of determining elastic properties of MPE alloys with five or fewer elements.

Virtual lattice approximation (VCA) is a method that allows the study of a crystal structure that its primitive unit cell can be periodically repeated, but it contains fictitious "virtual" atoms that interrupt between the atoms behaviors in the parent material [96]. By combining DFT and VCA methods, Tian et al. [97] calculated the elastic constants, ITS and ISS of single-phase TiVNbMo alloys. Their simulation results implied that the single-phase bcc TiVNbMo exhibits good ductility, but low tensile strength and shear strength [97]. Moreover, Mu et al. [59] used ab-initio EMTO and VCA methods to study the properties of refractory MPE alloys. They showed that the refractory MPE alloys consisting of Cr, Ti, Ta, Zr, V, Mo, Nb and W have stable single-phase bcc instead of fcc and hcp phases [59]. Also their studied alloys showed superior ductile and isotropic properties [59].

Table 1 summarizes the available Young's moduli of various MPE alloys in the literature. Despite the great progress that has been achieved by the aforementioned computational approaches, great amounts of experimental studies are still needed to validate the obtained computational results.

Table 1. Available Young's moduli (GPa) of MPE alloys in literature.

Alloys	E_{DFT}	E_{MD}	$E_{exp.}$
$Al_{0.1}CoCrFeNi$	-	199 [99]	203 [99]
CoCrFeNi	275.7 [94], 274.1 [89], 225 [100], 196 [100]	-	226 [101]
CoCrMnNi	265.6 [89]	-	171 [100]
CoFeMnNi	267.2 [95]	-	-
CoCrFeMnNi	262.4 [94], 279.7 [95], 207 [100]	-	215 [102], 137 [100]
$Cr_{10}Mn_{40}Fe_{40}Co_{10}$	328.1 [94]	-	-
TiZrVMoTaNb	71.9 [59]	-	-
TiZrVMoTaNbCr	130.9 [59]	-	-
TiZrVMoTaNbCrW	166.7 [59]	-	-
NbVTiZrAl	118.0 [98]	-	-
ZrNbHf	95.4 [93]	-	-
ZrVTiNb	95.1 [98], 117.5 [93]	-	80 [103], 101 [104]
ZrTiNbHf	88.9 [93]	-	-
ZrVTiNbHf	97.1 [93]	-	-
$TiZrNbMoV_x$	141.1 (x = 1) [90], 127.8 [96]	-	-
$Al_xMoNbTiV$	174.4 (x = 1) [92], 185.4 [96]	-	-
TaNbHfZrTi	185.4 [96]	-	78.5 [103], 87 [104]
NbTaTiWV	257.3 [96]	-	-
WNbMoTaV	218.0 [96]	-	-
MoNbTaTiV	130.5 [96]	-	-
MoTiZrNbHfTa	136.6 [96]	-	-

In addition to EMTO-CPA and VCA methods, several other theories and methods were developed to predict elastic properties of MPE alloys. The maximum entropy (MaxEnt) method [96], which is based on the first-principles, is used to study the effect of lattice distortions on the elasticity of several single-phase MPE alloys. Compared to EMTO-CPA, it has been shown that MaxEnt is a more reliable method in studying the elastic properties of MPE alloys, as well as studying the local atomic environment [96]. Moreover, the MaxEnt method can be potentially used in the MPE alloy systems with 4 to 10 elements, with different concentrations [96]. In addition to the discussed methods, first-principles approaches can be directly applied to study the mechanical properties of MPE alloys. For instance, Qiu et al. [98] examined the effect of Al on structure stability, electronic structure, chemical bonding and mechanical properties, such as strengthening mechanisms of bcc NbVTiZrAl alloys by directly using first-principles calculations.

Although, the majority of the studies on elastic properties of MPE alloys have used first-principles based approaches, some meso-scale computational techniques also have sparsely employed. For example, a finite element analysis (FEA) model was developed by Štamborsk et al. [105] to study the effect of anisotropic as-cast microstructure on high-temperature compression deformation of multiphase $Co_{24}Cr_{19}Fe_{24}Ni_{19}Al_8(Ti,Si,C)_6$ Compositionally Complex Alloys (CCA), which is derived from single phase CoCrFeNi-based MPE alloys. They qualitatively studied the local microstructure evolutions, such as the fragmentation of brittle phase and plastic deformation of the ductile phases, with respect to 3D numerical modelling of local strains and stresses [105]. Their approach was proposed to be helpful in the initial screening of the composition of alloy during the design of new MPE alloys [105].

3.1.2. Plastic Deformation

Similar to the other types of structural material, it is essential to study the plastic deformation behaviors and the underlying deformation mechanisms of MPE alloys [106].

In a study on equimolar MPE alloys, Ye et al. [107] identified a non-symmetric residual strain field using first-principles calculations with atomic scale fluctuations, which provided a further understanding of the plasticity enhancement in MPE alloys, such as dislocation strengthening. Besides, to calculate the essential residual strain in MPE alloys and other types of alloys, Ye et al. [108] developed a self-sufficient geometric model. This approach is expected to be useful in compositional selection for designing of new MPE alloys. But the chemistry effects were ignored in their calculations [108].

The stacking fault energy (SFE) of materials is one of the most important factors in determining the dominant plastic deformation mechanisms. By using a first-principles approach, Kivy and Asle Zaeem [109] recently calculated the generalized stacking fault energy (GSFE) of CoCrFeNi-based single phase MPE alloys to investigate the effect of Cu, Ti, Mo, Mn, Al elements on their plastic deformation mechanisms. With randomly distributed alloy systems, possible variations in formation energies were obtained. Uncertainties were calculated by first running five to nine simulations for each composition to determine the atomic positions within the DFT supercell that results in the most stable (lowest energy) configuration, then, the stacking fault plane was placed at different locations to take into account the composition effects at the stacking fault plane and its neighboring planes on GSFE calculations. Examples of calculating GSFE curves are shown in Figure 9. This study demonstrated that relatively high amounts of Al, or the presence of Cu and Mn, endorse martensitic transformation and dislocation mediated slip as plastic deformation mechanisms. Alternatively, mechanical twining and dislocation glide in alloys containing Mo or Ti, and dislocation gliding for alloys with a low amount of Al would be the plastic deformation mechanisms [109].

In another study, SFEs for a series of solid solution alloys (SSAs) were studied by first-principles calculations [110]. Their obtained results indicated that these SSAs exhibit low (even negative) SFEs depending on the alloying elements [110]. To determine the temperature dependence of the SFE in Fe-Cr-Co-Ni-Mn alloy, Huang et al. [111] analyzed the chemical, strain and magnetic effects using the first-principles method. They predicted very low SFE values with a large positive temperature factor at cryogenic conditions, which can be used to explain the observed twinning induced plasticity (TWIP) effect at temperatures lower than zero and transformation induced plasticity (TRIP) effect. These effects may also explain the observed combinations of superior ductility and strength of Fe-Cr-Co-Ni-Mn MPE alloys [111].

With a different approach, Pei et al. [112] used the algorithm of particle swarm optimization within Peierls–Nabarro model to obtain the dislocation structures and provided a pathway to efficiently determine their mobility and geometries.

In a recent study, Choudhuri et al. [113] investigated the effects of σ and B2 intermetallic phases on deformation twinning of $Al_{0.3}$CoCrFeN MPE alloy, using a combination of experiments and MD simulations. According to this study, σ and B2 intermetallic compounds endorsed deformation twining and strain hardening of the fcc matrix. They used ~900 K atoms with an EAM interatomic potential in their simulations [113]. In a three dimensional MD simulation study, Wang et al. [114] investigated the strengthening mechanism and plastic deformation behaviors of AlCrCuFeNi MPE alloys. They used 1.5 M atom simulations utilizing a combination of EAM and Morse potential, and determined the surface topography, friction coefficient, dislocation density and subsurface damaged structure during the process of nanoscale scratching, and the dynamic evolution of scratching forces, and compared them with those in pure metals [114]. In another study, the strain-induced transformation plasticity of single-crystal and nanocrystalline $Co_{25}Ni_{25}Fe_{25}Al_{7.5}Cu_{17.5}$ MPE alloy during fcc to bcc phase transition was investigated using MD simulations by Li et al. [115]. With 2.3M atom simulations using a normalized EAM potential, they concluded that the fcc to bcc phase transition provides an alternative approach in designing novel MPE alloys with improved strength and ductility [115]. Sharma and Balasubramanian [99] have investigated the deformation mechanisms of a single phase $Al_{0.1}$CoCrFeNi, under tension by employing MD simulations. They considered an EAM-LJ hybrid potential with 62,500 atoms. Their simulation results attempted to offer insights on the nucleation and dynamic evolution of defects, which cannot be achieved by experiments and not unfeasible by first-principles calculations [99]. Finally, by employing MD simulations focusing on dislocation motion behavior, and utilizing an EAM potential in 1.4M atom simulations, Smith et al. [116] demonstrated that the SFE in equiatomic CrMnFeCoNi MPE alloys should be considered as a spatially local property instead of global variable. In Table 2 the calculated SFE values of several MPE alloys by DFT, MD, and experiment are compared, demonstrating that calculations of SFE is mostly ignored by the most of previous MD studies, therefore one can't confidently utilize such MD simulations to study plastic deformation in MPE alloys.

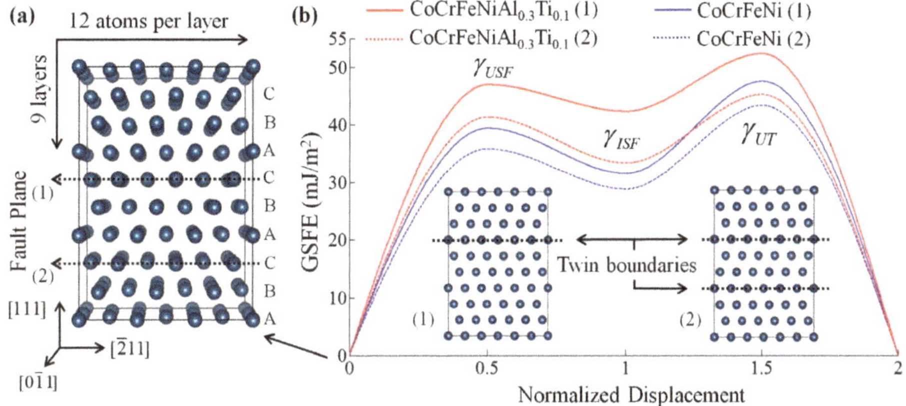

Figure 9. (**a**) fcc supercell structure used for calculating generalized stacking fault energy curves and surface energies of CoCrFeNi-based single phase MPE alloys. (**b**) Calculated GSFE curves for CoCrFeNi and CoCrFeNiAl0.3Ti0.1 by considering two different fault planes shown in (**a**); subset pictures show twin boundary formation for these two cases (this figure is from Reference [109]).

As discussed before, the main issue in utilizing MD simulations to study the MPE alloys, or multicomponent alloys in general, is unavailability of interatomic potentials for these alloys. The accuracy of the results in MD simulations is directly controlled by the interatomic potentials, therefore developing and benchmarking of interatomic potentials for MPE alloys are urgently desired.

As can be seen in Tables 1 and 2, the calculated results for Young's moduli and SFEs show considerable differences with respect to the experimental data. These discrepancies are the results of the configurational disorders of MPE alloys, discussed previously in Section 2 of the current review artilce. SFE results presented in Figure 9 shows how the distribution of atoms in fcc supercells and their relative positions to the stacking fault plane affect the GSFE curves. Although specific modeling methods, such as SQS have shown good results for some MPE systems, these methods still show large uncertainties for some other MEP alloys, and in general they can be computationally expensive.

Table 2. A summary of available SFE (mJ/m^2) of MPE alloys in literature.

Alloys	$\gamma_{SFE\text{-}DFT}$	$\gamma_{SFE\text{-}MD}$	$\gamma_{SFE\text{-}exp}$
FeCrCoNiMn	21 [111], 27.3 [100], 29.7 [109]	-	25 [117], 26.5 [118], 19 [100]
CoCrFeNi	31.6 [109], 31.7 [100]	-	27 [118]
$CoCrFeNiCu_{0.5}$	29.0 [109]	-	-
CoCrFeNiCu	27.5 [109]	-	-
$CoCrFeNiCuAl_{0.5}$	32.0 [109]	-	-
$CoCrFeNiCuTi_{0.5}$	37.4 [109]	-	-
$CoCrFeNiAl_{0.3}$	35.2 [109]	-	-
CoCrFeNi	31.6 [109]	-	26.8 [117]
$CoCrFeNiCu_{0.5}$	29.0 [109]	-	-
CoCrFeNiCu	27.5 [109]	-	49.0 [115]
$CoCrFeNiCuAl_{0.5}$	32.0 [109]	-	-
$Co_{20}Cr_{26}Fe_{20}Ni_{14}Mn_{20}$	-	-	3.5 [117]
$Co_{15}Cr_{20}Fe_{20}Ni_{25}Mn_{20}$	-	-	38 [118]
$Co_{26}Cr_{18.5}Fe_{18.5}Ni_{18.5}Mn_{18.5}$	-	-	9.7 [100]
$(CoCrFeNi)_{86}Mn_{14}$	-	-	29 [118]
$(CoCrFeNi)_{94}Mn_6$	-	-	28 [118]
FeCrNi	-	20 [114]	-

3.1.3. Solute Strengthening

Solute strengthening is an important characteristic of MPE alloys which causes improvement in the strength of these alloys compared to their constitutes [98]. It is well known that the strength of materials can be changed by adding more solute atoms with large differences in shear moduli and size of solvent and solute atoms [119–121].

First-principles calculations have shown good potential in elucidating the mechanisms of solute strengthening. For instance, the Labusch model was used to study the solute strengthening of fcc NiCoFeCr and NiCoFeCrMn MPE alloys by Varvenne et al. [122]. This model uses dislocation/solute interaction energies from first-principles calculations as inputs, and quantitatively analyzes the effects of composition, strain-rate and temperature on the yield strength and activation volume [122]. Their calculated results for temperature dependent yield strength for $77K \leq T \leq 700K$ showed good agreement with the experimental results available for these MPE alloys [122]. In another study, Varvenne et al. [123] also demonstrated that the additional solute strengthening of MPE alloys due to dilute additions of another solute could be predicted. Moreover, Toda-Caraballo et al. [124,125] applied the Labusch model to compute the hardening parameter in multicomponent alloys, and analyzed the capability of this model in predicting the solid solution hardening (SSH) effect. Wang et al. [80] developed an analytical model based on the Labusch formula to study the effects of the addition of Hf and the resulting lattice distortion on strength and solid solution strengthening mechanisms of bcc $TiNbTaZrHf_x$ MPE alloys. They have predicted that this theoretical model could be used in studying the yield stress, and also in the design of the other bcc MPE alloys.

Besides Labusch model, another general model was proposed by Walbrühl et al. [126] to study the SSH coefficients. Their model is based on an Integrated Computational Materials Engineering (ICME) framework. Different from Labusch [127] models, Walbrühl et al. model directly fits the experimental hardness data, but avoids modeling of misfit parameters. Prediction of SSH of MPE alloys by this model has ±13% overall accuracy [126]. Moreover, I. Toda-Caraballo et al. [128] proposed a methodology which can be employed to compute the distribution of interatomic distances in HEAs by using the unit cell parameter and bulk modulus of component elements. They applied the method for benchmarking the bcc MoNbTaVW alloy and its 5 sub-quaternary systems. The results obtained by this method shows mean variations in the range of 1–2 pm with respect to DFT simulations, which means it can be used as a better starting point for the time consuming DFT simulations and to quantify the SSH effects in MPE alloys [128].

3.2. Thermo-Chemical Properties

Due to the potential applications in extreme conditions, such as high temperatures or corrosive environments, the studies of thermo-chemical properties of MPE alloys are of great interest.

First-principles approaches were applied to investigate the magnetic and thermal properties of FeCoNiCu-based MPE alloys by Huang et al. [129], and $L1_2$ $(Co,Ni)_3(Al,Mo,Nb)$ phases by Yao et al. [71], providing data for the design of novel tungsten-free high-temperature Co based MPE alloys. In another study and by employing the EMTO-CPA method, Cao et al. [92] predicted that the addition of Al slightly decreased the thermodynamic stability of bcc $Al_xMoNbTiV$ MPE alloys [92]. Furthermore, Löffler et al. [130] assessed the heat capacity of the quaternary AlCuMgSi Q phase precipitation by the combination of experiment and first-principles calculations.

To study the heat-transfer behavior in a WTaMoNb refractory MPE alloy, Zhang et al. [131] developed a macro-grid and micro-grid nested model coupled with the finite difference-finite element (FD-FE) method. They have simulated the thermal stress-strain and temperature distributions in a continuous selective laser melting (SLM) process [131]. Moreover, by coupling finite element model (FEM) with the thermomechanical simulation using Gleeble 3800 thermo-mechanical simulator, Rahul et al. [132] studied the material flow pattern and strain field distribution at different conditions.

Although above efforts have provided some valuable data on thermal or chemical properties of some specific MPE alloys, such data for most of MPE alloys do not exist, and first-principles calculations or MD simulations can be utilized to provide databases for thermal or chemical properties of MPE alloys.

3.3. Magnetic Properties

In addition to the mechanical and thermo-chemical properties of MPE alloys, the magnetic performance of these alloys has attracted great attention from researchers. The magnetic behavior of an MPE alloy depends on the alloying elements and composition, and the crystal structures of the generated phase(s). In recent years, some computational studies, mostly based on DFT calculations, have been conducted to investigate the magnetic properties of some MPE alloys.

For example, a computational method based on DFT calculations and the small unit cell SQS was proposed by Zunger et al. [133] to derive defect formation enthalpies for MPE alloys. This model was designed to simulate the multisite correlation functions and some relevant near neighbor pairs of random substitutional alloys [134]. It was found that the vacancy significantly affects its surrounding local spin magnetic moment. In another study, DFT calculations and five different experimental methods were combined and utilized to determine the magnetic ordering in the CoCrFeMnNi MPE alloys [135]. Their first-principles calculations showed that interactions of Fe-located and/or Mn-located moments with the nearby magnetic structure may be responsible for the experimental macroscopic magnetization bias. Separately, systematic first-principle calculations were carried out in order to study the Curie temperature (T_C) of some equiatomic MPE alloys [136]. An integrated computational study of the mean field, and DFT calculations was also conducted to compute Curie temperatures of MPE alloys [137]. Some candidate MPE alloys with good magnetic properties were revealed, including $CoFeNiCrAg_{0.37}$, $CoFeNiCr_{0.8}Cu_{0.64}$ or $CoFeNiCrAu_{0.29}$. It was also concluded that the hypothetical T_C maps can be directly used in creating ferromagnetic MPE alloys with well-defined target magnetizations and T_C's. Figure 10 shows the calculated local magnetic moments in $Al_xCrMnFeCoNi$ ($0 \leq x \leq 5$) MPE alloys using a first principles approach [138].

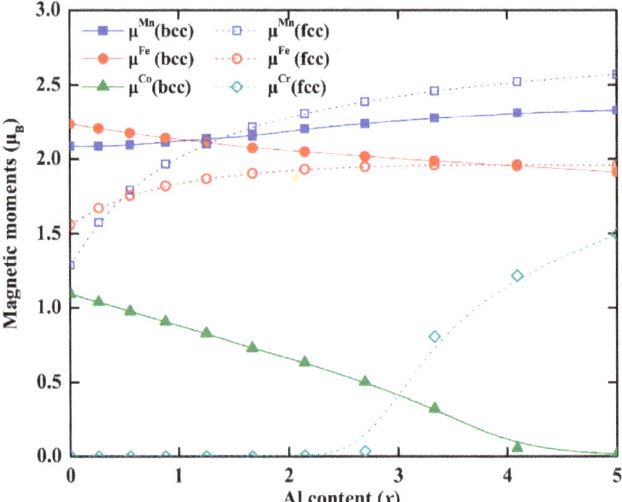

Figure 10. Theoretical local magnetic moments at 0 K for paramagnetic bcc (solid symbols) and fcc (hollowed symbols) $Al_xCrMnFeCoNi$ ($0 \leq x \leq 5$) MPE alloys as a function of Al content (this figure is from Reference [138]).

In some other work, the magnetic properties of CoFeMnNiX (X = Cr, Al, Sn and Ga) MPE alloys were investigated experimentally, and compared with DFT and AIMD simulation results [139,140]. DFT calculations predicted that the anti-ferromagnetic order related to Mn in the CoFeMnNi fcc phase can favor ferromagnetism in the CoFeMnNiAl bcc phase. While CoFeMnNiCr alloy is paramagnetic. By employing DFT calculations and Korringa–Kohn–Rostoker method with the coherent potential approximation (KKR-CPA), Calvo-Dahlborg et al. studied the effects of electronic structures on magnetic properties of some Fe, Co, Cr, and Ni based MPE alloys [141]. Moreover, Sun et al. have studied the dissimilar magnetic moment behaviors of the elements in different concentrations of doped Al in CrMnFeCoNi [138]. Further experimental, theoretical, and computational efforts are still desired in improving the understanding of the magnetic properties of MPE alloys reported in most of these computational studies. Specially to study the effects of grain boundaries and defects on magnetic properties in micro scale, meso-scale models, such as phase-field simulations can be utilized.

4. Summary

In this review article, we have provided a comprehensive overview of current computational modeling and simulation tools for the study and design of MPE alloys. In particular, we focused on the capabilities of different computational modeling tools to study, and predict the structures and properties of MPE alloys. Table 3 summarizes the computational modeling and simulation approaches reported in the literature to study the structures and properties of MPE alloys.

As it can be seen in this table, first-principles approaches have been commonly used to study MPE alloys, and there are many databases created for structures and properties (in particular mechanical properties) of MPE alloys. This can be due to the readiness and the high accuracy of these methods, however the first-principles calculations (e.g., DFT calculations and AIMD simulations) are limited in the model size (up to a few hundred of atoms) and simulation time.

In larger length and time scales, MD simulations were employed in few studies to investigate the structures and mechanical properties of some MPE alloys, but the applicability of such simulations relies on the availability of the interatomic potentials for MPE alloys. The reliability of results is also significantly dependent on the accuracy of interatomic potentials. Therefore, development of accurate interatomic potentials for MPE alloys is of great interest to study and predict structures and phases of these alloys by MD simulations; such models will also open the door to study nanostructural evolutions in MPE alloys.

To study microstructural evolutions and microstructure-dependent properties of MEP alloys, computational modeling and simulations are needed at the mesoscale (such as PFM simulations). Such studies and models are scarce in the literature, and also for verification and validation of such models high-throughput and in-situ experiments are required.

Table 3. The reviewed computational modeling methods used to study the structures and properties of MPE alloys.

Method		First-Principles	Monte-Carlo	MD	Microscale (e.g., PFM simulations)	FEM	Thermodynamics
Structures/ Phases		a	b	c	×	×	d
Properties	Mechanical	e	×	f	×	g	×
	Thermo-Chemical	h	×	×	×	i	×
	Magnetic	j	×	×	×	×	×

a: [52–65]; b: [74,75] [84,85]; c: [77,78]; d: [17,20,21] [13,22,32–37,39,40,42–45]; e: [87–90,92–95] [106–111] [122–125]; f: [116] [99] [114,115]; g: [105]; h: [71] [92] [129,130]; i: [131,132]; j: [133–141].

Author Contributions: M.B.K., Y.H., and M.A.Z. wrote the manuscript, and M.A.Z. coordinated the whole work.

Conflicts of Interest: The authors state no conflict of interest.

References

1. Miracle, D.; Senkov, O. A critical review of high entropy alloys and related concepts. *Acta Mater.* **2017**, *122*, 448–511. [CrossRef]
2. Murty, B.S.; Yeh, J.-W.; Ranganathan, S. *High-Entropy Alloys*; Butterworth-Heinemann: Oxford, UK, 2014.
3. Zhang, C.; Zhang, F.; Chen, S.; Cao, W. Computational thermodynamics aided high-entropy alloy design. *JOM* **2012**, *64*, 839–845. [CrossRef]
4. Yeh, J.-W. Alloy Design Strategies and Future Trends in High-Entropy Alloys. *JOM* **2013**, *65*, 1759–1771. [CrossRef]
5. Tsai, K.-Y.; Tsai, M.-H.; Yeh, J.-W. Sluggish diffusion in Co–Cr–Fe–Mn–Ni high-entropy alloys. *Acta Mater.* **2013**, *61*, 4887–4897. [CrossRef]
6. Wu, Z.; Parish, C.; Bei, H. Nano-twin mediated plasticity in carbon-containing FeNiCoCrMn high entropy alloys. *J. Alloy. Compd.* **2015**, *647*, 815–822. [CrossRef]
7. Wang, Z.; Baker, I. Interstitial strengthening of a fcc FeNiMnAlCr high entropy alloy. *Mater. Lett.* **2016**, *180*, 153–156. [CrossRef]
8. Li, Z.; Tasan, C.C.; Springer, H.; Gault, B.; Raabe, D. Interstitial atoms enable joint twinning and transformation induced plasticity in strong and ductile high-entropy alloys. *Sci. Rep.* **2017**, *7*, 40704. [CrossRef]
9. Laurent-Brocq, M.; Sauvage, X.; Akhatova, A.; Perrière, L.; Leroy, E.; Champion, Y. Precipitation and Hardness of Carbonitrides in a CrMnFeCoNi High Entropy Alloy. *Adv. Eng. Mater.* **2017**, *19*. [CrossRef]
10. Beyramali Kivy, M.; Kriewall, C.S.; Zaeem, M.A. Formation of chromium-iron carbide by carbon diffusion in Al X CoCrFeNiCu high-entropy alloys. *Mater. Res. Lett.* **2018**, *6*, 321–326. [CrossRef]
11. Fu, X.; Schuh, C.; Olivetti, E. Materials selection considerations for high entropy alloys. *Scr. Mater.* **2017**, *138*, 145–150. [CrossRef]
12. Yeh, J.W.; Chen, S.K.; Lin, S.J.; Gan, J.Y.; Chin, T.S.; Shun, T.T.; Tsau, C.H.; Chang, S.Y. Nanostructured high-entropy alloys with multiple principal elements: Novel alloy design concepts and outcomes. *Adv. Eng. Mater.* **2004**, *6*, 299–303. [CrossRef]
13. Ye, Y.; Wang, Q.; Zhao, Y.; He, Q.; Lu, J.; Yang, Y. Elemental segregation in solid-solution high-entropy alloys: Experiments and modeling. *J. Alloy. Compd.* **2016**, *681*, 167–174. [CrossRef]
14. Zhang, Y.; Yan, X.; Ma, J.; Lu, Z.; Zhao, Y. Compositional gradient films constructed by sputtering in a multicomponent Ti–Al–(Cr, Fe, Ni) system. *J. Mater. Res.* **2018**, *33*, 3330–3338. [CrossRef]

15. Sheng, G.; Liu, C.T. Phase stability in high entropy alloys: Formation of solid-solution phase or amorphous phase. *Prog. Natl. Sci. Mater. Int.* **2011**, *21*, 433–446.
16. Pickering, E.; Jones, N.G. High-entropy alloys: A critical assessment of their founding principles and future prospects. *Int. Mater. Rev.* **2016**, *61*, 183–202. [CrossRef]
17. Zhang, Y.; Zhou, Y.J.; Lin, J.P.; Chen, G.L.; Liaw, P.K. Solid-Solution Phase Formation Rules for Multi-component Alloys. *Adv. Eng. Mater.* **2008**, *10*, 534–538. [CrossRef]
18. Mizutani, U. *The Hume-Rothery Rules for Structurally Complex Alloy Phases, in Surface Properties and Engineering of Complex Intermetallics*; World Scientific: Singapore, 2010; pp. 323–399.
19. Guo, S.; Ng, C.; Lu, J.; Liu, C. Effect of valence electron concentration on stability of fcc or bcc phase in high entropy alloys. *J. Appl. Phys.* **2011**, *109*, 103505. [CrossRef]
20. Tsai, M.-H.; Tsai, K.-Y.; Tsai, C.-W.; Lee, C.; Juan, C.-C.; Yeh, J.-W. Criterion for sigma phase formation in Cr-and V-containing high-entropy alloys. *Mater. Res. Lett.* **2013**, *1*, 207–212. [CrossRef]
21. Tsai, M.-H.; Yeh, J.-W. High-entropy alloys: A critical review. *Mater. Res. Lett.* **2014**, *2*, 107–123. [CrossRef]
22. Morral, J.; Chen, S.-L. High Entropy Alloys, Miscibility Gaps and the Rose Geometry. *J. Phase Equilib. Diffus.* **2017**, *38*, 319–331. [CrossRef]
23. Kaufman, L.; Bernstein, H. *Computer Calculation of Phase Diagrams. With Special Reference to Refractory Metals*; Academic Press: Cambridge, MA, USA, 1970.
24. Kroupa, A. Modelling of phase diagrams and thermodynamic properties using Calphad method—Development of thermodynamic databases. *Comput. Mater. Sci.* **2013**, *66*, 3–13. [CrossRef]
25. Kattner, U.R. The thermodynamic modeling of multicomponent phase equilibria. *JOM* **1997**, *49*, 14–19. [CrossRef]
26. Bale, C.; Bélisle, E.; Chartrand, P.; Decterov, S.; Eriksson, G.; Hack, K.; Jung, I.-H.; Kang, Y.-B.; Melançon, J.; Pelton, A. FactSage thermochemical software and databases—Recent developments. *Calphad* **2009**, *33*, 295–311. [CrossRef]
27. Zhang, Y.; Zhang, J. First principles study of structural and thermodynamic properties of zirconia. *Mater. Today Proc.* **2014**, *1*, 44–54. [CrossRef]
28. Chen, H.-L.; Mao, H.; Chen, Q. Database development and Calphad calculations for high entropy alloys: Challenges, strategies, and tips. *Mater. Chem. Phys.* **2018**, *210*, 279–290. [CrossRef]
29. CompuTherm, L. *Pandat 8.0-Phase Diagram Calculation Software for Multi-Component Systems*; CompuTherm LLC: Madison, WI, USA, 2008; Volume 53719.
30. Idury, K.S.; Murty, B.; Bhatt, J. Thermodynamic modeling and composition design for the formation of Zr–Ti–Cu–Ni–Al high entropy bulk metallic glasses. *Intermetallics* **2015**, *65*, 42–50. [CrossRef]
31. Idury, K.S.; Murty, B.; Bhatt, J. Identifying non-equiatomic high entropy bulk metallic glass formers through thermodynamic approach: A theoretical perspective. *J. Non-Cryst. Solids* **2016**, *450*, 164–173. [CrossRef]
32. Tang, Z.; Gao, M.C.; Diao, H.; Yang, T.; Liu, J.; Zuo, T.; Zhang, Y.; Lu, Z.; Cheng, Y.; Zhang, Y. Aluminum alloying effects on lattice types, microstructures, and mechanical behavior of high-entropy alloys systems. *JOM* **2013**, *65*, 1848–1858. [CrossRef]
33. Sonkusare, R.; Janani, P.D.; Gurao, N.; Sarkar, S.; Sen, S.; Pradeep, K.; Biswas, K. Phase equilibria in equiatomic CoCuFeMnNi high entropy alloy. *Mater. Chem. Phys.* **2018**, *210*, 269–278. [CrossRef]
34. Saal, J.E.; Berglund, I.S.; Sebastian, J.T.; Liaw, P.K.; Olson, G.B. Equilibrium high entropy alloy phase stability from experiments and thermodynamic modeling. *Scr. Mater.* **2018**, *146*, 5–8. [CrossRef]
35. Yao, H.; Qiao, J.; Gao, M.; Hawk, J.; Ma, S.; Zhou, H.; Zhang, Y. NbTaV-(Ti, W) refractory high-entropy alloys: Experiments and modeling. *Mater. Sci. Eng. A* **2016**, *674*, 203–211. [CrossRef]
36. Feng, R.; Gao, M.C.; Lee, C.; Mathes, M.; Zuo, T.; Chen, S.; Hawk, J.A.; Zhang, Y.; Liaw, P.K. Design of light-weight high-entropy alloys. *Entropy* **2016**, *18*, 333. [CrossRef]
37. Stepanov, N.; Yurchenko, N.Y.; Skibin, D.; Tikhonovsky, M.; Salishchev, G. Structure and mechanical properties of the AlCrxNbTiV (x = 0, 0.5, 1, 1.5) high entropy alloys. *J. Alloy. Compd.* **2015**, *652*, 266–280. [CrossRef]
38. Gwalani, B.; Gorsse, S.; Choudhuri, D.; Styles, M.; Zheng, Y.; Mishra, R.S.; Banerjee, R. Modifying transformation pathways in high entropy alloys or complex concentrated alloys via thermo-mechanical processing. *Acta Mater.* **2018**, *153*, 169–185. [CrossRef]
39. Abu-Odeh, A.; Galvan, E.; Kirk, T.; Mao, H.; Chen, Q.; Mason, P.; Malak, R.; Arróyave, R. Efficient exploration of the High Entropy Alloy composition-phase space. *Acta Mater.* **2018**. [CrossRef]

40. Tancret, F.; Toda-Caraballo, I.; Menou, E.; Díaz-Del, P.E.J.R. Designing high entropy alloys employing thermodynamics and Gaussian process statistical analysis. *Mater. Des.* **2017**, *115*, 486–497. [CrossRef]
41. Gorsse, S.; Tancret, F. Current and emerging practices of CALPHAD toward the development of high entropy alloys and complex concentrated alloys. *J. Mater. Res.* **2018**, *33*, 2899–2923. [CrossRef]
42. Liu, Y.; Ma, S.; Gao, M.C.; Zhang, C.; Zhang, T.; Yang, H.; Wang, Z.; Qiao, J. Tribological properties of AlCrCuFeNi2 high-entropy alloy in different conditions. *Metall. Mater. Trans. A* **2016**, *47*, 3312–3321. [CrossRef]
43. Haase, C.; Tang, F.; Wilms, M.B.; Weisheit, A.; Hallstedt, B. Combining thermodynamic modeling and 3D printing of elemental powder blends for high-throughput investigation of high-entropy alloys–Towards rapid alloy screening and design. *Mater. Sci. Eng. A* **2017**, *688*, 180–189. [CrossRef]
44. Arslan, H.; Dogan, A. Thermodynamic investigations on the component dependences of high-entropy alloys. *Russ. J. Phys. Chem. A* **2016**, *90*, 2339–2345. [CrossRef]
45. Eshed, E.; Larianovsky, N.; Kovalevsky, A.; Popov, V., Jr.; Gorbachev, I.; Popov, V.; Katz-Demyanetz, A. Microstructural Evolution and Phase Formation in 2nd-Generation Refractory-Based High Entropy Alloys. *Materials* **2018**, *11*, 175. [CrossRef] [PubMed]
46. Beyramali Kivy, M.; Zaeem, M.A.; Lekakh, S. Investigating phase formations in cast AlFeCoNiCu high entropy alloys by combination of computational modeling and experiments. *Mater. Des.* **2017**, *127*, 224–232. [CrossRef]
47. Toda-Caraballo, I.; Wróbel, J.; Nguyen-Manh, D.; Perez, P.; Rivera-Díaz-del-Castillo, P. Simulation and modeling in high entropy alloys. *JOM* **2017**, *69*, 2137–2149. [CrossRef]
48. Kresse, G.; Marsman, O.; Furthmuller, J. VASP the Guide. Available online: http://cms.mpi.univie.ac.at/vasp/vasp.pdf (accessed on 24 April 2016).
49. Giannozzi, P.; Baroni, S.; Bonini, N.; Calandra, M.; Car, R.; Cavazzoni, C.; Ceresoli, D.; Chiarotti, G.L.; Cococcioni, M.; Dabo, I. QUANTUM ESPRESSO: A modular and open-source software project for quantum simulations of materials. *J. Phys. Condens. Matter* **2009**, *21*, 395502. [CrossRef] [PubMed]
50. Gonze, X.; Beuken, J.-M.; Caracas, R.; Detraux, F.; Fuchs, M.; Rignanese, G.-M.; Sindic, L.; Verstraete, M.; Zerah, G.; Jollet, F. First-principles computation of material properties: The ABINIT software project. *Comput. Mater. Sci.* **2002**, *25*, 478–492. [CrossRef]
51. Segall, M.; Lindan, P.J.; Probert, M.A.; Pickard, C.J.; Hasnip, P.J.; Clark, S.; Payne, M. First-principles simulation: Ideas, illustrations and the CASTEP code. *J. Phys. Condens. Matter* **2002**, *14*, 2717. [CrossRef]
52. Ma, D.; Grabowski, B.; Körmann, F.; Neugebauer, J.; Raabe, D. Ab initio thermodynamics of the CoCrFeMnNi high entropy alloy: Importance of entropy contributions beyond the configurational one. *Acta Mater.* **2015**, *100*, 90–97. [CrossRef]
53. Niu, C.; Zaddach, A.; Koch, C.; Irving, D. First principles exploration of near-equiatomic NiFeCrCo high entropy alloys. *J. Alloy. Compd.* **2016**, *672*, 510–520. [CrossRef]
54. Yalamanchili, K.; Wang, F.; Schramm, I.; Andersson, J.; Jöesaar, M.J.; Tasnadi, F.; Muecklich, F.; Ghafoor, N.; Odén, M. Exploring the high entropy alloy concept in (AlTiVNbCr) N. *Thin Solid Films* **2017**, *636*, 346–352. [CrossRef]
55. Jiang, C.; Uberuaga, B.P. Efficient ab initio modeling of random multicomponent alloys. *Phys. Rev. Lett.* **2016**, *116*, 105501. [CrossRef]
56. Tian, F.; Varga, L.K.; Vitos, L. Predicting single phase CrMoWX high entropy alloys from empirical relations in combination with first-principles calculations. *Intermetallics* **2017**, *83*, 9–16. [CrossRef]
57. Li, Z.; Körmann, F.; Grabowski, B.; Neugebauer, J.; Raabe, D. Ab initio assisted design of quinary dual-phase high-entropy alloys with transformation-induced plasticity. *Acta Mater.* **2017**, *136*, 262–270. [CrossRef]
58. Heidelmann, M.; Feuerbacher, M.; Ma, D.; Grabowski, B. Structural anomaly in the high-entropy alloy ZrNbTiTaHf. *Intermetallics* **2016**, *68*, 11–15. [CrossRef]
59. Mu, Y.; Liu, H.; Liu, Y.; Zhang, X.; Jiang, Y.; Dong, T. An ab initio and experimental studies of the structure, mechanical parameters and state density on the refractory high-entropy alloy systems. *J. Alloy. Compd.* **2017**, *714*, 668–680. [CrossRef]
60. Zhang, F.; Zhao, S.; Jin, K.; Bei, H.; Popov, D.; Park, C.; Neuefeind, J.C.; Weber, W.J.; Zhang, Y. Pressure-induced fcc to hcp phase transition in Ni-based high entropy solid solution alloys. *Appl. Phys. Lett.* **2017**, *110*, 011902. [CrossRef]

61. Tian, F.; Delczeg, L.; Chen, N.; Varga, L.K.; Shen, J.; Vitos, L. Structural stability of NiCoFeCrAl x high-entropy alloy from ab initio theory. *Phys. Rev. B* **2013**, *88*, 085128. [CrossRef]
62. Middleburgh, S.; King, D.; Lumpkin, G.; Cortie, M.; Edwards, L. Segregation and migration of species in the CrCoFeNi high entropy alloy. *J. Alloy. Compd.* **2014**, *599*, 179–182. [CrossRef]
63. Takaki, T.; Ohno, M.; Shibuta, Y.; Sakane, S.; Shimokawabe, T.; Aoki, T. Two-dimensional phase-field study of competitive grain growth during directional solidification of polycrystalline binary alloy. *J. Cryst. Growth* **2016**, *442*, 14–24. [CrossRef]
64. Leong, Z.; Wróbel, J.S.; Dudarev, S.L.; Goodall, R.; Todd, I.; Nguyen-Manh, D. The effect of electronic structure on the phases present in high entropy alloys. *Sci. Rep.* **2017**, *7*, 39803. [CrossRef] [PubMed]
65. Gutierrez, M.; Rodriguez, G.; Bozzolo, G.; Mosca, H. Melting temperature of CoCrFeNiMn high-entropy alloys. *Comput. Mater. Sci.* **2018**, *148*, 69–75. [CrossRef]
66. Zhang, Y.; Stocks, G.M.; Jin, K.; Lu, C.; Bei, H.; Sales, B.C.; Wang, L.; Béland, L.K.; Stoller, R.E.; Samolyuk, G.D. Influence of chemical disorder on energy dissipation and defect evolution in concentrated solid solution alloys. *Nat. Commun.* **2015**, *6*, 8736. [CrossRef] [PubMed]
67. Pielnhofer, F.; Schöneich, M.; Lorenz, T.; Yan, W.; Nilges, T.; Weihrich, R.; Schmidt, P. A Rational Approach to IrPTe–DFT and CalPhaD Studies on Phase Stability, Formation, and Structure of IrPTe. *Z. Anorg. Allg. Chem.* **2015**, *641*, 1099–1105. [CrossRef]
68. Mathieu, R.; Dupin, N.; Crivello, J.-C.; Yaqoob, K.; Breidi, A.; Fiorani, J.-M.; David, N.; Joubert, J.-M. CALPHAD description of the Mo–Re system focused on the sigma phase modeling. *Calphad* **2013**, *43*, 18–31. [CrossRef]
69. Bigdeli, S. *Developing the Third Generation of Calphad Databases: What Can ab-Initio Contribute?* KTH Royal Institute of Technology: Stockholm, Sweden, 2017.
70. Ikeda, Y.; Grabowski, B.; Körmann, F. Ab initio phase stabilities and mechanical properties of multicomponent alloys: A comprehensive review for high entropy alloys and compositionally complex alloys. *Mater. Charact.* **2018**, *147*, 464–511. [CrossRef]
71. Yao, Q.; Shang, S.-L.; Wang, K.; Liu, F.; Wang, Y.; Wang, Q.; Lu, T.; Liu, Z.-K. Phase stability, elastic, and thermodynamic properties of the L1 2 (Co, Ni) 3 (Al, Mo, Nb) phase from first-principles calculations. *J. Mater. Res.* **2017**, *32*, 2100–2108. [CrossRef]
72. Gao, M.C.; Zhang, B.; Guo, S.; Qiao, J.; Hawk, J. High-entropy alloys in hexagonal close-packed structure. *Metall. Mater. Trans. A* **2016**, *47*, 3322–3332. [CrossRef]
73. Gao, M.C.; Zhang, B.; Yang, S.; Guo, S. Senary refractory high-entropy alloy HfNbTaTiVZr. *Metall. Mater. Trans. A* **2016**, *47*, 3333–3345. [CrossRef]
74. Choi, W.-M.; Jo, Y.H.; Sohn, S.S.; Lee, S.; Lee, B.-J. Understanding the physical metallurgy of the CoCrFeMnNi high-entropy alloy: An atomistic simulation study. *npj Comput. Mater.* **2018**, *4*, 1. [CrossRef]
75. Sharma, A.; Deshmukh, S.A.; Liaw, P.K.; Balasubramanian, G. Crystallization kinetics in AlxCrCoFeNi ($0 \leq x \leq 40$) high-entropy alloys. *Scr. Mater.* **2017**, *141*, 54–57. [CrossRef]
76. Mooney, C.Z. *Monte Carlo Simulation*; Sage Publications: Thousand Oaks, CA, USA, 1997; Volume 116.
77. Anzorena, M.S.; Bertolo, A.; Gagetti, L.; Kreiner, A.; Mosca, H.; Bozzolo, G.; del Grosso, M. Characterization and modeling of a MoTaVWZr high entropy alloy. *Mater. Des.* **2016**, *111*, 382–388. [CrossRef]
78. Feng, W.Q.; Zheng, S.M.; Qi, Y.; Wang, S.Q. Periodic Maximum Entropy Random Structure Models for High-Entropy Alloys. *Mater. Sci. Forum* **2017**, *898*, 611–621. [CrossRef]
79. Widom, M. Modeling the structure and thermodynamics of high-entropy alloys. *J. Mater. Res.* **2018**, *33*, 2881–2898. [CrossRef]
80. Wang, Z.; Fang, Q.; Li, J.; Liu, B.; Liu, Y. Effect of lattice distortion on solid solution strengthening of BCC high-entropy alloys. *J. Mater. Sci. Technol.* **2018**, *34*, 349–354. [CrossRef]
81. del Grosso, M.; Bozzolo, G.; Mosca, H. Modeling of high entropy alloys of refractory elements. *Phys. B Condens. Matter* **2012**, *407*, 3285–3287. [CrossRef]
82. Toda-Caraballo, I.; Rivera-Díaz-del-Castillo, P.E. Modelling and design of magnesium and high entropy alloys through combining statistical and physical models. *JOM* **2015**, *67*, 108–117. [CrossRef]
83. Kucza, W.; Dąbrowa, J.; Cieślak, G.; Berent, K.; Kulik, T.; Danielewski, M. Studies of "sluggish diffusion" effect in Co-Cr-Fe-Mn-Ni, Co-Cr-Fe-Ni and Co-Fe-Mn-Ni high entropy alloys; determination of tracer diffusivities by combinatorial approach. *J. Alloy. Compd.* **2018**, *731*, 920–928. [CrossRef]

84. Fernández-Caballero, A.; Fedorov, M.; Wróbel, J.S.; Mummery, P.M.; Nguyen-Manh, D. Configurational Entropy in Multicomponent Alloys: Matrix Formulation from Ab Initio Based Hamiltonian and Application to the FCC Cr-Fe-Mn-Ni System. *Entropy* **2019**, *21*, 68. [CrossRef]
85. Fernandez-Caballero, A.; Wróbel, J.; Mummery, P.; Nguyen-Manh, D. Short-range order in high entropy alloys: Theoretical formulation and application to Mo-Nb-Ta-VW system. *J. Phase Equilib. Diffus.* **2017**, *38*, 391–403. [CrossRef]
86. Gludovatz, B.; Hohenwarter, A.; Catoor, D.; Chang, E.H.; George, E.P.; Ritchie, R.O. A fracture-resistant high-entropy alloy for cryogenic applications. *Science* **2014**, *345*, 1153–1158. [CrossRef]
87. Andersen, O.; Jepsen, O.; Krier, G. Exact Muffin-Tin Orbital Theory. In *Lectures on Methods of Electronic Structure Calculations*; World Scientific: Singapore, 1994; pp. 63–124.
88. Vitos, L.; Skriver, H.L.; Johansson, B.; Kollár, J. Application of the exact muffin-tin orbitals theory: The spherical cell approximation. *Comput. Mater. Sci.* **2000**, *18*, 24–38. [CrossRef]
89. Tian, F.; Varga, L.K.; Shen, J.; Vitos, L. Calculating elastic constants in high-entropy alloys using the coherent potential approximation: Current issues and errors. *Comput. Mater. Sci.* **2016**, *111*, 350–358. [CrossRef]
90. Tian, F.; Varga, L.K.; Chen, N.; Shen, J.; Vitos, L. Ab initio design of elastically isotropic TiZrNbMoVx high-entropy alloys. *J. Alloy. Compd.* **2014**, *599*, 19–25. [CrossRef]
91. Li, X.; Irving, D.L.; Vitos, L. First-principles investigation of the micromechanical properties of fcc-hcp polymorphic high-entropy alloys. *Sci. Rep.* **2018**, *8*, 11196. [CrossRef] [PubMed]
92. Cao, P.; Ni, X.; Tian, F.; Varga, L.K.; Vitos, L. Ab initio study of AlxMoNbTiV high-entropy alloys. *J. Phys. Condens. Matter* **2015**, *27*, 075401. [CrossRef] [PubMed]
93. Li, X.; Tian, F.; Schönecker, S.; Zhao, J.; Vitos, L. Ab initio-predicted micro-mechanical performance of refractory high-entropy alloys. *Sci. Rep.* **2015**, *5*, 12334. [CrossRef] [PubMed]
94. Li, X.; Schönecker, S.; Li, W.; Varga, L.K.; Irving, D.L.; Vitos, L. Tensile and shear loading of four fcc high-entropy alloys: A first-principles study. *Phys. Rev. B* **2018**, *97*, 094102. [CrossRef]
95. Ge, H.; Song, H.; Shen, J.; Tian, F. Effect of alloying on the thermal-elastic properties of 3d high-entropy alloys. *Mater. Chem. Phys.* **2018**, *210*, 320–326. [CrossRef]
96. Zheng, S.-M.; Feng, W.-Q.; Wang, S.-Q. Elastic properties of high entropy alloys by MaxEnt approach. *Comput. Mater. Sci.* **2018**, *142*, 332–337. [CrossRef]
97. Tian, F.; Wang, D.; Shen, J.; Wang, Y. An ab initio investgation of ideal tensile and shear strength of TiVNbMo high-entropy alloy. *Mater. Lett.* **2016**, *166*, 271–275. [CrossRef]
98. Qiu, S.; Miao, N.; Zhou, J.; Guo, Z.; Sun, Z. Strengthening mechanism of aluminum on elastic properties of NbVTiZr high-entropy alloys. *Intermetallics* **2018**, *92*, 7–14. [CrossRef]
99. Sharma, A.; Balasubramanian, G. Dislocation dynamics in Al0. 1CoCrFeNi high-entropy alloy under tensile loading. *Intermetallics* **2017**, *91*, 31–34. [CrossRef]
100. Zaddach, A.; Niu, C.; Koch, C.; Irving, D. Mechanical properties and stacking fault energies of NiFeCrCoMn high-entropy alloy. *JOM* **2013**, *65*, 1780–1789. [CrossRef]
101. Senkov, O.; Miller, J.; Miracle, D.; Woodward, C. Accelerated exploration of multi-principal element alloys with solid solution phases. *Nat. Commun.* **2015**, *6*, 6529. [CrossRef] [PubMed]
102. Lucas, M.; Belyea, D.; Bauer, C.; Bryant, N.; Michel, E.; Turgut, Z.; Leontsev, S.; Horwath, J.; Semiatin, S.; McHenry, M. Thermomagnetic analysis of FeCoCr x Ni alloys: Magnetic entropy of high-entropy alloys. *J. Appl. Phys.* **2013**, *113*, 17A923. [CrossRef]
103. Dirras, G.; Lilensten, L.; Djemia, P.; Laurent-Brocq, M.; Tingaud, D.; Couzinié, J.-P.; Perrière, L.; Chauveau, T.; Guillot, I. Elastic and plastic properties of as-cast equimolar TiHfZrTaNb high-entropy alloy. *Mater. Sci. Eng. A* **2016**, *654*, 30–38. [CrossRef]
104. Dirras, G.; Gubicza, J.; Heczel, A.; Lilensten, L.; Couzinié, J.-P.; Perrière, L.; Guillot, I.; Hocini, A. Microstructural investigation of plastically deformed Ti20Zr20Hf20Nb20Ta20 high entropy alloy by X-ray diffraction and transmission electron microscopy. *Mater. Charact.* **2015**, *108*, 1–7. [CrossRef]
105. Štamborská, M.; Lapin, J. Effect of anisotropic microstructure on high-temperature compression deformation of CoCrFeNi based complex concentrated alloy. *Kov. Mater.* **2017**, *55*, 369–378.
106. Lu, Y.; Dong, Y.; Jiang, L.; Wang, T.; Li, T.; Zhang, Y. A criterion for topological close-packed phase formation in high entropy alloys. *Entropy* **2015**, *17*, 2355–2366. [CrossRef]
107. Ye, Y.; Zhang, Y.; He, Q.; Zhuang, Y.; Wang, S.; Shi, S.; Hu, A.; Fan, J.; Yang, Y. Atomic-scale distorted lattice in chemically disordered equimolar complex alloys. *Acta Mater.* **2018**, *150*, 182–194. [CrossRef]

108. Ye, Y.; Liu, C.; Yang, Y. A geometric model for intrinsic residual strain and phase stability in high entropy alloys. *Acta Mater.* **2015**, *94*, 152–161. [CrossRef]
109. Kivy, M.B.; Asle Zaeem, M. Generalized stacking fault energies, ductilities, and twinnabilities of CoCrFeNi-based face-centered cubic high entropy alloys. *Scr. Mater.* **2017**, *139*, 83–86. [CrossRef]
110. Zhao, S.; Stocks, G.M.; Zhang, Y. Stacking fault energies of face-centered cubic concentrated solid solution alloys. *Acta Mater.* **2017**, *134*, 334–345. [CrossRef]
111. Huang, S.; Li, W.; Lu, S.; Tian, F.; Shen, J.; Holmström, E.; Vitos, L. Temperature dependent stacking fault energy of FeCrCoNiMn high entropy alloy. *Scr. Mater.* **2015**, *108*, 44–47. [CrossRef]
112. Pei, Z.; Eisenbach, M. Acceleration of the Particle Swarm Optimization for Peierls–Nabarro modeling of dislocations in conventional and high-entropy alloys. *Comput. Phys. Commun.* **2017**, *215*, 7–12. [CrossRef]
113. Choudhuri, D.; Gwalani, B.; Gorsse, S.; Komarasamy, M.; Mantri, S.A.; Srinivasan, S.G.; Mishra, R.S.; Banerjee, R. Enhancing strength and strain hardenability via deformation twinning in fcc-based high entropy alloys reinforced with intermetallic compounds. *Acta Mater.* **2019**, *165*, 420–430. [CrossRef]
114. Wang, Z.; Li, J.; Fang, Q.; Liu, B.; Zhang, L. Investigation into nanoscratching mechanical response of AlCrCuFeNi high-entropy alloys using atomic simulations. *Appl. Surf. Sci.* **2017**, *416*, 470–481. [CrossRef]
115. Li, J.; Fang, Q.; Liu, B.; Liu, Y. Transformation induced softening and plasticity in high entropy alloys. *Acta Mater.* **2018**, *147*, 35–41. [CrossRef]
116. Smith, T.; Hooshmand, M.; Esser, B.; Otto, F.; McComb, D.; George, E.; Ghazisaeidi, M.; Mills, M. Atomic-scale characterization and modeling of 60 dislocations in a high-entropy alloy. *Acta Mater.* **2016**, *110*, 352–363. [CrossRef]
117. Zaddach, A.; Scattergood, R.; Koch, C. Tensile properties of low-stacking fault energy high-entropy alloys. *Mater. Sci. Eng. A* **2015**, *636*, 373–378. [CrossRef]
118. Liu, S.; Wu, Y.; Wang, H.; He, J.; Liu, J.; Chen, C.; Liu, X.; Wang, H.; Lu, Z. Stacking fault energy of face-centered-cubic high entropy alloys. *Intermetallics* **2018**, *93*, 269–273. [CrossRef]
119. Ventelon, L.; Lüthi, B.; Clouet, E.; Proville, L.; Legrand, B.; Rodney, D.; Willaime, F. Dislocation core reconstruction induced by carbon segregation in bcc iron. *Phys. Rev. B* **2015**, *91*, 220102. [CrossRef]
120. Tsuru, T.; Chrzan, D. Effect of solute atoms on dislocation motion in Mg: An electronic structure perspective. *Sci. Rep.* **2015**, *5*, 8793. [CrossRef] [PubMed]
121. Yu, Q.; Qi, L.; Tsuru, T.; Traylor, R.; Rugg, D.; Morris, J.; Asta, M.; Chrzan, D.; Minor, A.M. Origin of dramatic oxygen solute strengthening effect in titanium. *Science* **2015**, *347*, 635–639. [CrossRef] [PubMed]
122. Varvenne, C.; Leyson, G.; Ghazisaeidi, M.; Curtin, W. Solute strengthening in random alloys. *Acta Mater.* **2017**, *124*, 660–683. [CrossRef]
123. Varvenne, C.; Curtin, W.A. Strengthening of high entropy alloys by dilute solute additions: CoCrFeNiAlx and CoCrFeNiMnAlx alloys. *Scr. Mater.* **2017**, *138*, 92–95. [CrossRef]
124. Toda-Caraballo, I.; Rivera-Díaz-del-Castillo, P.E. Modelling solid solution hardening in high entropy alloys. *Acta Mater.* **2015**, *85*, 14–23. [CrossRef]
125. Toda-Caraballo, I. A general formulation for solid solution hardening effect in multicomponent alloys. *Scr. Mater.* **2017**, *127*, 113–117. [CrossRef]
126. Walbrühl, M.; Linder, D.; Ågren, J.; Borgenstam, A. Modelling of solid solution strengthening in multicomponent alloys. *Mater. Sci. Eng. A* **2017**, *700*, 301–311. [CrossRef]
127. Labusch, R. A statistical theory of solid solution hardening. *Phys. Status Solidi* **1970**, *41*, 659–669. [CrossRef]
128. Toda-Caraballo, I.; Wróbel, J.; Dudarev, S.; Nguyen-Manh, D.; Rivera-Díaz-del-Castillo, P. Interatomic spacing distribution in multicomponent alloys. *Acta Mater.* **2015**, *97*, 156–169. [CrossRef]
129. Huang, S.; Vida, Á.; Heczel, A.; Holmström, E.; Vitos, L. Thermal Expansion, Elastic and Magnetic Properties of FeCoNiCu-Based High-Entropy Alloys Using First-Principle Theory. *JOM* **2017**, *69*, 2107–2112. [CrossRef]
130. Löffler, A.; Zendegani, A.; Gröbner, J.; Hampl, M.; Schmid-Fetzer, R.; Engelhardt, H.; Rettenmayr, M.; Körmann, F.; Hickel, T.; Neugebauer, J. Quaternary Al-Cu-Mg-Si Q Phase: Sample Preparation, Heat Capacity Measurement and First-Principles Calculations. *J. Phase Equilib. Diffus.* **2016**, *37*, 119–126. [CrossRef]
131. Zhang, H.; Xu, W.; Xu, Y.; Lu, Z.; Li, D. The thermal-mechanical behavior of WTaMoNb high-entropy alloy via selective laser melting (SLM): Experiment and simulation. *Int. J. Adv. Manuf. Technol.* **2018**, *96*, 461–474. [CrossRef]

132. Rahul, M.; Samal, S.; Venugopal, S.; Phanikumar, G. Experimental and finite element simulation studies on hot deformation behaviour of AlCoCrFeNi2. 1 eutectic high entropy alloy. *J. Alloy. Compd.* **2018**, *749*, 1115–1127. [CrossRef]
133. Zunger, A.; Wei, S.-H.; Ferreira, L.; Bernard, J.E. Special quasirandom structures. *Phys. Rev. Lett.* **1990**, *65*, 353. [CrossRef] [PubMed]
134. Chen, W.; Ding, X.; Feng, Y.; Liu, X.; Liu, K.; Lu, Z.; Li, D.; Li, Y.; Liu, C.; Chen, X.-Q. Vacancy formation enthalpies of high-entropy FeCoCrNi alloy via first-principles calculations and possible implications to its superior radiation tolerance. *J. Mater. Sci. Technol.* **2018**, *34*, 355–364. [CrossRef]
135. Schneeweiss, O.; Friák, M.; Dudová, M.; Holec, D.; Šob, M.; Kriegner, D.; Holý, V.; Beran, P.; George, E.P.; Neugebauer, J. Magnetic properties of the CrMnFeCoNi high-entropy alloy. *Phys. Rev. B* **2017**, *96*, 014437. [CrossRef]
136. Huang, S.; Holmström, E.; Eriksson, O.; Vitos, L. Mapping the magnetic transition temperatures for medium-and high-entropy alloys. *Intermetallics* **2018**, *95*, 80–84. [CrossRef]
137. Körmann, F.; Ma, D.; Belyea, D.D.; Lucas, M.S.; Miller, C.W.; Grabowski, B.; Sluiter, M.H. "Treasure maps" for magnetic high-entropy-alloys from theory and experiment. *Appl. Phys. Lett.* **2015**, *107*, 142404. [CrossRef]
138. Sun, X.; Zhang, H.; Lu, S.; Ding, X.; Wang, Y.; Vitos, L. Phase selection rule for Al-doped CrMnFeCoNi high-entropy alloys from first-principles. *Acta Mater.* **2017**, *140*, 366–374. [CrossRef]
139. Ma, S.; Zhang, S.; Gao, M.; Liaw, P.; Zhang, Y. A Successful Synthesis of the CoCrFeNiAl0. 3 Single-Crystal, High-Entropy Alloy by Bridgman Solidification. *JOM* **2013**, *65*, 1751–1758. [CrossRef]
140. Zuo, T.; Gao, M.C.; Ouyang, L.; Yang, X.; Cheng, Y.; Feng, R.; Chen, S.; Liaw, P.K.; Hawk, J.A.; Zhang, Y. Tailoring magnetic behavior of CoFeMnNiX (X = Al, Cr, Ga, and Sn) high entropy alloys by metal doping. *Acta Mater.* **2017**, *130*, 10–18. [CrossRef]
141. Calvo-Dahlborg, M.; Cornide, J.; Tobola, J.; Nguyen-Manh, D.; Wróbel, J.; Juraszek, J.; Jouen, S.; Dahlborg, U. Interplay of electronic, structural and magnetic properties as the driving feature of high-entropy CoCrFeNiPd alloys. *J. Phys. D Appl. Phys.* **2017**, *50*, 185002. [CrossRef]

Review

A Review of the Serrated-Flow Phenomenon and Its Role in the Deformation Behavior of High-Entropy Alloys

Jamieson Brechtl [1], Shuying Chen [2], Chanho Lee [2], Yunzhu Shi [3], Rui Feng [2], Xie Xie [2], David Hamblin [2], Anne M. Coleman [2], Bradley Straka [2], Hugh Shortt [2], R. Jackson Spurling [2] and Peter K. Liaw [2,*]

1 The Bredesen Center for Interdisciplinary Research and Graduate Education, The University of Tennessee, Knoxville, TN 37996, USA; brechtljm@ornl.gov

2 Department of Materials Science and Engineering, The University of Tennessee, Knoxville, TN 37996, USA; sychen2014@gmail.com (S.C.); clee70@vols.utk.edu (C.L.); fengruisjtu@gmail.com (R.F.); xie.xie@fcagroup.com (X.X.); dhamblin@vols.utk.edu (D.H.); amcolema@vols.utk.edu (A.M.C.); bstraka@vols.utk.edu (B.S.); hshortt@vols.utk.edu (H.S.); rspurlin@vols.utk.edu (R.J.S.)

3 State Key Laboratory for Advanced Metals and Materials, University of Science and Technology Beijing, Beijing 100083, China; yz.shi1@siat.ac.cn

* Correspondence: pliaw@utk.edu; Tel.: +1-865-974-6356

Received: 10 July 2020; Accepted: 11 August 2020; Published: 13 August 2020

Abstract: High-entropy alloys (HEAs) are a novel class of alloys that have many desirable properties. The serrated flow that occurs in high-entropy alloys during mechanical deformation is an important phenomenon since it can lead to significant changes in the microstructure of the alloy. In this article, we review the recent findings on the serration behavior in a variety of high-entropy alloys. Relationships among the serrated flow behavior, composition, microstructure, and testing condition are explored. Importantly, the mechanical-testing type (compression/tension), testing temperature, applied strain rate, and serration type for certain high-entropy alloys are summarized. The literature reveals that the serrated flow can be affected by experimental conditions such as the strain rate and test temperature. Furthermore, this type of phenomenon has been successfully modeled and analyzed, using several different types of analytical methods, including the mean-field theory formalism and the complexity-analysis technique. Importantly, the results of the analyses show that the serrated flow in HEAs consists of complex dynamical behavior. It is anticipated that this review will provide some useful and clarifying information regarding the serrated-flow mechanisms in this material system. Finally, suggestions for future research directions in this field are proposed, such as the effects of irradiation, additives (such as C and Al), the presence of nanoparticles, and twinning on the serrated flow behavior in HEAs.

Keywords: high-entropy alloys; serrated flow; microstructure; data analysis

---

1. Introduction

1.1. High-Entropy Alloys

High-entropy alloys (HEAs) are an important class of materials that emerged in the earlier part of this century [1,2]. Background investigations on HEAs officially began circa 1996, which resulted in five publications on the subject in 2004 [1–5]. HEAs are a type of complex concentrated alloy that typically consists of five or more (some HEAs with four elements have also been reported) principal elements with atomic concentrations that vary from 5 to 35 atomic percent (at.%) [1,2,6–8]. This combination of elements leads to a high mixing entropy that favors disordered solid solutions at elevated temperatures [9,10]. An HEA is also defined as having a configurational entropy (ΔS_{conf}) of 1.5R or more, where R is the ideal gas constant [11]. The configurational entropy of an alloy can be calculated with Equation (1) [12]:

$$\Delta S_{conf} = -R \sum_{i}^{n} \frac{1}{n_i} \text{Ln} \frac{1}{n_i} \tag{1}$$

where R is the ideal gas constant, and $1/n_i$ is the atomic fraction of the ith element in the alloy. In the case of an equiatomic alloy, $1/n_i$ can be written as X_i, where X_i equals the atomic percentage of the element, which gives $\Delta S_{conf} = RLnX_i$. Therefore, the primary elements in HEAs, in principle, can produce an alloy that has more than twice the configurational entropy, as compared to conventional alloys (0–0.7R) [13]. This relatively high entropy, as well as not having one primary element, results in several effects that account for the unique properties exhibited by HEAs, such as the high-entropy effect, severe lattice distortion, sluggish diffusion, and the cocktail effect [13,14].

It is commonly accepted that the high-entropy effect explains why HEAs often are observed to form only one or two primary phases, while the Gibbs's phase rule indicates that they could form five or more phases [13–16]. This effect also corresponds to a relatively lower Gibbs free energy of mixing for the solid-solution phases in the alloy, as determined by the following equation [13–15]:

$$\Delta G_{mix} = \Delta H_{mix} - T\Delta S_{mix} \tag{2}$$

where ΔG_{mix}, ΔH_{mix}, T, and ΔS_{mix} are the Gibbs free energy of mixing, the enthalpy of mixing, the temperature, and the entropy of mixing, respectively. Since the intermetallics consist of highly ordered structures, the increase in ΔS_{mix} is often sufficient to lower the ΔG_{mix} of the solid-solution phases to below that of the ΔG_{mix} of the corresponding intermetallic phases in HEAs [13,15]. Consequently, solid-solution phases are more energetically favorable than intermetallic phases in many HEAs. However, there is still controversy regarding the true effect of high entropy on the solid-solution stability of HEAs. For instance, recent research suggests that the mutual solubility of constituent elements in the primary structure of the HEAs is the contributing factor for only one or two primary phases in the matrix [15]. Therefore, more investigative work is needed to determine the true mechanism for the high solid-solution stability of HEAs.

Another factor contributing to the unique properties of HEAs is known as the lattice distortion effect, which corresponds to the random distribution of elements in the matrix [17,18]. This random distribution leads to a large amount of lattice distortion due to a lower adherence to the Hume-Rothery rules than conventional alloys [13]. Figure 1 compares the lattice structure of a monoatomic metal and a multiatomic alloy with an HEA [14]. It is apparent from the figure that the lattice of the HEA is significantly more distorted, as compared to the other metals. The severe lattice distortion is thought to enhance the strength, hardness, and thermal and electrical resistance of HEAs by disrupting dislocation motion and increasing the mean free path of electrons and phonons [13,14,17].

Furthermore, the random distribution of elements (and lattice-potential energy) in the alloy may also be a contributing factor to the sluggish diffusion of atoms and vacancies to lattice sites of lower energies [13,19]. This sluggish diffusion effect results in the potential beneficial properties of slower grain growth [2], exceptional elevated-temperature stability [20], high-temperature strength [21], and a higher recrystallization temperature. However, there is still some debate regarding the validity of the sluggish-diffusion concept [18,22–25].

The last effect is the cocktail effect, which states that inter-element interactions give rise to unusual behaviors, as well as average composite properties (rule-of-mixtures) [26]. It should also be mentioned that the composite properties arise from the mutual interactions among the different elements, as well as the severe lattice distortion [27]. This phenomenon has been described as an overall effect that arises from the composition, structure, and microstructure that can result in enhanced properties [26,28]. For example, the cocktail effect has been cited as a contributing factor for the high magnetization and electrical resistance, high strength, and excellent plasticity of a $FeCoNi(AlSi)_{0.2}$ HEA [29].

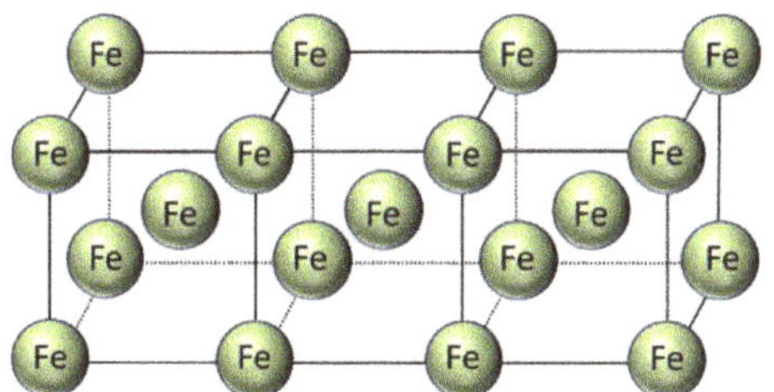

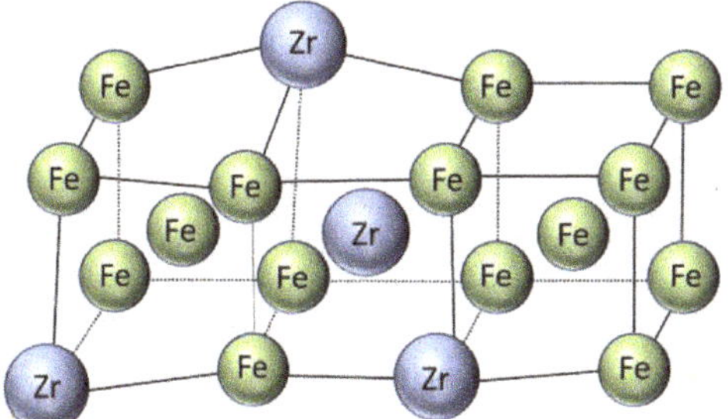

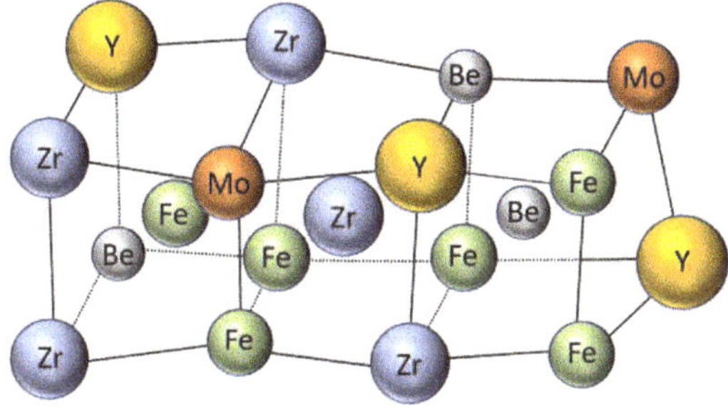

Figure 1. An illustration of the lattice distortion in an element, a conventional alloy, and a high-entropy alloy (adapted from Reference [14]).

In terms of their properties, HEAs exhibit good corrosion resistance [30–35], high fatigue resistance [36–41], high strength and fracture toughness [17,42–47], notable resistance to shear failure [48], decent irradiation stability [49–58], biocompatibility [59–61], good cryogenic tensile and fracture properties [46,62–65], and excellent wear resistance [66,67]. Figure 2 presents an Ashby plot that compares the fracture toughness and yield strength of HEAs with several other material systems [46,47]. As can be seen, the HEAs have fracture-toughness values that exceed almost every material on the graph. Furthermore, this alloy has comparable yield strength values with those of stainless steels. Figure 3 displays the endurance limit and ultimate tensile strength for HEAs and conventional alloys [68]. Except for the bulk metallic glass, the HEA exhibits comparatively higher values, as compared to the other alloy systems.

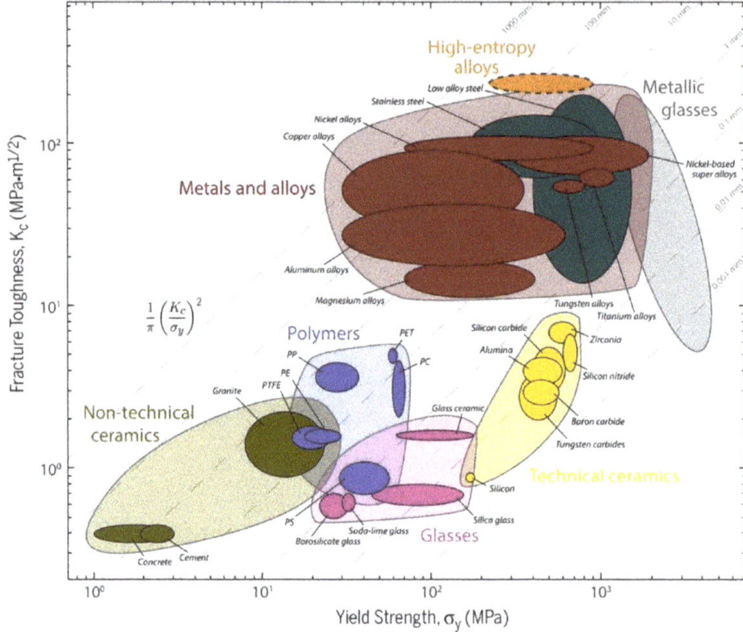

Figure 2. An Ashby plot that compares the fracture toughness and yield strength of HEAs with several other materials and bulk metallic glasses (reproduced from Reference [46] with permission).

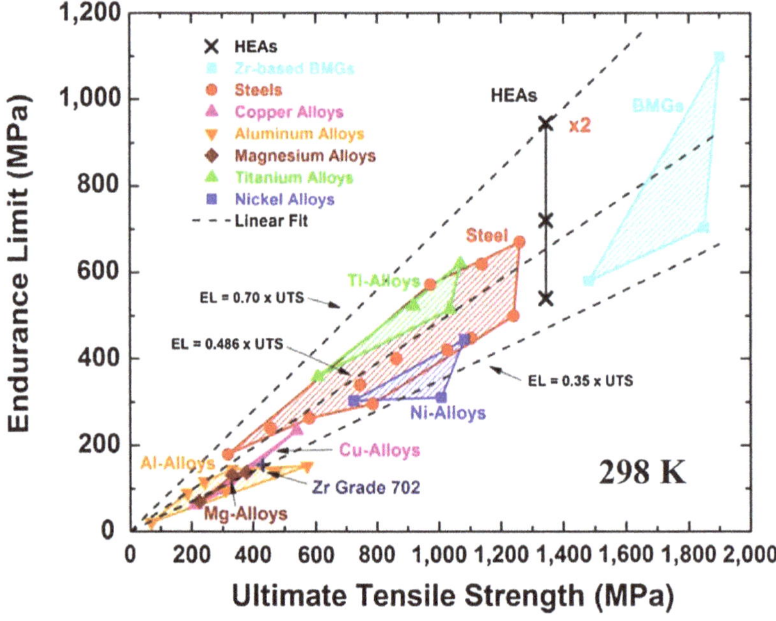

Figure 3. Endurance limits of the $Al_{0.5}$CoCrCuFeNi HEA as a function of the ultimate tensile strength, compared with other structural materials and bulk metallic glasses (BMGs) (reproduced from Reference [68] with permission).

1.2. The Serrated-Flow Phenomenon

The *Concise Oxford English Dictionary* defines a serration as "a tooth or point of a serrated edge or surface" [69]. A typical morphology of serrations is displayed in Figure 4a,b, which features the high-magnification light micrographs of the teeth for two species of shark [70]. The figure displays the two main types of serrations: large primary serrations and small, interspersed secondary serrations. Serrations have also been observed in different phenomena, such as the Barkhausen noise in magnetic materials [8,71–74] (see Figure 5), crackling noise during earthquakes [75–78], discrete strain bursts in nanocrystals [76], various economic indices [79–83], and neuronal avalanches that occur during the operation of the brain's network (see Figure 6) [84–92].

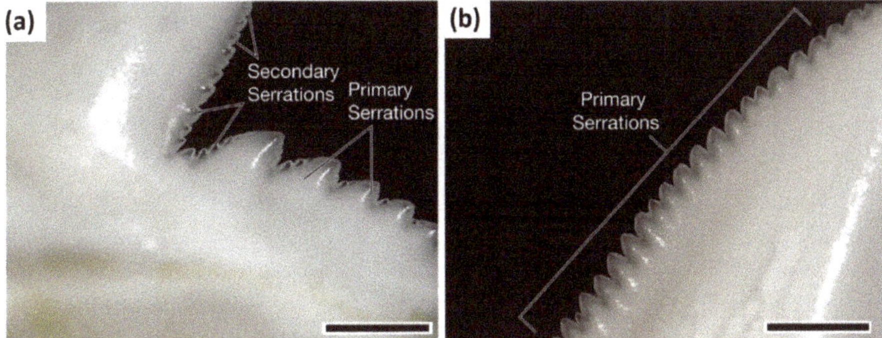

Figure 4. High-magnification light micrographs of the (**a**) primary and secondary serrations of the distal heel of the *Galeocerdo cuvier* shark tooth (scale bar of 2 mm), and (**b**) the primary serrations of the cutting edge of the *Prionace glauca* shark tooth (scale bar of 2 mm) (reproduced from Reference [70] with permission).

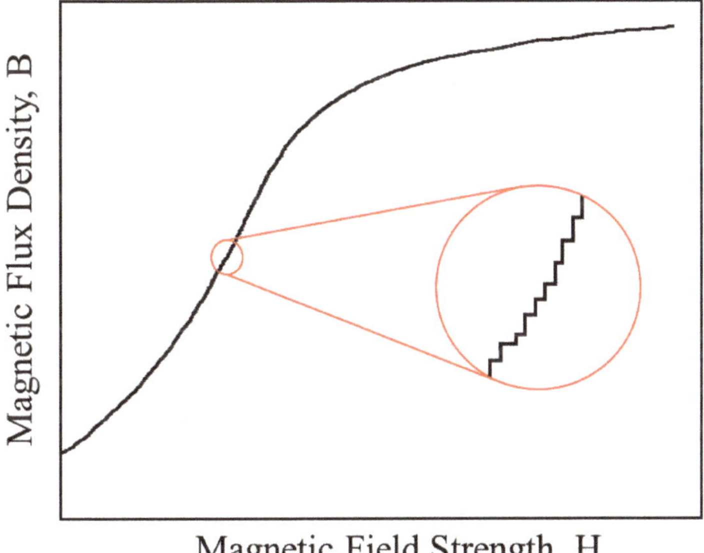

Figure 5. Magnetic Barkhausen noise, as illustrated by irreversible discontinuities in the magnetization, M, as the alternative current magnetic field, H, is varied (adapted from Reference [71]).

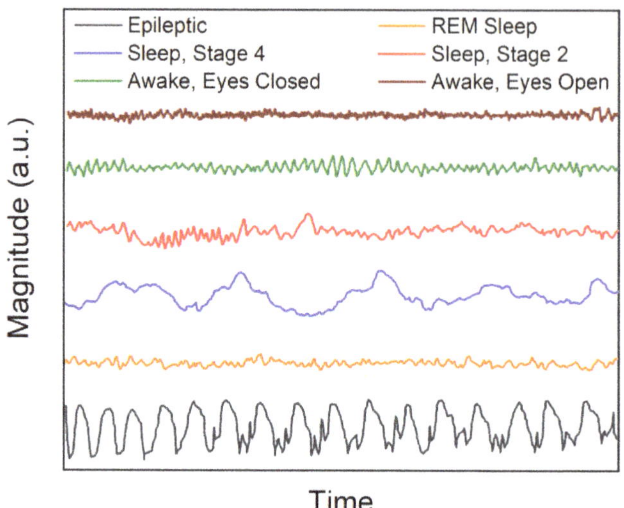

Figure 6. Electroencephalographic recordings during different brain states in humans, which include awake with eyes open, awake with eyes closed, sleep stage 2, sleep stage 4, rapid eye movement (REM) sleep state, and epileptic seizure (adapted from Reference [92]).

In addition to the above phenomena, serrations (or the serrated flow) can occur during the dynamic strain aging (DSA) of a material while undergoing mechanical testing. The serrated flow is of great engineering significance since it corresponds to substantial changes in the microstructure and can have deleterious effects on the mechanical behavior of a material [8,9,93–95]. For example, the plastic instabilities that occur during the serrated flow can lead to a loss in the ductility of the material, as well as an unacceptable surface quality [95]. The study of this type of behavior is important for the production of metal-alloy parts with smooth surfaces and elucidates the fundamental understanding of microscale plastic deformation [96]. The serrated-flow behavior that occurs during material deformation has been reported as far back as the 1970s [97]. With respect to the compression and tension testing, serrations are typically characterized by the fluctuations in the corresponding stress–strain graph [8]. Additionally, the serrated flow has also been observed in the form of pop-ins during nanoindentation testing, which are characterized by displacement bursts in the load vs. displacement graph. A variety of material systems have exhibited this type of behavior [8,9,98,99], including single and polycrystalline metals [100–103], Al alloys [8,104–115], Cu alloys [116], V alloys [117,118], steels [8,119–131], granular systems [132], bulk metallic glasses (BMGs) [8,97,132–151], and medium-entropy and high-entropy alloys [8,9,68,94,152–181]. It is thought that the standard interpretations can be applied to HEAs since a similar weakening effect is observed during the DSA, where solute atoms diffuse and lock dislocations, thereby immobilizing them [153,180].

A graph listing the number of journal publications that feature the serrated flow phenomenon in HEAs for years 2011 through July 2020 is displayed in Figure 7. The results featured in the graph were found via the Web of Science database. As can be seen, the number of articles continually increased per year. Before moving on, it should be mentioned that the numbers for 2020 are still ongoing.

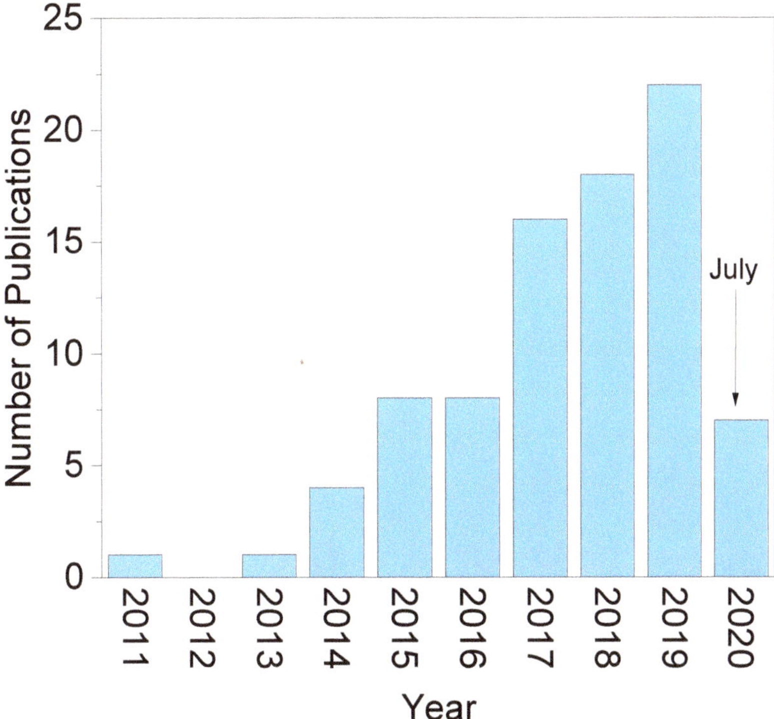

Figure 7. The number of publications that discuss the serrated-flow phenomenon in high-entropy alloys for years ranging from 2011 to July 2020, as determined from the Web of Science publication database.

1.3. Mechanisms of Serrated Flow

Several mechanisms can contribute to the serrated flow in materials during mechanical deformation, including the dislocation pinning by solute atoms, mechanical twinning, grain boundaries, order–disorder phase transformations, phase transformations induced by stress and strain, and yielding across fracture surfaces in brittle materials [182–186]. In terms of dislocation pinning by solute atoms, both interstitial and substitutional solutes can participate in the dislocation-pinning process. With respect to interstitial solute pinning, Cottrell and Bilby provided one of the earliest theoretical explanations for this phenomenon [187]. In their work, they suggested that at sufficiently high temperatures interstitial solute atoms (such as C or N) diffuse toward dislocation cores, where they impede their motion [186]. As mentioned above, substitutional solutes can also participate in dislocation pinning. Here, substitutional solutes diffuse via a vacancy mechanism during plastic deformation [188]. Solute atoms that are larger than the parent metal diffuse below the dislocation line, where the strains are tensile [189]. On the other hand, solutes that are smaller than the matrix atoms migrate to sites above the dislocation line, where compressive strains occur. Consequently, these strain fields pin the dislocation in place. Once the stress becomes sufficiently large, the dislocation can break free and then migrate until it is pinned again. If the pinning and unpinning processes are repeated cyclically, it leads to a DSA effect [190].

Figure 8 presents an illustration of the pinning–unpinning cycle [152]. Dislocations are pinned by diffusion solute atoms (labeled I in the figure). As the stress increases, the dislocation remains pinned (section II) until a critical stress is reached, where there is sufficient energy for the dislocation to break free. The escape of the dislocation corresponds to the stress drop in the stress–strain curve (marked by III in the figure). After escaping the solute atoms, the dislocation moves until it is pinned again by a migrating solute atom (designated IV in the figure). This pinning is accompanied by an increase in the stress, and the process then repeats.

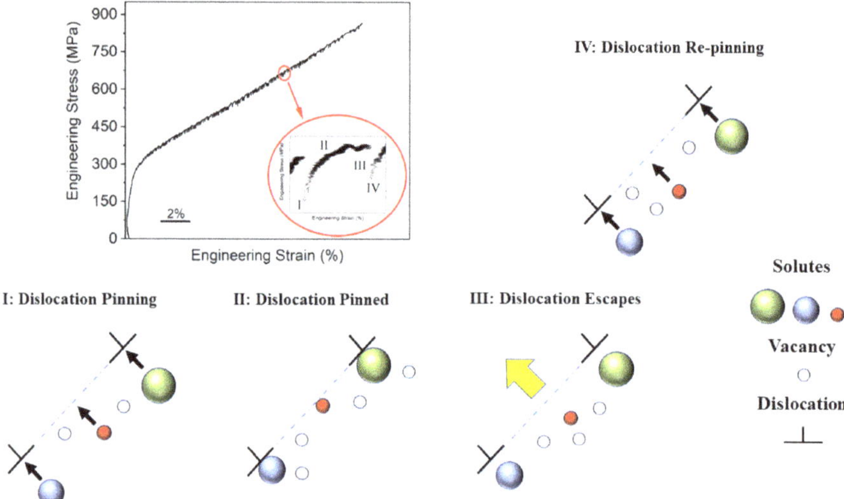

Figure 8. An illustration of the serration process that occurs due to the pinning and unpinning of solute atoms by dislocations. Each numbered section corresponds to the different interactions between the solutes and dislocations during the serration event as follows: (**I**) the solute atoms pin a moving dislocation, which corresponds to the onset of the serration; (**II**) the dislocation remains pinned as the stress increases; (**III**) enough stress builds up, and the dislocation escapes; and (**IV**) the solute atoms catch and re-pin the dislocation (adapted from Reference [152]).

It is widely accepted that, in HEAs, any atom can act as a solute atom [1]. However, there are cases in which only certain atoms in the matrix can participate in dislocation locking. In fact, some investigations have revealed that only certain additives such as Al may hinder dislocation motion in HEAs during deformation [94,180,191]. For instance, Yasuda et al. compared the mechanical behavior of $Al_{0.3}$CoCrFeNi and CoCrFeNi HEAs during DSA [191]. It was reported that the Al-containing HEA exhibited serrations, whereas the latter specimen did not. It was hypothesized that the creation of Al-containing solute atmospheres near a moving dislocation core results in an increase of the frictional stress on the dislocations, thus resulting in serrations. Niu et al. also found a similar trend when investigating the DSA in $Al_{0.5}$CoCrFeNi and CoCrFeNi HEAs [94]. In a study of carbon-doped CoCrFeMnNi HEAs, serrated flows at room temperature (RT) occurred due to interactions between carbon impurities and dislocations, as well as an increase in the stacking fault (SF) energy due to the presence of the carbon [192].

In addition to solute atoms, nanoparticles and phase structures may also participate in the dislocation-locking process during the serrated flow in HEAs. For instance, Chen et al. reported that $L1_2$ particles could provide an obstacle for the mobile dislocations during compression in an $Al_{0.5}$CoCrCuFeNi HEA [155]. Transmission electron microscopy (TEM) characterization revealed that $L1_2$ nanoparticles were present in the matrix for the samples tested at 500 and 600 °C. In another study [193], it has been reported that, in CoCrFeNiMnV_x HEAs, sigma-phase particles and dendrite boundaries can provide a barrier to dislocation motion, resulting in serrations. Twinning, which occurs at cryogenic temperatures, is another mechanism that can lead to the serrated flow in HEAs by hindering dislocation motion at twin boundaries [8,9,122,123,159]. In Reference [159], it was found that, in the CoCrFeNi HEA, twinning at 4.2 K was accompanied by a phase transformation from face-centered cubic (FCC) to hexagonal close-packed (HCP) structures. In other studies [194,195], the lack of observed serrations during mechanical testing of FeCrCoNiMnV and FeNiCoCuV HEAs indicates that the presence of secondary phase precipitates within the solid solution grains may be required for the serrated flow to occur. However, another investigation on an FeCoNiCuCr HEA found that serrations did not occur, even though there were intermetallic compound precipitates present in the matrix [196].

In contrast to crystalline alloys such as HEAs and steels, the serrated flow in BMGs occurs via different mechanisms. Table 1 compares some of the different mechanisms responsible for the serrated flow in crystalline and amorphous alloys. These mechanisms include the flow defect unit and liquid-like site agglomeration, and shear band formation and propagation [197–201]. The activation and percolation of the flow units are typically caused by an applied stress or elevated temperature [202]. In terms of the shear bands, there is a general agreement that their initiation is caused by local structural softening, owing to free volume generation [203]. After the initiation of a shear band, it propagates until the stored elastic energy is sufficiently released, leading to its arrest [198].

Table 1. Some of the underlying mechanisms of the serrated flow in crystalline and amorphous alloys.

Material System	Serration Mechanism
Crystalline Alloys (including HEAs)	Dislocation pinning by solute atoms or nanoparticles, mechanical twinning, order–disorder transformations, phase transformations, yielding across fracture surfaces.
Amorphous Alloys (BMGs)	Excess free volume generation, flow defect and liquid-like site agglomeration, shear band initiation and propagation.

Figure 9 displays an illustration of the serration process that occurs due to the flow defect accumulation and subsequent shear-band initiation and propagation in a BMG [197]. At the beginning of a serration event, the applied stress leads to stress concentrations that arise from the modulus difference between the flow units and the glassy matrix. Consequently, these stress concentrations activate the flow units (labeled I in the figure). With an increase in the applied stress, the liquid-like regions that are composed of the flow defects grow and coalesce with adjacent regions (labeled II in the figure). Once the applied stress reaches a critical value, which corresponds to the local peak in the stress (labeled III in the graph), a shear band is initiated in the coalesced liquid-like region. Consequently, the shear band activates and propagates, leading to a drop in the stress that corresponds to a dissipation of the stored elastic energy. Once a sufficient amount of elastic energy is dissipated, the shear band is arrested. This arrest is accompanied by a recovery of the liquid-like layer and subsequent restoration of the solid-like (glassy) matrix (labeled IV in the figure) in the shear plane [198]. Once the glassy matrix is fully restored, the process repeats.

1.4. Types of Serrated Flow

Serrations have been categorized into five distinct types, which have been labeled as A, B, C, D, and E [8,9,153,182,204]. Figure 10 displays the five types of serrations, and as can be observed, each type of serrations has distinct characteristics. Type-A serrations occur in a periodic fashion in which the curve rises above the general level of stress values before there is a sharp decrease in the stress. Type-B serrations typically fluctuate above the general level of stress with a higher frequency, as compared to Type-A serrations. Type-C serrations are characterized by yield drops that occur below the general level of stress. Furthermore, the elapsed time between stress drops is randomly distributed. In contrast to the serration types previously described, Type-D serrations consist of a stair-stepping pattern in the stress vs. strain graph. Finally, Type-E serrations are characterized by fluctuations that display an irregular pattern and occur with little or no working hardening during band propagation [182]. It should also be noted that combinations of serration types can occur during the serrated flow, such as A + B and B + C [204]. As is discussed later, in Section 1.5, the serration type is typically dependent on the applied strain rate or test temperature during either tension or compression testing [8,9,133,153]. For instance, serrations have been observed to transition from Type-A to Type-C behavior when the testing temperature is increased [180].

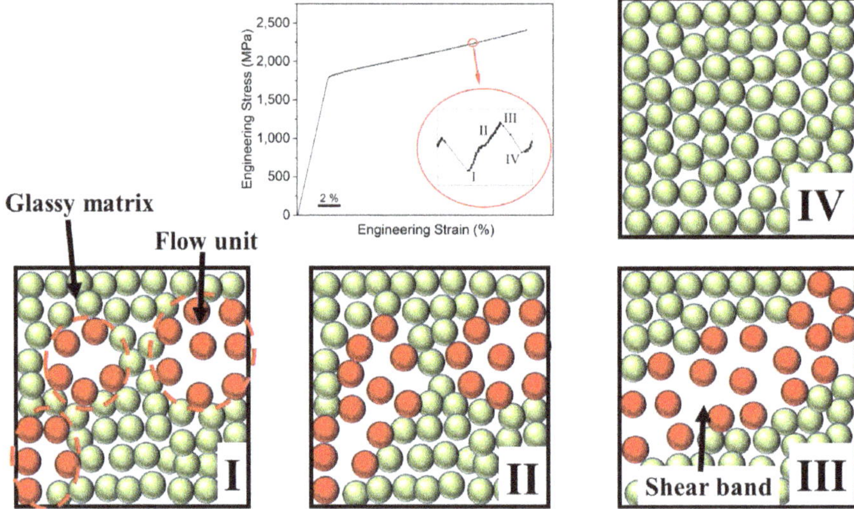

Figure 9. An illustration of the serration process that occurs due to the flow defect accumulation in a BMG during compression. Each numbered section corresponds to the evolution of the flow unit and glassy matrix during the serration event as follows: (**I**) activation of flow units in liquid-like regions under an applied stress, (**II**) growth and coalescence of the liquid-like cites with adjacent cites with an increase in stress, (**III**) initiation of a shear band when a critical stress is reached, and (**IV**) a drop in the stress that is accompanied by shear-band arrest and subsequent restoration of the glassy matrix in the shear plane (adapted from Reference [197]).

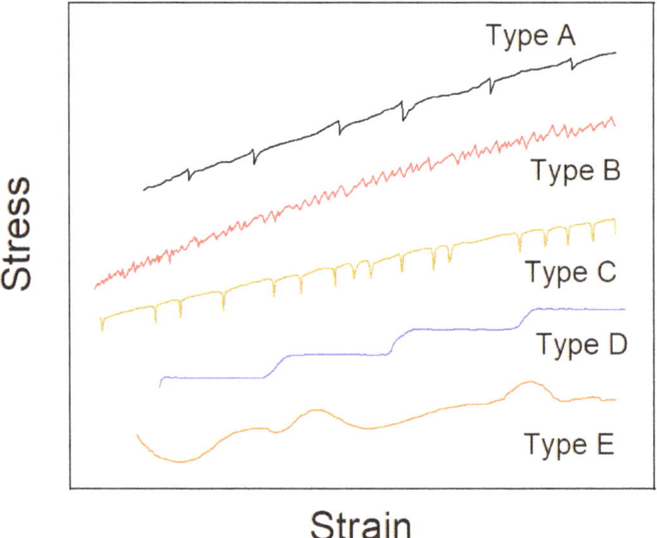

Figure 10. The five types of serrations that can occur during mechanical testing (adapted from References [180,182].

1.5. Factors Affecting Serrated Flow

The serration behavior can be affected by factors such as the composition, temperature, irradiation damage, and strain rate. For example, Cai et al. [205] observed that, in Ni-based superalloys, the γ' precipitates could lead to the aggregation of solute atoms around moving dislocations, thereby enhancing serrations in the material. Similarly, Chandravathi et al. found that the serration behavior in a 9Cr-1Mo steel was also born out of DSA, as the serrated flow was observed in all variations of microstructures, except in certain conditions where increased soaking temperatures above 1423 K led to elevated hardness due to the larger austenitic grain size [206]. Bayramin et al. determined that microstructure played a primary role in the serration behavior of a dual-phase low-carbon steel [207]. They found that the DSA was more pronounced in a sample with a higher concentration of the ferrite phase present. From the results, they hypothesized that the pronounced DSA was attributed to a dislocation density, which increased with increasing strain, such that the dislocation pinning was more prevalent in the ferrite microstructures. Such behavior has been observed in prior studies [207–209].

Temperature has also been reported to play an important role in the serration behavior of materials [8,117,153,210–214]. With an increase in temperature, the mobility of diffusing solute atoms increases, allowing them to pin dislocations at a faster rate. However, once a limiting temperature is exceeded, serrations do not occur due to factors such as a prohibitively high critical strain, smaller critical plastic strain for the termination of serrations, or exceedingly large atomic thermal vibrations, which prevent solutes from settling down the effective locking [8,153]. Conversely, a decrease in the temperature reduces the migration speed of the solute atoms, thereby diminishing their ability to lock dislocations. Once the temperature becomes low enough, the solutes can no longer catch and pin dislocations [215].

The increase in the pinning rate of dislocations (due to an increase in temperature) results in a significant change of the dynamical behavior of the stress and corresponding serrated flow curves (and serration type). For instance, Sakthivel et al. observed that when Hastelloy X underwent tension testing, the serration type changed with an increase in temperature [211]. More specifically, the serrated flow evolved from Type-A at 400 °C to Type-A + Type-B at 500–550 °C, followed by Type-B at 600 °C, and then finally to Type-C at 650 °C. It was also reported that the height of serrations in the stress–strain curve increased with an increase in the temperature. The critical plastic strain, which is the strain associated with the onset of serrations, was found to decrease with an increase in the test temperature.

It has also been reported that the combinations of temperature and irradiation damage can affect the serrated-flow behavior. Rowcliffe et al. reported that temperature and neutron irradiation led to a change in the serrated-flow behavior of the V-4Cr-4Ti alloy during tension [117]. Figure 11a,b presents the engineering stress vs. strain behavior for the unirradiated and irradiated alloys during tension at temperatures ranging from 20 to 850 °C and a strain rate of 10^{-3} s^{-1}. For the irradiation conditions, samples were bombarded by neutrons to doses of 0.5 displacements per atom (dpa) at temperatures ranging from 160 to 420 °C, and also to a dose of 0.1 dpa at 505 °C. It was observed that, for the unirradiated samples, pronounced serrations occurred for temperatures of 300–700 °C. Below 300 °C, serrations were not observed, since the interstitial solutes were relatively immobile. For temperatures above 700 °C, the magnitude and frequency of the serrations significantly declined, while at 800 °C, serrations were restricted to the Lüders region. At 850 °C, there were no observable serrations. As for the irradiated samples, the serrated flow was apparent for temperatures ≥420 °C. Furthermore, as compared to the unirradiated samples, there were significantly fewer and less pronounced serrations in the irradiated specimens. This reduction in the serrated flow may have been caused by a loss of pinning solute atoms that become trapped in either small dislocation loops (T ≤ 400 °C) or fairly coarse {0 0 1} defects (T > 420 °C) that formed during irradiation.

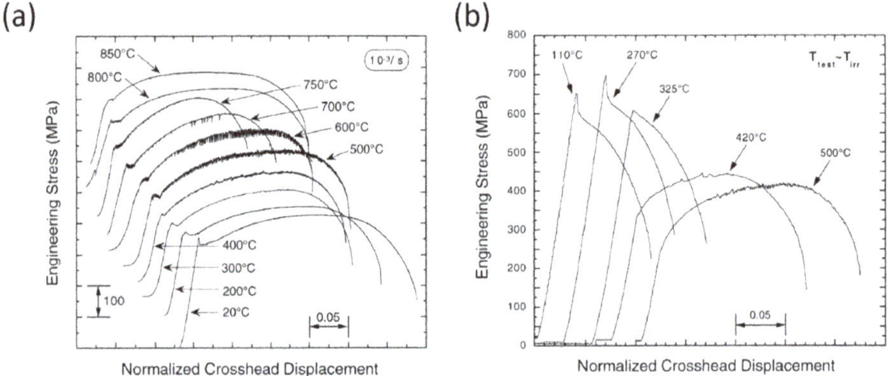

Figure 11. (**a**) Tensile curves for the unirradiated annealed V-4Cr-4Ti at a strain rate of 10^{-3} s^{-1} and (**b**) tensile curves for V-4Cr-4Ti irradiated to 0.5 dpa at 110–420 °C and to 0.1 dpa at 505 °C. The curves were offset on the strain axis for clarity (reproduced from Reference [117] with permission).

In addition to temperature and irradiation, strain rate can also affect the serrated-flow behavior [8,211,212,216,217]. Typically, the critical strain for the onset of serrations increases with an increase in the strain rate [8]. However, there are cases where the critical strain will also increase with a decreasing strain rate [217]. In this scenario, the increase in the critical strain is due to an increase in the strength of solute obstacles, which reduces the ability of solutes to migrate and pin dislocations. Robinson et al. reported the dependence of the serrated flow during Lüders-band formation on the strain rate (as well as temperature) in an Al-5Mg (at.%) alloy [212]. The results of this study can be observed in Figure 12, and as can be seen, Type-C serrations tended to occur at higher temperatures, while Types A and B were found at lower temperatures. Furthermore, the strain rate at which a transition in the serration type occurred, i.e., Type-A to Type-A + Type-B transition increased with an increase in the temperature.

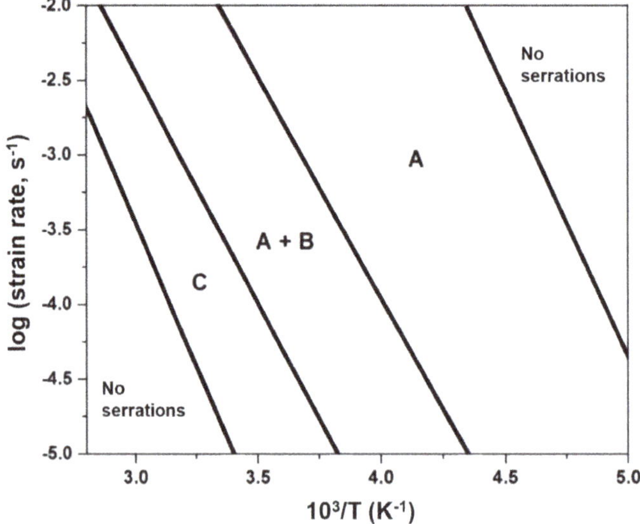

Figure 12. Temperature and strain rate dependence of the serrated flow in the Al-5Mg (at.%) alloy that occurs during Lüders-band formation (reproduced with permission from Reference [8,212].

Similar to conventional crystalline alloys, the serrated flow in HEAs can be affected by factors such as temperature and strain rate [152,153,155,180,193]. For example, Type-C serrations have been typically observed at relatively lower strain rates and higher temperatures, while the reverse is true for Type-A serrations [152,153,155,180]. In an $Al_{0.5}CoCrCuFeNi$ HEA, for example, the transition from Types A to C serrations with increasing temperature was attributed to the increase in the migration speed of the solute atoms that pin dislocations [153]. In this scenario, the increased speed allows the solutes to immediately pin the dislocation after it breaks free. After pinning, the stress increase is characterized by a downward bending curve that is typical of Type-C serrations. Furthermore, the formation of ordered $L1_2$ nanoprecipitates and the emergence of the B2 phase in the alloy during testing at 600 °C were thought to play a role in altering the characteristic behavior of the serrated flow in this HEA [180]. Tsai et al. hypothesized that increasing the number of solute atoms in an HEA can increase the abundance and the atomic-size differences that enhances the pinning force in the lattice [181]. Such a change results in an increase in the range of temperatures in which the serrated flow can occur.

In BMGs, factors such as the strain rate, temperature, and pre-annealing can also affect the serrated-flow dynamics [133,145,150,151,218]. It has been reported that, during compression, an increase in the strain rate leads to the creation of free volumes [218]. The creation of free volumes is accompanied by the formation of shear bands that can better accommodate the applied strain during compression. Jiang et al. observed that, with a decrease in the temperature, the time interval between stress peaks during serrations significantly decreased [145]. It has been observed that, in a $Zr_{52.5}Cu_{17.9}Ni_{14.6}Al_{10}Ti_5$ BMG, pre-annealing at 300 °C for one week led to more irregular serration behavior during compression testing [150]. The increase in the irregularity of the serrated flow was thought to be due to the annihilation of excess free volumes (and associated defects) during the thermally induced structural relaxation of the alloy [219,220].

2. Modeling and Analytical Techniques

In the past, several different types of modeling and analytical techniques have been used to help quantify the serration behavior in a great amount of alloy systems. These techniques include statistical [8,9,76,105,133–135,152–155,221,222], multifractal [180,216,221,223,224], and complexity [104,119,131,180,225] methods. The above methods have been used to model and analyze the serrated flow, since it is believed that serration curves contain scaling laws, self-similarity, and complex dynamics [180].

2.1. Complexity Modeling and Analysis

Entropy-based complexity measurements have been employed to analyze a host of different fluctuating phenomena, including financial-time series, biological signals, and time-dependent serrated-flow behavior during mechanical testing [119,226–228]. In the present context, entropy increases with the degree of disorder (or irregularity) and is the maximum for completely random systems [226]. With respect to the serration behavior in HEAs, higher values are attributed to more complex dynamical behavior during the serrated flow. For biological systems, the complexity of physiological signals is related to the ability of a living system to adjust to an ever-changing environment [227]. For instance, a relatively lower complexity measure has been attributed to pathological conditions, such as the congestive heart failure [226,227,229]. Applying the above ideas to material systems, the complexity of the serrated flow may be a measure of a material's ability to adapt to an applied load [151].

2.1.1. Approximate Entropy Algorithm

One such method that has been used to gauge the complexity of a time series is known as the approximate entropy (ApEn) technique [230]. Lower ApEn values are assigned to more regular time series, while higher ApEn values are assigned to more irregular, less predictable time series [227].

This method has been widely used in physiology and medicine [226,231]. However, there are issues with this method, namely that the technique inherently includes a bias toward regularity (or regularity bias), as it will count a self-match of vectors [232].

The procedure used to calculate the ApEn is discussed below. Given a time series of size, N, one constructs the dataset, $X(i) = [x(i), x(i+1), \ldots, x(i+m-1)]$ (m denotes the embedding dimension), consisting of N data samples from a given time series with $1 \leq i \leq N - m + 1$ [233]. Next, one determines the number of given vectors, $n_i^m(r)$, which are within a distance, $d[X(i), X(j)] < r = 0.2\sigma$, where σ is the standard deviation of X [154]. This distance is defined as follows [230]:

$$d[X(i), X(j)] = \max_{k=1,2,\ldots,m} \left\{ \left| x(i+k-1) - x(j+k-1) \right| \right\} \tag{3}$$

Next we define the following:

$$\Phi^m(r) = \frac{1}{N-m+1} \sum_{i=1}^{N-m+1} C_i^m(r) \tag{4}$$

where $C_i^m(r) = n_i^m(r)/(N - m + 1)$ [230], and $\Phi^m(r)$ represents the average degree of self-correlation [153]. The approximate entropy, ApEn, can now be defined for a fixed m and r as follows:

$$ApEn(m, r, N) = \Phi^m(r) - \Phi^{m+1}(r) \tag{5}$$

where $\Phi^m(r) - \Phi^{m+1}(r)$ signifies the degree of randomness of the sequences, $X(i)$ ($i = 1, 2, \ldots$).

2.1.2. Refined Composite Multiscale Entropy Methods

A technique that does not suffer from the issue of regularity bias is known as the multiscale-entropy (MSE) algorithm [227]. Similar to the ApEn technique, this method has been used to derive the representations of a system's dynamics on different time scales [234]. The MSE method has been used to analyze the complexity of the serration behavior in different material systems, including the low-carbon steel, Al-Mg alloy, and HT-9 steel [104,119,225].

Previous studies have observed that the microstructural composition can affect the complexity of serrations. For example, Sarkar et al. compared the serrated-flow behavior of a low-carbon steel and an Al-Mg alloy [225]. The results of the MSE analysis revealed that the carbon steel produced serrations that had significantly larger complexity values than the Al-Mg alloy. The authors surmised that the complexity values were related to the type of solute atoms in the matrix. In the present context, the carbon solutes in the steel are able to interact with both screw and edge dislocations, whereas the Al-substitutional solutes could only interact with edge dislocations [225]. As such, the higher variety of interactions exhibited by the low-carbon steel resulted in more intricate behavior during the serrated flow and, hence, higher complexity values as compared to the Al-Mg alloy. It has also been reported that an increase in the concentration of carbon impurities in steels led to an increase in the complexity of the serrated flow during tension [131]. In addition to the microstructure, the complexity of the serrated flow has also been linked to different serration types [104,180]. For instance, Type-A and Type-B serrations have been associated with a higher degree of complexity, as compared to Type-C [104].

Due to issues regarding the accuracy of the MSE algorithm when applied to smaller datasets ($N < 750$), alternative techniques were later developed [227,234,235]. One such algorithm includes the refined composite entropy multiscale entropy (RCMSE) method [235], which has been used to investigate various phenomena, such as the serrated flow, chaos, and noise, and cognitive tasks [131,151,178,180,235–237]. The advantage of the RCMSE technique was illustrated in Reference [235], where the RCMSE and MSE algorithms were applied to model and analyze the $1/f$ noise (f is the frequency of the generated noise). It was reported that the MSE technique had a non-zero probability of inducing undefined entropy values, whereas the RCMSE technique had negligible probability.

To begin the RCMSE analysis, the underlying trend of the stress vs. time data in the strain-hardening regime is first eliminated [119]. To perform this task, a moving average or a third-order polynomial is used to fit the data [238]. The fit is then subtracted from the original data. Subsequently, the coarse-grained time series, $y^{\tau}_{k,j}$, is constructed by using the following equation [131]:

$$y^{\tau}_{k,j} = \frac{1}{\tau}\sum_{i=(j-1)\tau+k}^{j\tau+k-1} x_i \quad ; \; 1 \le j \le \frac{N}{\tau} \qquad 1 \le k \le \tau \tag{6}$$

where x_i is the ith point from the detrended time-series data, N is the total number of data points from the detrended time series, τ is the scale factor, and k is an indexing factor, which denotes at what data point to begin the averaging. To give the reader perspective, setting τ equal to 1 yields the original detrended time series. Once $y^{\tau}_{k,j}$ is determined, construct the time series, y^{τ}_{k}, which is represented as a vector for each τ [235]:

$$y^{\tau}_{k} = \left\{ y^{\tau}_{k,1} \; y^{\tau}_{k,2} \; \cdots \; y^{\tau}_{k,M} \right\} \tag{7}$$

where each $y^{\tau}_{k,j}$ is determined from Equation (6), and M equals the integer below N/τ. Now determine the template vectors of dimension, m:

$$y^{\tau,m}_{k,i} = \left\{ y^{\tau}_{k,i} \; y^{\tau}_{k,i+1} \; \cdots \; y^{\tau}_{k,i+m-1} \right\} \; ; \quad 1 \le i \le N-m \, ; 1 \le k \le \tau \tag{8}$$

Next, find n-matching sets of distinct template vectors for each k, using the following equation:

$$d^{\tau,m}_{jl} = \| y^{\tau,m}_{j} - y^{\tau,m}_{l} \|_{\infty} = \max\left\{ \left| y^{\tau}_{1,j} - y^{\tau}_{1,l} \right| \cdots \left| y^{\tau}_{i+m-1,j} - y^{\tau}_{i+m-1,l} \right| \right\} < r \tag{9}$$

where $d^{\tau,m}_{jl}$ is the distance between two vectors, as calculated by the infinity norm [239], and r is typically chosen as 0.15 times the standard deviation of the data, which ensures that the sample entropy does not depend on the variance of the data [226,227,240]. As indicated by Equation (9), two vectors match when $d^{\tau,m}_{jl}$ is less than r. Now repeat the above process for template vectors of size $m + 1$. Next, determine the total number of matching vectors, $n^{m}_{k,\tau}$, for m and $m + 1$ by summing from $k = 1$ to τ. Finally, the RCMSE (or sample entropy) value for the detrended time-series data can be determined by taking the ratio of the natural log for these two sums, using the following equation [235]:

$$RCMSE(X,\, \tau,\, m,\, r) = Ln\left(\frac{\sum_{k=1}^{\tau} n^{m}_{k,\tau}}{\sum_{k=1}^{\tau} n^{m+1}_{k,\tau}} \right) \tag{10}$$

Here, it should be mentioned that the sample entropy is only undefined when $n^{m}_{k,\tau}$ or $n^{m+1}_{k,\tau}$, are zero for all k.

To illustrate the RCMSE method, the $1/f^{\beta}$ noise was analyzed, using the RCMSE method in a similar fashion as was done in Reference [131]. This process was performed for $\beta = 0$, 1, and 2, which correspond to white, pink, and brown noise, respectively. Figure 13 features a graph of the mean Sample En. (sample entropy) vs. scale factor for the three types of noise plotted. As can be seen, the white and brown noise displayed decreasing and increasing trends, respectively. The sample-entropy values for the brown noise were greater than those for the white noise when $\tau > 22$. This result indicates that the brown noise exhibits more complex behavior as compared to the white noise.

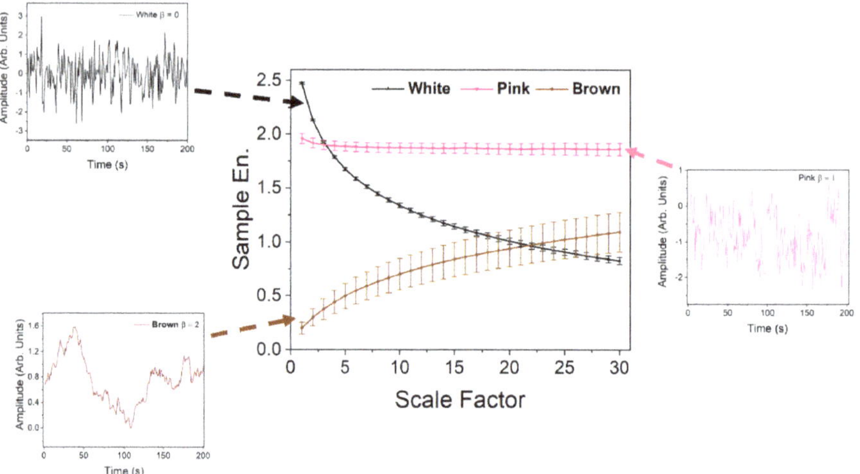

Figure 13. Mean sample entropy of the colored noise data and $1 \leq \tau \leq 30$ plotted with the colored noise plots for brown ($\beta = 2$), pink (or $1/f$ with $\beta = 1$), and white ($\beta = 0$) (adapted from Reference [131]).

The above findings also suggest that a random-walk signal with persistent behavior, such as the brown noise, contains higher spatiotemporal correlations, as compared to a noise-like signal that is inherent in the white noise [241]. As compared to the white and brown noise, the pink noise exhibits the greatest sample-entropy values for $\tau > 3$. This result supports the theory that, as compared to the brown and white noise, the $1/f$ noise exhibits fluctuations that contain a greater degree of long-range correlations and are the most complex in nature [227]. Furthermore, the relatively higher complexity of the pink noise may also be attributed to its inherent mix of random-walk and noise-like characters.

2.2. Multifractal Modeling and Analysis

In contrast to the fractal approach, which can only characterize global scaling properties, the multifractal formalism can analyze not only the underlying fractal geometry but also the distribution of the underlying physical properties on the fractal support [180,242]. The multifractal method has been used to model and analyze the serrated flow in different material systems, including Al-Mg alloys [216,221,223,224,243–245], bulk-metallic glasses [246,247], and HEAs [180]. Specifically, this method has been used to examine the relationship between the serration type and its corresponding multifractal spectra. For instance, Lebyodkin et al. reported that the multifractal characteristics of the serrated flow in an Al-Mg alloy were affected by the phase composition of the material [242]. They also observed that the serrated deformation curves with an underlying multifractal and hierarchical structure results from the self-organization of dislocation motion. In another investigation, it was observed that the transition from Type-B to Type-A serrations was associated with a sharp increase in the multifractality of the stress-burst behavior [216].

In a similar fashion as the RCMSE method, the underlying trend associated with the strain hardening is omitted from the original signal. As discussed earlier, one can accomplish this task by subtracting away the moving average of the polynomial fit from the data. Next, determine the magnitude of the burst in the plastic activity, $\beta_i \approx \left|\frac{d\sigma}{dt}\right|_{t=t_i}$. This step is accomplished by taking the absolute value of the numerical time derivative of the stress at each point in time. Now partition the detrended dataset into non-overlapping windows of size $\Delta\tau$ [244,245] and let the size of the window vary as $(\Delta\tau)^{2j}$ for $j = 1 \ldots n$. For each window, calculate the normalized amplitude of the probability measure for the plastic bursts:

$$p_i(\Delta\tau) = \frac{\left(\sum_{j=1}^{M} \beta_j\right)_i}{\sum_{k=1}^{P} \beta_k} \tag{11}$$

where M is the total number of points in the ith window. Now compute the measure, $\mu_i(\Delta\tau, q)$:

$$\mu_i(\Delta\tau, q) = \frac{p_i^q(\Delta\tau)}{\sum_j p_j^q(\Delta\tau)} \tag{12}$$

where q is a real number. From μ_i, determine the local singularity strength, α [216,248]:

$$\alpha = \lim_{\Delta\tau \to 0} \frac{\sum_i \mu_i(\Delta\tau, q) Ln[p_i(\Delta\tau)]}{Ln(\Delta\tau)} \tag{13}$$

Since $p_i(\Delta\tau) \sim \Delta\tau^{\alpha}$, the singularity strength is an important quantity, since it can give a measure of the scaling relationship between the interval size and the probability measure of the dataset. Next, one can solve for the fractal dimension $f(\alpha)$ of the subset of intervals characterized by the singularity strength [180]:

$$f(\alpha) = \lim_{\Delta\tau \to 0} \frac{\sum_i \mu_i(\Delta\tau, q) Ln\mu_i(\Delta\tau)}{Ln(\Delta\tau)} \tag{14}$$

It is important to note that, typically, the limits from Equations (13) and (14) cannot be calculated directly, and instead a linear-regression analysis must be implemented. A plot of $f(\alpha)$ vs. α is known as the multifractal spectra and is typically characterized by a parabolic shape. The width of the multifractal spectrum, which is defined here as $\Delta = \alpha_{max} - \alpha_{min}$, measures the signal heterogeneity and is associated with the dynamical heterogeneity of the serrated flow [244].

2.3. Mean-Field Theory and the Mean-Field Interaction Model

During deformation, several materials, such as BMGs, HEAs, densely packed granular materials, rocks, and other composite materials, can exhibit jerky flows that occur via slip avalanches [76,134,153,249–252]. These slip avalanches typically exhibit a broad distribution of slip sizes, which can be described by relatively simple probability distribution function (pdf) models [253]. These mean-field interaction models, which are based on the mean-field theory (MFT), assume that a solid material has weak spots that are elastically coupled [132]. The type of weak spots will also depend on the kind of material. For instance, in crystalline materials, weak spots may comprise regions with dislocations that can undergo dislocation slip. In non-crystalline materials, on the other hand, weak spots may be regions that consist of shear bands, shear-transformation zones, or other relatively weak regions in the material [133,254,255]. This type of long-range interaction terminates once the stress at every point in the system is below its current failure stress [256].

The MFT model assumes that, during an applied shear stress, a weak spot can trigger other weak spots to slip, resulting in a slip avalanche [133]. Furthermore, the model assumes that the long length-scale behavior of the slip statistics is independent of the microscopic structural details of the material. The power of the MFT model lies in its ability to predict many statistical distributions and quantities. Typically, these models involve a power-law distribution of slip sizes multiplied with an exponentially decaying cutoff function [132–134,140,251,252,256]. As an example, the underlying equation for the MFT model has been written as follows [133]:

$$D(S, q) = S^{-\kappa} D'\left(Sq^{\lambda}\right) \tag{15}$$

where S is the avalanche-slip size; q is an experimentally tunable parameter, such as the applied stress or strain rate; $D'(y)$ is a universal scaling function; and κ and λ are universal power-law exponents.

It is important to note that the maximum avalanche-slip size, S_{max}, is proportional to q as q^{λ}. For a more comprehensive discussion on the MFT model, please see Reference [76].

Another important quantity is the corresponding complementary cumulative distribution function (CCDF), $C(S, q)$, which gives the probability of observing an avalanche of size greater than S, is useful for systems with low numbers of avalanches [133,180]:

$$C(S,q) = \int_{S}^{\infty} D(S',q)dS' \tag{16}$$

Importantly, this type of analysis is used by experimentalists and theorists to model the stress-drop behavior in multiple systems, including BMGs [133] and HEAs [8,9,32,153,180]. If we substitute Equation (15) into Equation (16) and integrate, we obtain the following:

$$C(S,q) = q^{\lambda(\kappa-1)}C'\left(Sq^{\lambda}\right) \tag{17}$$

where $C'(Sq^{\lambda})$ is some other scaling function. Next, plot $C(S, q)q^{-\lambda(\kappa - 1)}$ vs. Sq^{λ} and then tune the universal exponents until the curves lie on top of each other. This procedure will yield the correct values of the critical exponents.

This type of modeling and analysis has been proven to be a useful tool in the analysis of serrated flows in HEAs. Zhang et al. applied the MFT-modeling technique to examine the effect of temperature on the weak spot (dislocations) avalanche dynamics in HEAs [8]. Carroll et al. employed the MFT model to analyze the serrations that were exhibited by low-, medium-, and high-entropy alloys [153]. Results indicated that the serration-type change from Type-A to Type-B to Type-C with an increase in the temperature. It was also reported that an increase in the strain rate corresponded to a transition from Type-C to Type-B to Type-A serrations in the HEAs.

In addition to analyzing the stress-drop behavior of materials undergoing tension and compression testing, statistical analysis of the nanoindentation pop-in behavior has also been conducted. For this type of analysis, the cumulative probability distribution of the size of the displacement bursts, S, is analyzed. A displacement burst is characterized as a sudden displacement at a constant load during a plastic-deformation event. One such equation that can be used to analyze this type of behavior is written as follows [257]:

$$P(>S) = AS^{-\beta}e^{-\frac{S}{S_c}} \tag{18}$$

where A is a normalization constant, β is a scaling exponent, and S_c is the cutoff value of S. It should be mentioned that, for pop-in sizes smaller than S_c, their behavior follows a power-law distribution. Once S exceeds the cutoff value, the magnitudes of the pop-ins decrease in an exponential fashion.

2.4. Chaos Analysis

Chaos, which is the exponential sensitivity to small perturbations, is a ubiquitous phenomenon in nature [258,259]. Furthermore, chaos is related to how a deterministic dynamical system can be potentially unpredictable due to an extreme sensitivity to initial conditions, which is also known as the "Butterfly Effect" [259]. Chaos has been observed in many different phenomena, including electric circuits [260–262], weather [263–265], physiological systems [266–268], financial markets [269–271], complex networks [272], the serrated flow in alloys [116,216,273], and ecological systems [274–276].

One of the earliest attempts to apply a chaotic mathematical model to the repeated yielding behavior was done by Ananthakrishna et al. [277]. Similar to the Lorenz system [278], their model consisted of a set of four time-dependent differential equations. The equations were based on the time-dependent behavior of the stress, the mobile density, the immobile density, and the density of structures with clouds of solute atoms. The results of the analysis indicated that the dynamical behavior of this system exhibited an infinite sequence of period-doubling bifurcations that eventually led to chaos. Subsequent works confirmed the presence of chaos in the serrated flow of CuAl and AlMg alloys during tension testing [116,216,273]. For the studies involving the AlMg alloy, the chaotic behavior was observed for strain rates below ~3×10^{-4} s^{-1} and corresponded to Type-B serrations [216,273].

One such way to determine whether a dynamical system is exhibiting the chaotic behavior is to determine the largest Lyapunov exponent [279] for the given time series. A positive exponent signifies that the underlying dynamical behavior is chaotic. On the other hand, a negative value indicates that the system will evolve into a stable state [159]. The following discussion gives a basic recipe on how to calculate this value for the time-dependent serration data from an experiment. From a given stress-time series, one can define the reconstructed attractor $Y(t, m) = [\sigma(t), \sigma(t + \tau), \ldots, \sigma(t + [m - 1]\tau)]$, where σ is the stress value at time $t = [1, 2, \ldots N - (m - 1)\tau]$, τ is an arbitrarily chosen delay time, m is the embedding dimension, and N is the number of data points in the time series [154,280].

To begin the analysis, one locates the nearest neighbor (in a Euclidean sense [280]) to the initial point, $Y_0(t_0, m)$, and determines the distance between them. This distance is defined as $L(t_0) = |Y(t_0, m) - Y_0(t_0, m)|$. At some later time, which is denoted as t_1, the initial length will become $L'(t_1) = |Y(t_1, m) - Y_0(t_1, m)|$. However, if it is determined that $Y_0(t_1, m)$ is not a nearest neighbor point of $Y(t_1, m)$, then a new point, $Y_1(t_1, m)$, will be chosen. Next, determine $L(t_1) = |Y(t_1, m) - Y_1(t_1, m)|$. For each step, the angle between $L(t_i)$ and $L(t_{i+1})$ is minimized to reduce the influence on the orbit evolution when a nearest neighbor point is chosen [159]. This process is repeated until the m-dimensional vector, $Y(t_i)$, reaches the end of the time series [154]. From the results, one can determine the largest Lyapunov exponent, λ_1 [280]:

$$\lambda_1 = \frac{1}{t_M - t_0} \sum_{k=1}^{M} Log_2 \frac{L'(t_k)}{L(t_{k-1})} \quad (19)$$

where M is defined as the total number of the replacement steps, and t_M is the final time point.

3. Serration Studies in HEAs

3.1. $Al_{0.5}CoCrCuFeNi$ HEA

Chen et al. performed multiple investigations on the serrated-flow behavior in an $Al_{0.5}CoCrCuFeNi$ HEA [152,154,155,157]. In Reference [152], they examined the effects of temperature on the serrated flow of the above alloy during compression tests. Here, tests were performed at temperatures of 400, 500, and 600 °C, using a strain rate of $5 \times 10^{-5} \cdot s^{-1}$. Synchrotron X-ray diffraction (XRD) (Advanced Photon Source (APS), Argonne National Laboratory, Chicago, IL, USA) revealed that the alloy contained only an FCC phase after testing at 400 and 500 °C, whereas both FCC and body-centered-cubic (BCC) phases were present in the sample tested at 600 °C. Furthermore, it was reported that, with an increase in the test temperature, the critical strain for the onset of serrations decreased, whereas the Young's modulus increased. The increase in the Young's modulus with temperature was attributed to the presence of the BCC phase in the sample. It was thought that the decrease in the critical strain with increasing temperature was due to the increased mobility of solute atoms that can catch and pin dislocations, reducing the amount of strains that can occur before the onset of serrations.

In a subsequent study, the serration behavior of the above alloy was examined after being compressed at strain rates of 5×10^{-5} s^{-1}, 2×10^{-4} s^{-1}, and 2×10^{-3} s^{-1} and temperatures of 400, 500, and 600 °C [180]. Figure 14 displays the stress vs. strain curves of the experiments. Table 2 presents a list of the serration type for each of the experimental parameters, as defined above. The serration type was dependent on the test temperature, and to a smaller extent, the strain rate. For instance, Type-A serrations were primarily observed at 400 °C, Types A and B were seen at 500 °C, and Type-C serrations occurred at 600 °C.

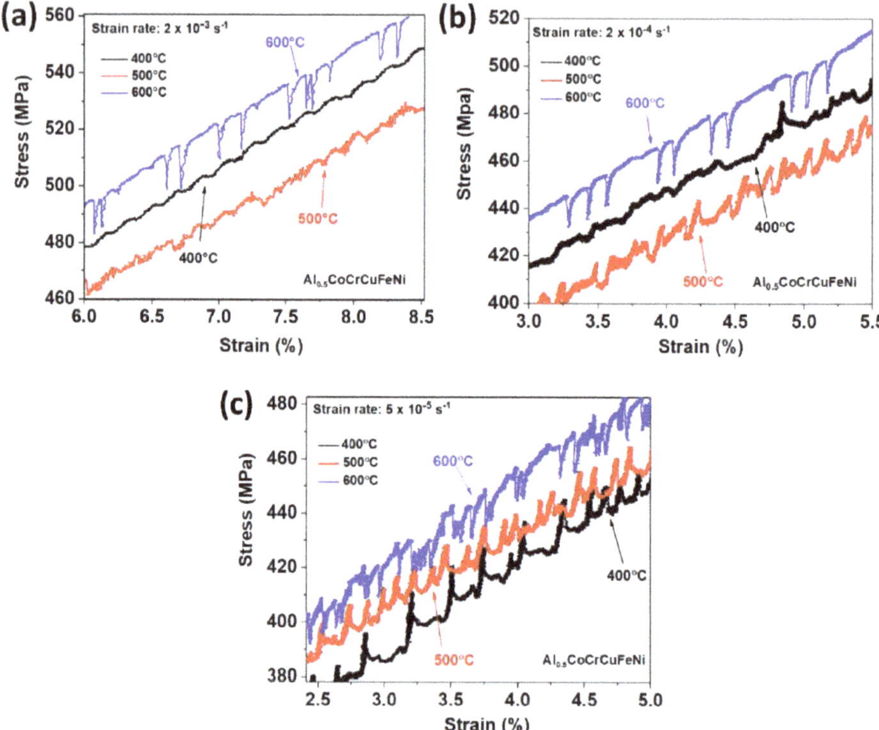

Figure 14. Stress vs. strain graph for the $Al_{0.5}CoCrCuFeNi$ HEA samples tested at strain rates of (**a**) 2×10^{-3} s^{-1}, (**b**) 2×10^{-4} s^{-1}, and (**c**) 5×10^{-5} s^{-1} and temperatures of 400–600 °C (reproduced from Reference [180] with permission).

Table 2. Strain rates, test temperatures, and serration types for the $Al_{0.5}CoCrCuFeNi$ HEA that underwent compression testing (from Reference [180]).

Strain Rate (s^{-1})	Temperature (°C)	Serration Type
	400	A
5×10^{-5}	500	B
	600	C
	400	A
2×10^{-4}	500	B
	600	C
	400	A
2×10^{-3}	500	A
	600	C

The serration data were analyzed by using different analytical techniques, such as the RCMSE (complexity) and multifractal methods. The results of the complexity analysis can be observed in Figure 15a–c. For all of the strain rates, the sample-entropy values were the highest for the samples compressed at 500 °C, while they were the lowest for those compressed at 600 °C. This result indicates that, at 500 °C, the serrations exhibited the most complex dynamical behavior, while the opposite was true for the specimens compressed at 600 °C. Using the results from Table 2, we can surmise that, as compared to the Type-A and Type-B serrations, the Type-C serrations exhibited less complex behavior. This lower complexity was attributed to the repeated pinning and unpinning of dislocations [152], which is characteristic of simple behavior. In contrast, the more complex behavior inherent in the Type-A and Type-B serrations corresponded to a couple of factors. Firstly, the relatively greater complexity of the serration behavior could be attributed to the presence of the deterministic chaos that is inherent in Type-B serrations. Secondly, the increased complexity could be related to the greater variety of defect interactions that occur during the serrated flow. These interactions include the solute atom–dislocation line, dislocation line–dislocation line, dislocation line–precipitate interactions, as well as the solute atom–solute atom interactions [180].

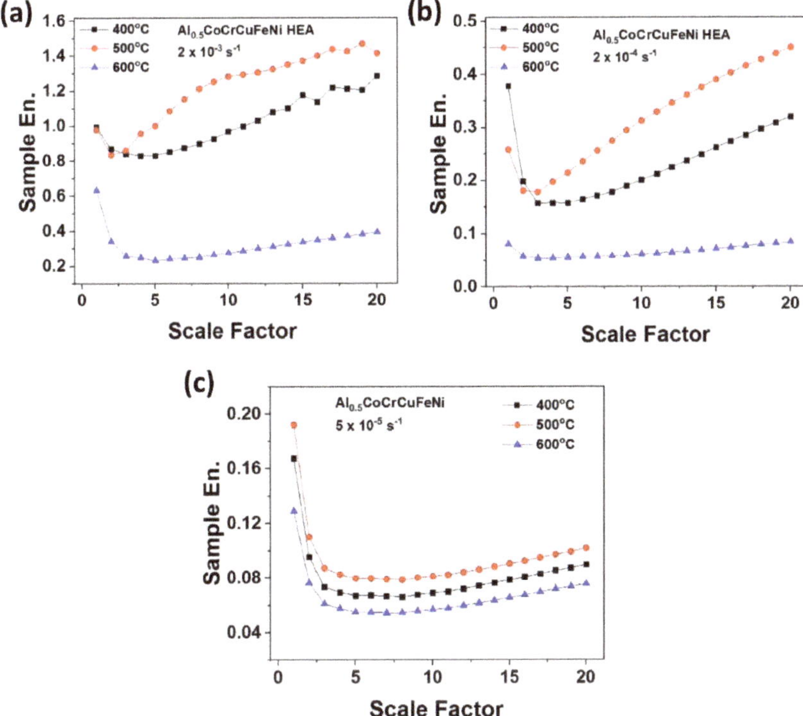

Figure 15. Sample entropy vs. scale factor for the $Al_{0.5}CoCrCuFeNi$ HEA samples tested at strain rates of (**a**) 2×10^{-3} s^{-1}, (**b**) 2×10^{-4} s^{-1}, and (**c**) 5×10^{-5} s^{-1} and temperatures of 400–600 °C, where the sample entropy was plotted for scale factors ranging from 1 to 20 (reproduced from Reference [180] with permission).

Figure 16a–c displays the multifractal spectra for the specimens tested at the prescribed temperatures and strain rates. As can be observed, all the spectra exhibited a parabolic shape for all the experimental conditions. Furthermore, the multifractality was generally found to increase with an increase in the temperature for each strain rate. Consequently, the Type-C serrations were accompanied by the widest multifractal spectra (greatest multifractality), as compared to the other serration types. There was also a sharp increase in the multifractality, and hence the dynamical heterogeneity, at 600 °C for a strain rate of 2×10^{-3} s^{-1}. This burst in the multifractality may represent a transition point between different serration types or dynamical behavior [244]. It should also be mentioned that, for the samples tested at 400 and 500 °C, where Type-A and Type-B serrations were observed, the multifractality of the serrated flow was markedly lower. The relatively lower multifractality of the serrations at these temperatures may correspond to serration dynamics that exhibit the self-organized criticality [216].

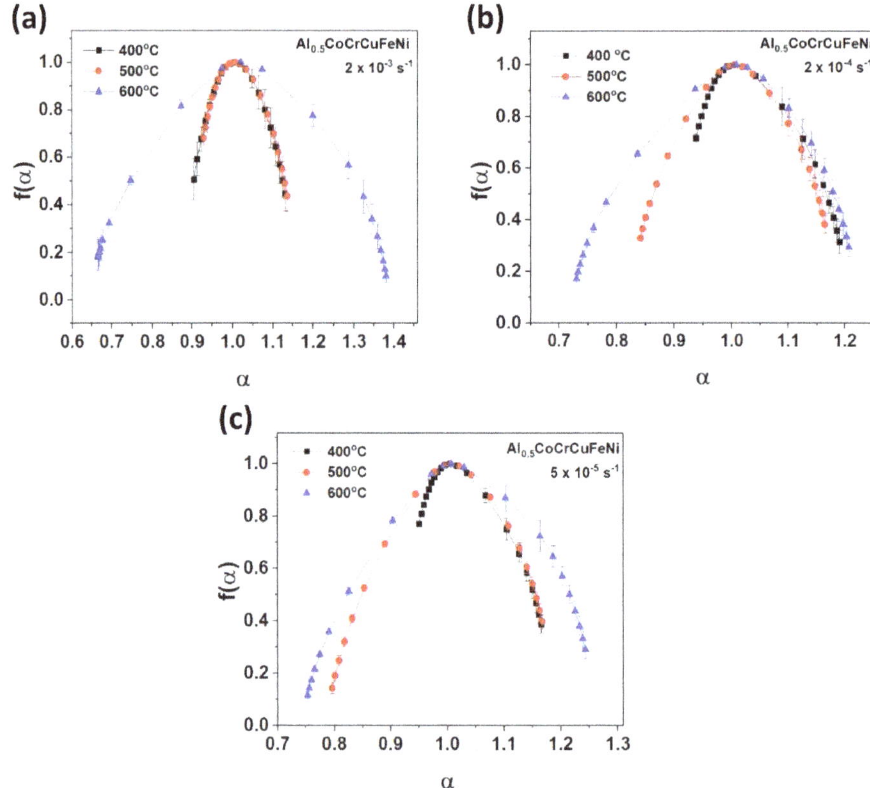

Figure 16. The multifractal spectra (f(α) vs. α) for the $Al_{0.5}$CoCrCuFeNi HEA samples tested at strain rates of (**a**) 2×10^{-3} s^{-1}, (**b**) 2×10^{-4} s^{-1}, and (**c**) 5×10^{-5} s^{-1} and temperatures of 400–600 °C, where the sample entropy was plotted for scale factors ranging from 1 to 20. (Results reproduced from Reference [180] with permission.)

In another study [155], the $Al_{0.5}$CoCrCuFeNi HEA was compression tested at temperatures ranging from RT to 700 °C and a strain rate of 5×10^{-5} s^{-1}. The subsequent TEM (see Figure 17a–d) and XRD characterization revealed that, after testing at a temperature of 600 °C, the matrix contained dislocations, fully ordered $L1_2$ particles, and both BCC and FCC phase structures in the alloy. It was also determined that, at 500 °C, there were the partially ordered $L1_2$ particles and an FCC phase present in the matrix. It was hypothesized that, during compression, the $L1_2$ particles can act as an obstacle for the moving dislocations [155]. This obstruction reduces the mobility of the dislocations enough such that mobile solute atoms can catch and pin them, resulting in the serrated flow. With regards to the serration dynamics, the less complex serrations (Type-C) that were observed at 600 °C may somehow be related to the presence of the fully ordered $L1_2$ nanoparticles and BCC phase in the matrix. On the other hand, the more complex serrations observed at 500 °C (Types A and B) were apparently associated with a matrix that contains partially ordered nanoparticles and a single FCC phase. However, the exact relation between the nanoparticles and serration dynamics are still not well understood and should therefore be the focus of future investigations.

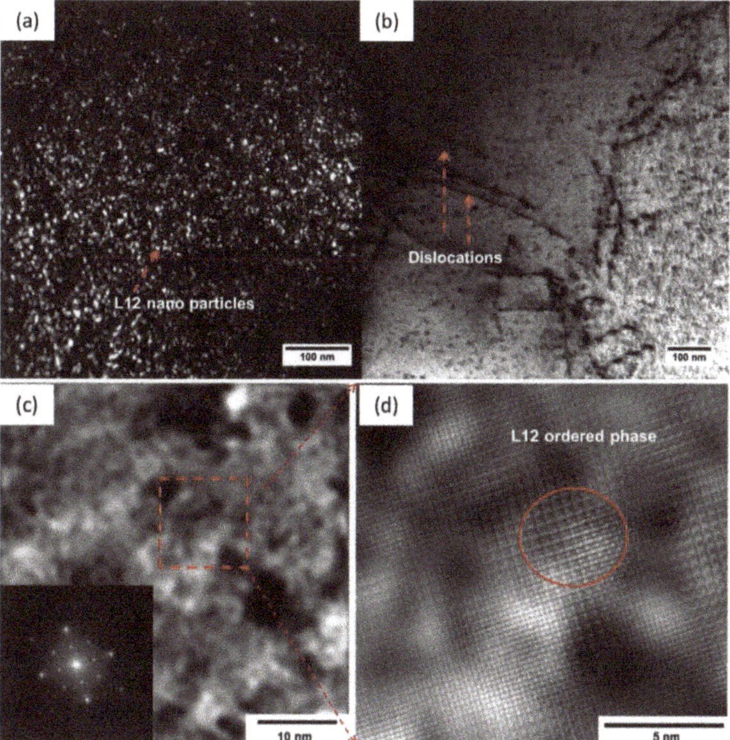

Figure 17. (**a**,**b**) The dark and bright field TEM image of the $Al_{0.5}CoCrCuFeNi$ HEA deformed at 600 °C. (**c**,**d**) The high-resolution TEM (HRTEM) images orientating in the (001) zone axis and the corresponding electron-diffraction pattern. Zoom-in images show the structure of the ordered $L1_2$ phase (reproduced from Reference [155] with permission).

In addition to compression and tension testing, the nanoindentation [281] serration behavior of the $Al_{0.5}CoCrCuFeNi$ HEA has also been examined [157]. For their work, nanoindentations were performed, using a NanoTest Vantage (Micro Materials Ltd., Wrexham, UK) equipped with a diamond Berkovich indenter. For the indentations, the samples were indented to a maximum load of 100 mN, using a loading rate of 10 mN/s. To study the effect of temperature on the indentation deformation, the experiments were performed at RT and 200 °C. Figure 18 displays the nanoindentation load as a function of displacement. The magnified insets show that there was a noticeable stair-step pattern for both the RT and 200 °C conditions, which correspond to the serrated flow (displacement bursts). These serrations were thought to be caused by several factors, including the breakaway of dislocations from various obstacles, such as atoms or precipitates, dislocation multiplication, or the evolution of dislocation cells or tangles [282].

Furthermore, it was observed that, for the 200 °C nanoindentation, the sample exhibited larger displacement bursts, as compared to the RT condition. These relatively larger displacement bursts indicate that, at higher temperatures, there is a greater concentration of dislocations that can be activated during the nanoindentation deformation. It should also be noted that, as compared to the RT condition, the tip penetrated the sample at greater depths during nanoindentation at 200 °C. This result suggests that the sample exhibited softening with an increase in the temperature and was most likely a direct result of the promoted thermal activation of dislocations during deformation at higher temperatures.

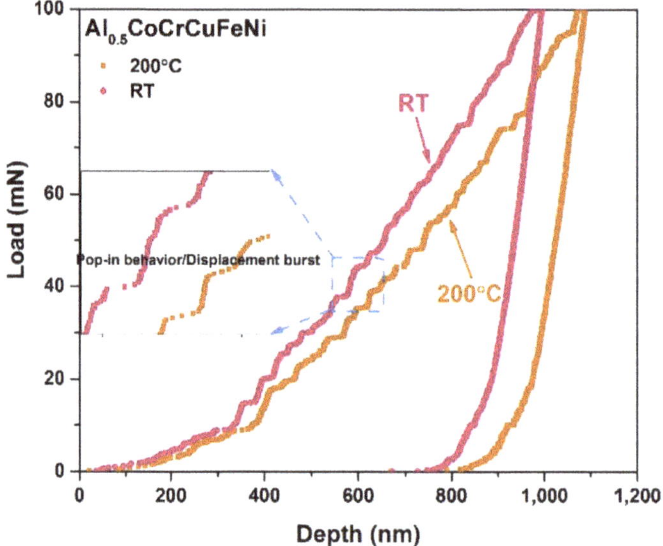

Figure 18. Nanoindentation pop-in behavior of the $Al_{0.5}CoCrCuFeNi$ HEA tested at room temperature (RT) and 200 °C (reproduced from Reference [157] with permission).

In a later study, Chen et al. analyzed the nanoindentation-serration behavior of the $Al_{0.5}CoCrCuFeNi$ HEA, using multiple analytical techniques, including the chaos analysis and ApEn methods [154]. Here, indentations were performed, using a Berkovich tip at RT and 200 °C to a peak load of 100 mN. For each temperature condition, three different holding times, namely 5, 10, and 20 s, at the peak load were used. Figure 19 displays the associated depth vs. time data for the nanoindentation experiments with the prescribed temperatures and loading times. As can be observed, there was serrated flow in the holding regime of the data.

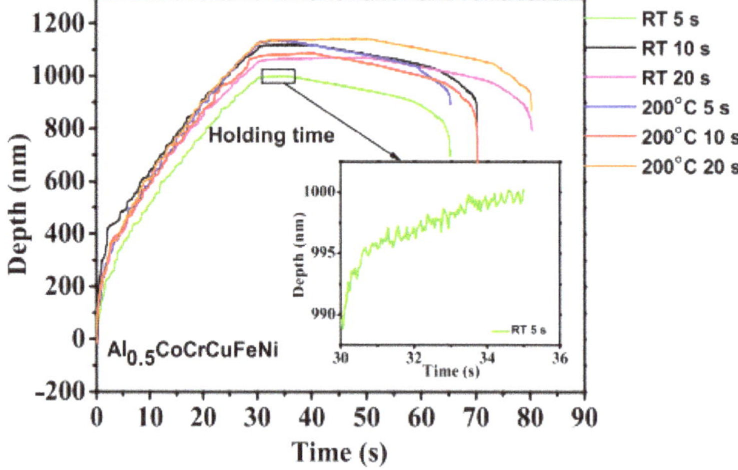

Figure 19. Depth-time curves for the nanoindentation of the $Al_{0.5}CoCrCuFeNi$ HEA with the holding times of 5, 10, and 20 s at RT and 200 °C (from Reference [154]).

The data corresponding to the serrated flow that occurred during the holding periods were analyzed by using methods such as the chaos and ApEn analytical techniques. Figure 20 displays the largest Lyapunov exponent, λ_1, and the ApEn results as a function of holding time for times ranging from 5 to 20 s. For both the RT and 200 °C conditions, the λ_1 values are positive for all the loading times. This finding signifies that for all the experimental conditions, the serrated flow is associated with slip-band dynamics that exhibit chaotic behavior. The result also means that the serration dynamics are sensitive to initial conditions. Furthermore, the largest Lyapunov exponent increased with an increase in the temperature (for a given holding time), indicating that the serrated flow exhibited a greater degree of chaotic behavior at 200 °C. This increase in the chaotic dynamics at the elevated temperature was attributed to a greater mobility of dislocations. It was also observed that, for both temperatures, λ_1 attained a minimum at a holding time of 10 s, which indicates that the serrated-flow behavior was the least affected by initial conditions, as compared to the other holding times. In terms of the ApEn values, they exhibited a similar trend as λ_1 for the sample tested at 200 °C. Furthermore, both values attained a maximum for a holding time of 5 s. The authors hypothesized that the maximum values for the lower holding time corresponded to the increase in the number of interactions among the slip bands that are indicative of a more complex process. As for the room-temperature nanoindentation, a holding time of 10 s was accompanied by both a minimum λ_1 value and a maximum value for the ApEn. This result indicates that the serration behavior exhibited dynamical behavior that has both a relatively low sensitivity to initial conditions and also a greater degree of freedom [154].

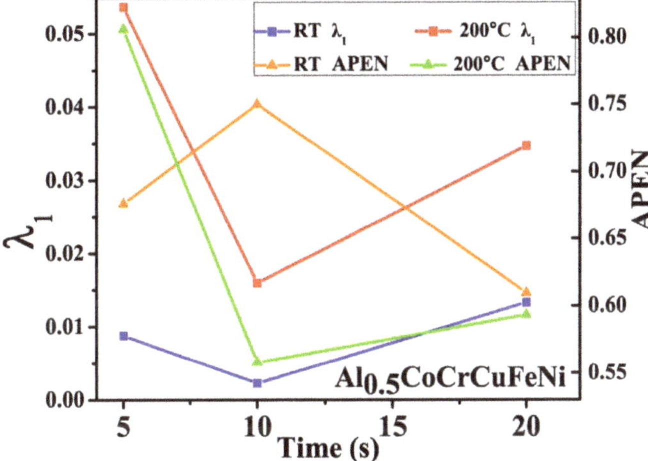

Figure 20. The largest Lyapunov exponent, λ_1, and approximate entropy, ApEn, for the holding times of 5, 10, and 20 s at RT and 200 °C (reproduced from Reference [154]).

Antonaglia et al. applied the mean-field theory to analyze the slip-avalanche statistics for the investigated conditions [167]. For the experiments, specimens with a length of 4 mm and diameter of 2 mm underwent uniaxial-compression tests at a strain rate of 4×10^{-4} s^{-1}, at temperatures of 7, 7.5, and 9 K. The test temperature was controlled by using liquid He. Figure 21a shows the stress vs. strain data resulting from the compression experiments, and the results of the analysis are displayed in Figure 21b. As can be seen, serrations were observed at all three test temperatures. Furthermore, it was found that the magnitude of the serrations significantly increases with an increase in the compression strain. It was also noted that a similar trend has previously been observed in BMGs [133,136,283]. The largest stress-drop size decreases as the temperature increases from 7 to 9 K. This decrease in the slip size for the above temperature range is in agreement with the model prediction, which states that the magnitude of the slip avalanche will be reduced as the temperature increases [284]. The reduction in the avalanche size was related to the twinning phenomenon in the HEA at cryogenic temperatures, where an increase in the temperature increased the difficulty in which deformation twinning could be induced during compression.

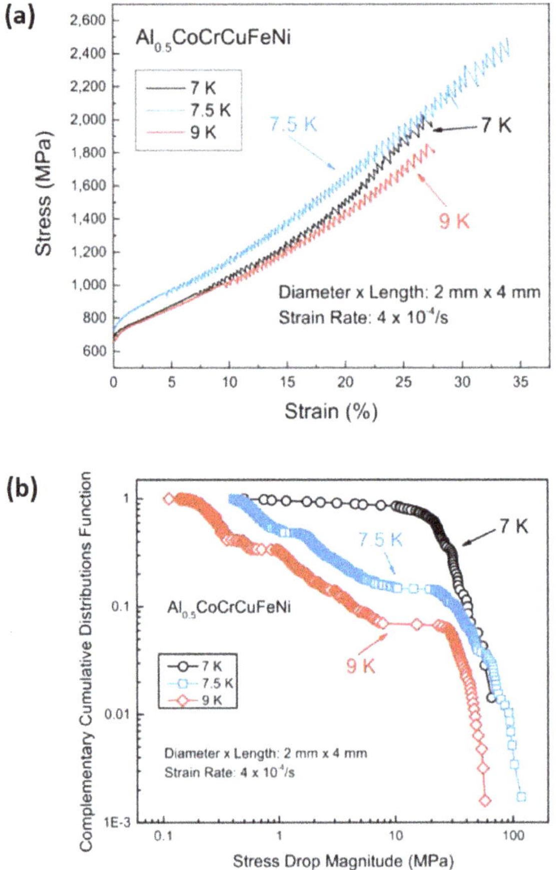

Figure 21. (**a**) Compressive stress–strain curves for the $Al_{0.5}CoCrCuFeNi$ HEA at temperatures ranging from 7 to 9 K, at a strain rate of 4×10^{-4} s^{-1}, and (**b**) the CCDF for the $Al_{0.5}CoCrCuFeNi$ HEA (reproduced from Reference [167] with permission).

In a similar study, the serrated flow during uniaxial compression was analyzed by using the chaos formalism [165]. Here, samples were compressed at a strain rate of 4×10^{-4} s^{-1} and temperatures ranging from 4.2 to 9 K. It was found that the serrated flow exhibited similar behavior (7.5 and 9 K) to that observed by Antonaglia et al. [167]. The results of the analysis revealed that the largest Lyapunov exponent was negative at all the test temperatures. This result indicates that the serrated flow did not exhibit chaotic behavior and indicates that the slip dynamics are stable. It is interesting to note that these results are in contrast with those from Chen et al, where it was determined that the serrated flow during nanoindentation at RT and 200 °C exhibited chaotic behavior [154]. However, it is important to note that the deformation mechanisms associated with twinning at cryogenic temperatures are completely different from those which occur during nanoindentation pop-ins at temperatures greater than or equal to RT. Therefore, these findings suggest that the sensitivity of the serration behavior to initial conditions may depend on the underlying deformation mechanisms.

3.2. $Al_{0.1}CoCrFeNi$ HEA

Xia et al. studied the deformation mechanisms of an $Al_{0.1}CoCrFeNi$ HEA that was produced by vacuum-levitation methods [285]. For the experiment, specimens underwent uniaxial-compressive tests where samples were exposed to a strain of 2×10^{-4} s^{-1} and temperatures of 77 K, 200 K, and 298 K. Results showed that the serrated flow was only observed in the sample compressed at 77 K.

To gain a better understanding of the underlying deformation mechanisms, TEM characterization was performed on the samples after testing (see Figure 22a–d). Figure 22a,b displays the microstructures of the specimens that were tested at 200 and 298 K, respectively. As can be seen, the sample that was compressed at 200 K had a greater dislocation density, as compared to the sample that was tested at 298 K. The results of the Burgers-vector analysis revealed that at these temperatures, the plastic deformation in the HEA occurred mainly by the planar slip of 1/2 < 110> type dislocations on {111}-type planes. This result indicates that deformation at these temperatures occurs only by the dislocation glide. Importantly, this finding is consistent with what is observed in other FCC solid solutions [286]. Figure 22c,d shows the TEM imaging results for the sample tested at 77 K. Here, the microstructure consisted of narrow deformation twins that had widths on the order of tens of nanometers. Furthermore, the TEM characterization indicated that nanoscale-deformation twins leading to the {111} <110> primary slip system were the primary deformation mechanisms responsible for the serrated flow during compression.

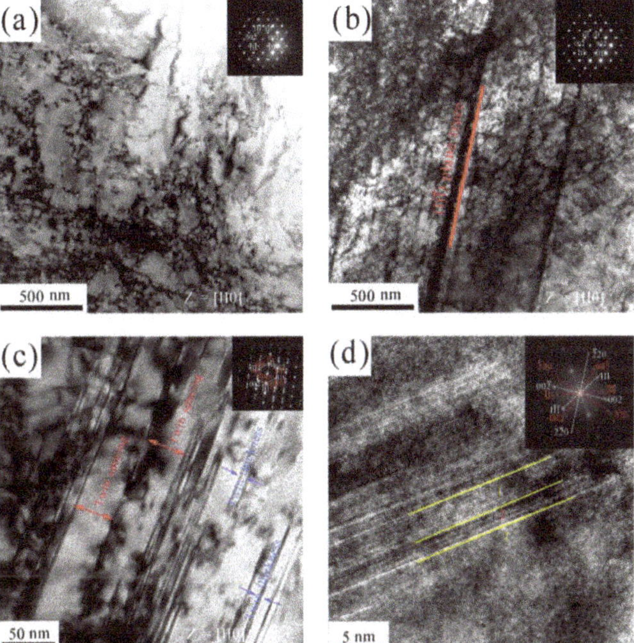

Figure 22. TEM bright field images of the $Al_{0.1}CoCrFeNi$ alloy tested at (**a**) 25 °C (298 K), which indicates that there is a tendency toward the dislocation-cell formation when the strain reaches 50%, (**b**) −73 °C (200 K), with the (111)-type slip plane, (**c**) −196 °C (77 K), which displays nanotwins, and (**d**) the corresponding high-resolution TEM image for (**c**) that displays the region where two deformation twins with a thickness of about 2 nm are located. The selected area electron diffraction patterns and fast Fourier transform (zone axis (110)) are also displayed (reproduced from Reference [285] with permission).

Hu et al. examined the serrations in the compressed nanopillars composed of the $Al_{0.1}CoCrFeNi$ HEA [160]. For the experiments, pillars with diameters of 500–700 nm underwent in situ compression tests, using direct electron imaging. The compression tests were performed by using a strain rate of ~1×10^{-3} s^{-1} in a Hysitron PI95 picoindenter equipped with a 2 µm flat punch diamond indenter in a JEOL 2010 LaB6 TEM (JEOL USA Inc., Boston, MA, USA) with an operating energy of 200 keV. The mechanical-deformation data were accumulated, using a data-acquisition rate of 500 Hz.

Figure 23 presents the stress vs. strain data for the in situ TEM compression test. As can be observed, there were multiple slip events that occurred during the experiment. Furthermore, the authors divided the deformation process into three stages that consisted of (1) little or no stress drops, (2) medium-sized stress-drops, and (3) relatively larger stress-drops. A comparison of the stress-time data and the three corresponding stages of the stress-drop behavior, as observed by the TEM and SEM, can be seen in Figure 24a–e. With respect to stage I, the stress drops corresponded to the wave-like propagation of dislocations from the top of the pillar to the bottom. As for Stage-II stress drops, they were attributed to small dislocation avalanches. Finally, stage III stress drops corresponded to dislocation avalanches that led to large crystal slips.

The CCDF analysis was performed on the serration behavior, and the results can be seen in Figure 25a,b. The data were modeled according to Equations (15) and (16), for which $D(S, q) = S^{-(\kappa - 1)}g[S(\sigma - \sigma_c)^{1/\beta}]$, where κ and β are critical exponents, σ is the applied stress, σ_c is the failure stress, and g is a universal scaling function defined as $x^{\kappa-1}\int_x^{\infty} e^{-At}t^{-\kappa}dt$ [222] with $A = 1.2$. These data were for samples with diameters ranging from ~512 to 655 nm. For Figure 25a, the CCDF, as a function of the stress level over the maximum stress, was plotted with respect to the slip size. Figure 25b displays the rescaled CCDFs in which the curves have collapsed onto one another. The critical exponents, κ and β, were determined by tuning them until the curves collapsed, and were found to be consistent with the predicted MFT values of 1.5 and 0.5, respectively [222].

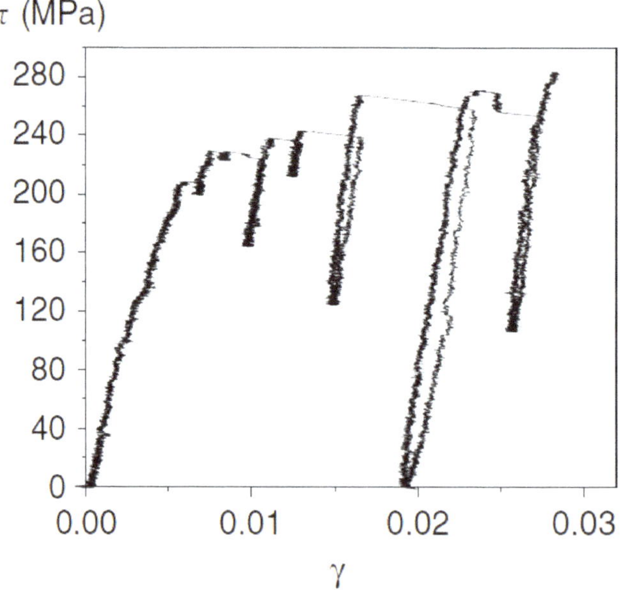

Figure 23. The stress vs. strain curve for the in situ TEM nanopillar compression tests for the $Al_{0.1}CoCrFeNi$ HEA (reproduced from Reference [160]).

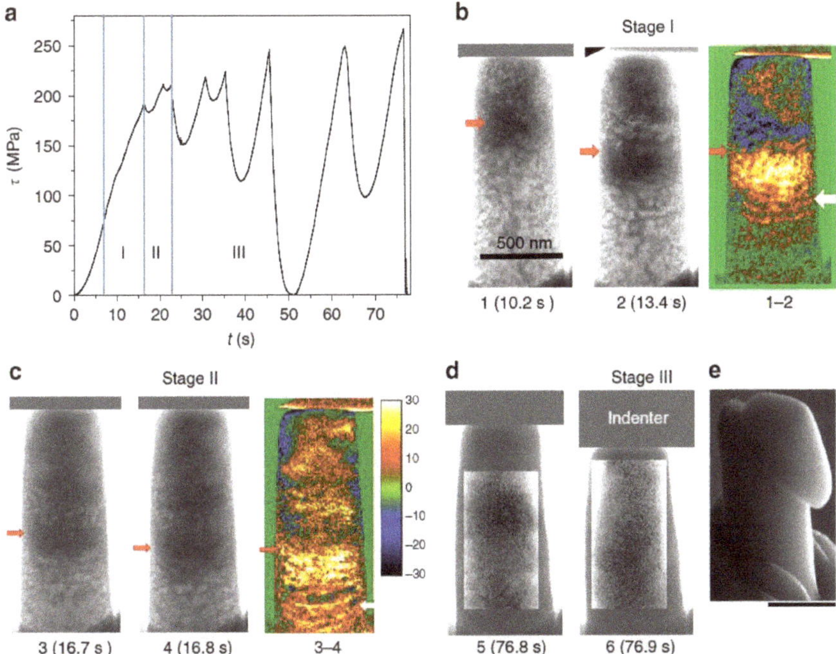

Figure 24. The in situ nanomechanical indentation and TEM results for the $Al_{0.1}CoCrFeNi$ HEA in terms of the (**a**) stress vs. time data (stages I, II, and III labeled in the figure), the TEM imaging results for the serrations that occurred during (**b**) stage I, (**c**) stage II, and (**d**) stage III (with color-coded deformation stages), and (**e**) the SEM image of the deformed nanopillar that confirms that large crystal-slip ensued during the experiment (reproduced from Reference [160]).

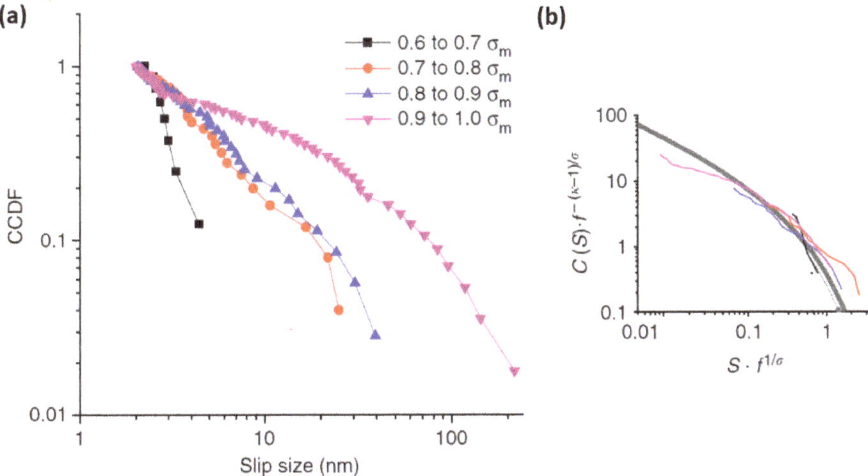

Figure 25. (**a**) The plots of the stress-binned CCDF of slip sizes as a function of the stress level over the maximum stress samples with diameters ranging from ~512 to 655 nm, compressed at a strain rate of 1×10^{-3} s^{-1} and with similar load–displacement characteristic. (**b**) The scaling collapse of the same data with the predicted scaling function (reproduced from Reference [160]).

The results of the analysis led to the following important conclusions. Firstly, the distribution used in their analysis is predicted by a simple, coarse-grained model [222,256]. Secondly, the avalanche mechanism responsible for the observed serration behavior in the samples is based not only on the interactions between the dislocation bands and the dislocation pileups, but also the pileup and the dislocation pinning centers, as well. Thirdly, the applied stress is a critical tuning parameter due to the dependence of the slip avalanche on the stress level. Finally, the deformation behavior exhibited by the $Al_{0.1}CoCrFeNi$ HEA is characteristic of the tuned critical behavior rather than the self-organized criticality, indicating that the corresponding avalanche distribution is universal in nature.

3.3. *$Al_{0.3}CoCrFeNi$ HEA*

The effect of Al on the serrated flow has also been observed in an investigation involving the $Al_{0.3}CoCrFeNi$ HEA [191]. Here, Yasuda et al. examined the deformation behavior of the single-crystal CoCrFeNi and $Al_{0.3}CoCrFeNi$ HEAs during compression. For their experiment, they compressed samples at a strain rate of 1.7×10^{-4} s^{-1} for temperatures ranging from −180 to 1000 °C (93–1273 K). Figure 26 displays the results, and, as can be observed, the $Al_{0.3}CoCrFeNi$ HEA exhibited the serrated flow, while the other specimen did not when tested at 600 °C. The serrated flow that was observed in the DSA regime was thought to be related to Al-containing solute atmospheres that were created near a moving dislocation core, thus leading to an increase of the frictional stress on the dislocations. Similar to the findings reported by Niu et al. [94], this result indicates that Al atoms play an important role in the serrated-flow behavior that occurs in the HEA during DSA. This finding also suggests that not every atom can act as a solute that participates in dislocation locking that results in the serrated flow.

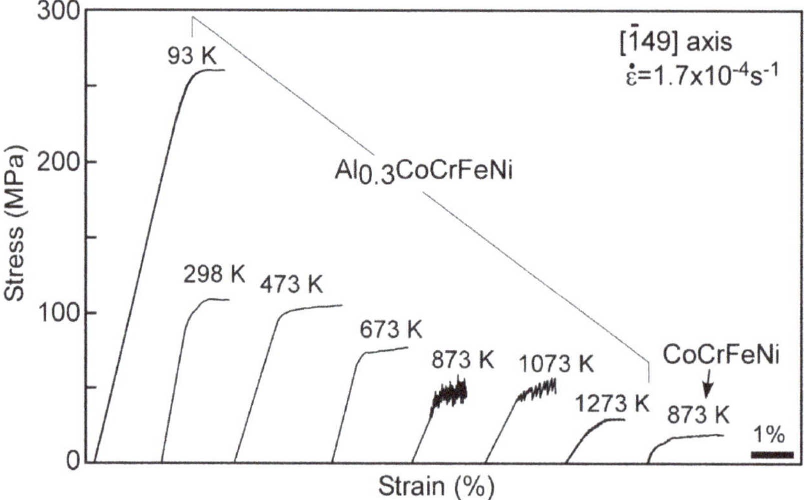

Figure 26. Stress vs. strain curves for the $Al_{0.3}CoCrFeNi$ and CoCrFeNi single crystals that were compressed at temperatures ranging from 93 to 1273 K (from −180 to 1000 °C) (reproduced from Reference [191] with permission).

Zhang et al. also examined the compression behavior of the $Al_{0.3}CoCrFeNi$ HEA [169]. The compression tests were performed on a Gleeble 3500 thermo-mechanical simulator (Dynamic Systems Inc., Poestenkill, NY, USA) at temperatures ranging from 400 to 900 °C at a strain rate of 10^{-3} s^{-1}. Figure 27a–d shows the true stress vs. true strain behavior for the samples tested at the above conditions. Serrations were observed in all of the deformation curves, and it was determined that the serration type was dependent on the test temperature. A summary of the serration type for each of the experimental conditions is displayed in Table 3. As can be seen in the table, Type-C serrations were observed at 400, 500, and 600 °C. At 700 °C, the serrated flow exhibited Type-B + Type-C serrations, whereas only Type-B serrations could be observed in the deformation curve at 800–900 °C. The authors reported that the number of relatively larger stress drops decreases with respect to the test temperature. Furthermore, it was also found that Type-A serrations were not observed in any of the deformation curves. These results appear to contradict those of other studies [8] where serrations have been observed to evolve from Type-A to Type-B, and then to Type-C with increasing temperature. Importantly, it was hypothesized that Type-C serrations were related to the nucleation and growth of twins, while Type-B serrations corresponded to the pinning (by mobile solute atoms) and unpinning of moving dislocations that is inherent in the DSA model [94,191].

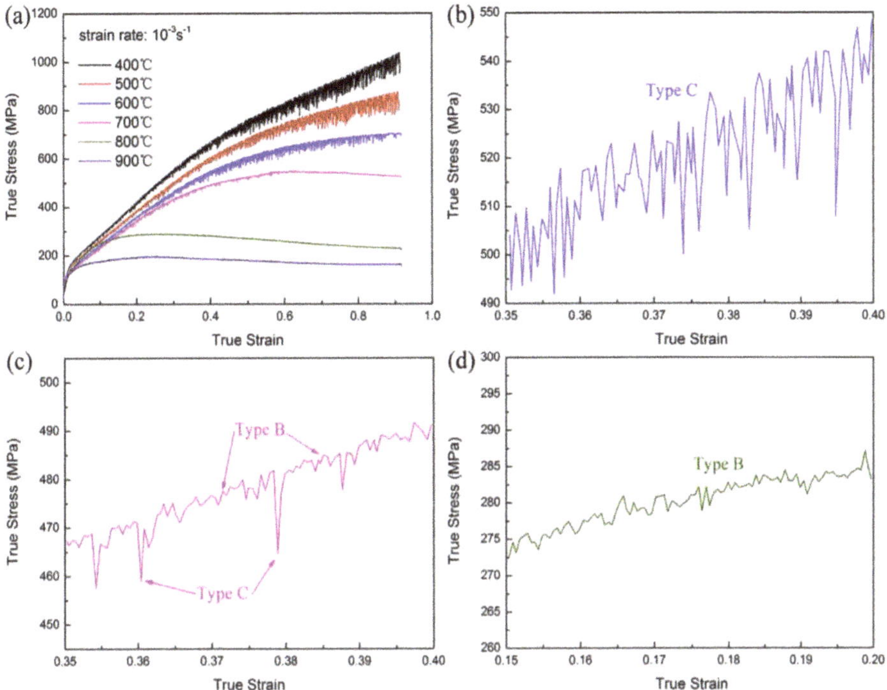

Figure 27. (**a**) True stress–strain curves of the $Al_{0.3}CoCrFeNi$ HEA tested at temperatures ranging from 400 to 900 °C and a strain rate of 10^{-3} s^{-1}, and the magnified regions of the serrations at temperatures of (**b**) 600 °C, (**c**) 700 °C, and (**d**) 800 °C (reproduced from Reference [169] with permission).

Table 3. Summary of the serration type exhibited by the $Al_{0.3}CoCrFeNi$ HEA during tension at a strain rate of 1×10^{-3} s^{-1} and temperatures of 400–800 °C (from Reference [169]).

Strain Rate (s^{-1})	Temperature (°C)	Serration Type
	400	C
	500	C
1×10^{-3}	600	C
	700	B + C
	800	B

Qiang et al. investigated the nanoindentation-serration behavior of as-cast and torsionally deformed (nano-grained) $Al_{0.3}CrCoFeNi$ HEA samples [287]. Room-temperature nanoindentations were performed by using a Hysitron Triboindenter TI950 (Bruker Corp., Santa Barbara, CA, USA) equipped with a Berkovich indenter. For the deformed samples, they underwent a compressive pressure of 10 GPa (RT) for 1, 3, and 10 rotations at one revolution per minute. For the nanoindentations, 15 indents were performed in which a loading rate of 250 μN/s was used. The samples were also examined, using different methods, such as XRD and Vickers-hardness tests. The XRD characterization revealed that all the samples contained a single-phase FCC structure. Vickers-hardness tests revealed that torsion led to a significant hardening of the alloy. The results also showed that the grain size was reduced from hundreds of microns in the as-cast state to tens of nanometers after 10 revolutions. It was also found that nanotwins and SFs were observed in the nanograins, despite the reported high SF energy.

In terms of mechanical behavior, it was surmised that, in the deformed nanograined HEA, the emission of Shockley partial dislocations at the grain boundaries and/or grain-boundary sliding may be the dominant deformation mechanism in the alloy. It was also reported that the as-cast HEA showed the slip-avalanche behavior, while the deformed samples did not exhibit any pronounced pop-ins (see Figure 28a). The lack of pop-ins exhibited by the deformed samples were attributed to the presence of affluent grain boundaries that act as extra mediators for plastic deformation and barriers for dislocation motion. The pop-ins that occurred in the as-cast sample exhibited relatively smaller pop-ins, as compared to alloys with a BCC lattice that displays only one pronounced pop-in [288,289]. This difference in the magnitude and the number of pop-ins was likely due to different dislocation-nucleation processes in the alloys, such as perfect dislocations in BCC crystals and partial dislocations in FCC crystals [289].

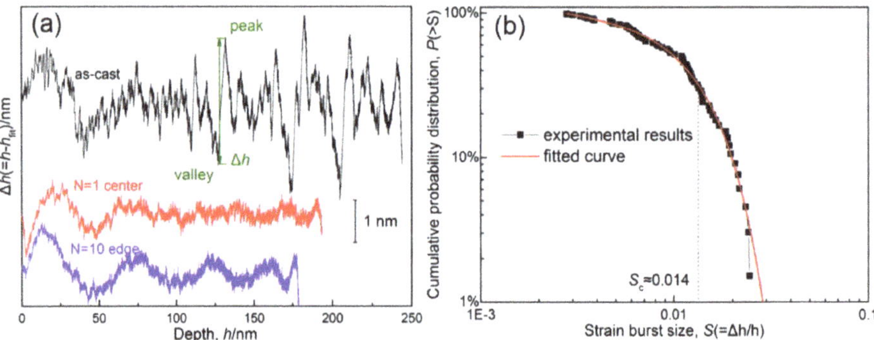

Figure 28. (**a**) Depth change of each pop-in event (Δh) as a function of indentation depth (h) and (**b**) the cumulative-probability distribution of the strain-burst size of the as-cast $Al_{0.3}$CrCoFeNi HEA (reproduced from Reference [287] with permission).

Figure 28b presents the cumulative-probability distribution of the displacement burst size, $S[P(>S) = AS^{-\beta}e^{-\frac{S}{S_c}}]$, for the as-cast $Al_{0.3}$CrCoFeNi HEA sample. As previously discussed, the distribution represents the percentage of pop-in events with the displacement burst size being larger than a given value, S. As can be seen in the figure, the probability of observing a pop-in of a size, S, significantly decreases with an increase in the size, S. It was also reported that the parameters S_c and β were approximately 0.014 ± 0.000 and 0.15 ± 0.02, respectively, which indicates that slip avalanches did indeed occur in the as-cast sample during the nanoindentation deformation.

3.4. $Al_{0.5}$CoCrFeNi HEA

Niu et al. examined the tension behavior of $Al_{0.5}$CoCrFeNi and CoCrFeNi alloys [94]. For their investigation, samples underwent tension at strain rates of 10^{-3} s^{-1}, 5×10^{-4} s^{-1}, and 10^{-4} s^{-1}, in addition to temperatures of 200–500 °C. Figure 29a–c presents the stress vs. strain curves for the

$Al_{0.5}CoCrFeNi$ HEA samples tested in the above conditions. As can be observed, the serrated flow consisted of Type-A, Type-B, Type-C, Type-A + Type-B, and Type B + Type-C serrations, which have been tabulated in Table 4. The results indicate that the serrations evolve from Type-A to Type-A + Type-B, and then to Type-B + Type-C, with a decrease in the strain rate and increase in the temperature. They also found that the serrated flow was a consequence of the DSA effect, and not strain-induced transformations or twinning. Statistical analysis was also performed on the serration behavior, and it was found that the distribution of the stress-drop magnitudes primarily exhibited the power-law behavior.

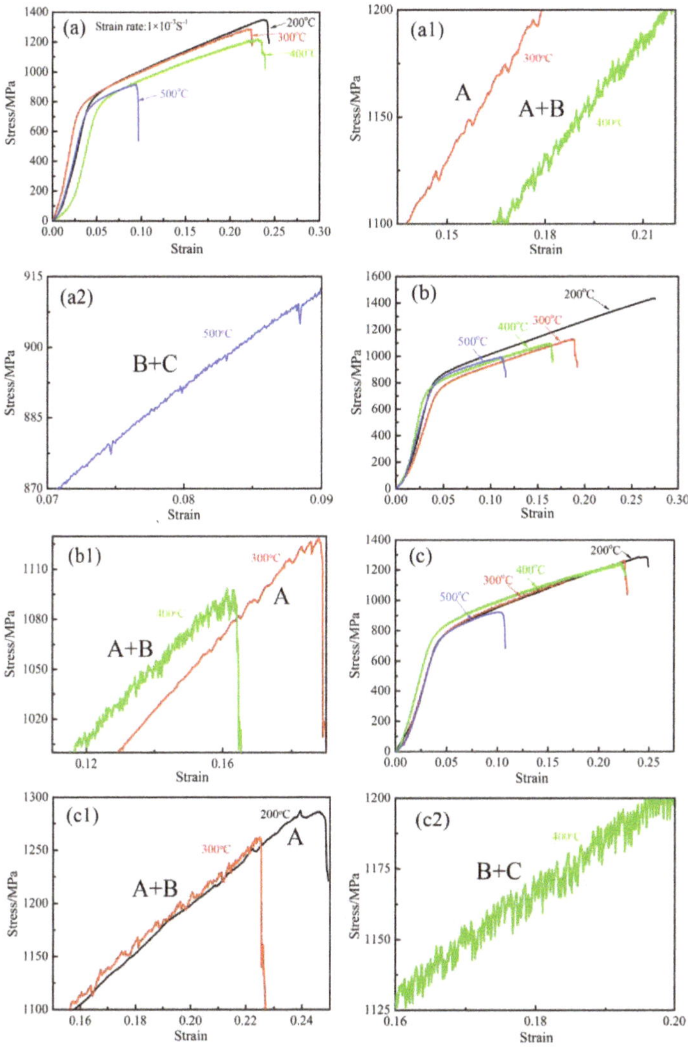

Figure 29. The tensile true stress–strain curves for the $Al_{0.5}CoCrFeNi$ HEA tested at temperatures of 200–500 °C and strain rates of (**a**) 10^{-3} s^{-1} where selected regions of the corresponding dynamic strain aging (DSA) curve are in (**a1**), (**a2**), and (**b**) 5×10^{-4} s^{-1} where selected regions of the corresponding DSA curve are in (**b1**), and (**c**) 10^{-4} s^{-1} where selected regions of the corresponding DSA curve are in (**c1**) and (**c2**) (reproduced from Reference [94] with permission).

Table 4. Strain rates, test temperatures, and serration types for the $Al_{0.5}CoCrCuFeNi$ HEA during tension testing (from Reference [94]).

Strain Rate (s^{-1})	Temperature (°C)	Serration Type
1×10^{-4}	200	A
	300	A + B
	400	B + C
5×10^{-4}	300	A
	400	A + B
1×10^{-3}	300	A
	400	A + B
	500	B + C

Figure 30a,b shows a comparison of the stress vs. strain behavior for the $Al_{0.5}CoCrFeNi$ and CoCrFeNi alloys. As can be seen, the Al-containing HEA exhibited significantly more pronounced serrations, as compared to the CoCrFeNi alloy. From this result, the authors surmised that the Al played a significant role in the DSA phenomenon in the HEA and was attributed to the atomic-size mismatch between the Al and the rest of the matrix atoms. In addition to the above finding, the authors reported some other interesting results. For example, it was observed that the tensile strength and elongation of the material both decreased with increasing temperature, signifying an embrittlement of the alloy. They noted that similar trends were reported in other Al-containing HEAs [290,291], and thus concluded that the trend of embrittlement may be related to the DSA phenomenon in HEAs.

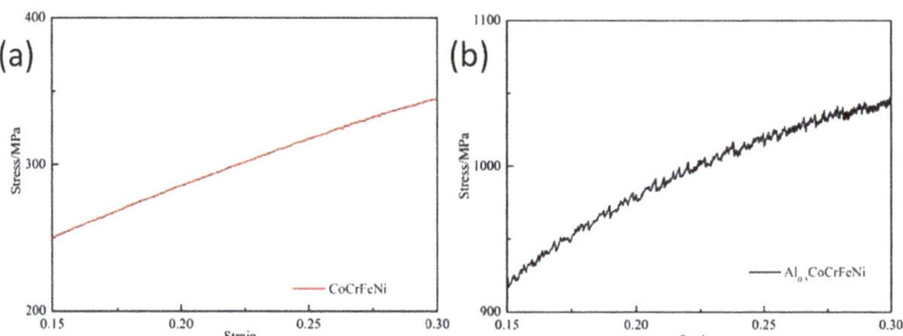

Figure 30. The engineering stress–strain curves for the (**a**) CoCrFeNi and (**b**) $Al_{0.5}CoCrFeNi$ HEAs during compression at a strain rate 10^{-3} s^{-1} and temperature of 400 °C (reproduced from Reference [94] with permission).

3.5. $Al_{0.7}CoCrFeNi$ HEA

Basu et al. investigated the nanoindentation-deformation behavior in a multi-phase $Al_{0.7}CrFeCoNi$ HEA that consisted of both BCC and FCC phases [292]. The energy-dispersive spectroscopy (EDS) characterization revealed that the BCC phases were composed of the (Ni, Al)-rich ordered B2 phase and (Fe, Cr)-rich disordered A2 phase. Figure 31a presents the load vs. displacement curve for four different indents that were performed inside the BCC HEA grain. Figure 31b presents a magnification of the indents, which features the elastic to plastic transitions, signifying a clear transition from Hertzian behavior [293]. The plastic regime of the nanoindentation curve that is displayed in Figure 31b consists of multiple pop-in events that are characteristic of the serrated flow in the alloy. The serrated flow was thought to be a consequence of the resistance provided by the interfacial-strengthening mechanisms between the soft A2 phase and the elastically stiffer B2 matrix against the dislocations.

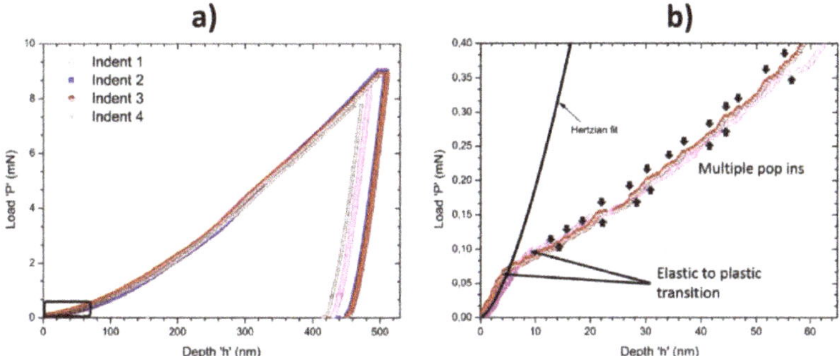

Figure 31. The (**a**) load vs. displacement curves for four different indents that were made inside the body-centered-cubic (BCC) grain in the HEA over indentation depths of 0 to 500 nm, and the (**b**) highlighted region, which shows the data for two randomly selected indent curves. Here, the graph features the elastic to plastic transition and contains multiple pop-ins (reproduced from Reference [292] with permission).

3.6. $Al_xNbTiMoV$ HEA

Chen et al. investigated the compressive behavior of a BCC $Al_xNbTiMoV$ HEA [179]. Here, the molar ratio, x, of Al ranged from 0 to 1.5. For their experiments, samples underwent compression tests at RT and strain rates of 5×10^{-5} s^{-1}, 5×10^{-4} s^{-1}, 5×10^{-3} s^{-1}, and 5×10^{-2} s^{-1}. Figure 32 displays the compressive engineering stress–strain curves for the alloy that was compressed at a strain rate of 5×10^{-4} s^{-1}. As can be seen, there were observable serrations for x = 0, 0.25, and 0.5 around the yielding point. The authors surmised that the serrated flow was a consequence of the interactions between moving dislocations and Cottrell atmospheres [294]. In this context, the serrations occur due to the drag effect caused by Cottrell atmospheres that surround a dislocation and lock it in place. It was found that the yield strength of the alloy increased with an increase in x for this range of molar ratios. Furthermore, these values of x corresponded to atomic-size differences and lattice parameters that were greater than 3.22% and 3.191 Å, respectively. This result suggests that in the Al-containing HEA, there is an atomic size disparity and lattice volume, which, when exceeded, prohibits dislocation locking from occurring. As for the samples that did not contain Al (x = 0), serrations were observed for all the given strain rates. Furthermore, the amplitudes of the serrations were relatively larger for the intermediate strain rates of 5×10^{-4} s^{-1} and 5×10^{-3} s^{-1}, while they were barely noticeable for the lowest strain rate of 5×10^{-5} s^{-1}.

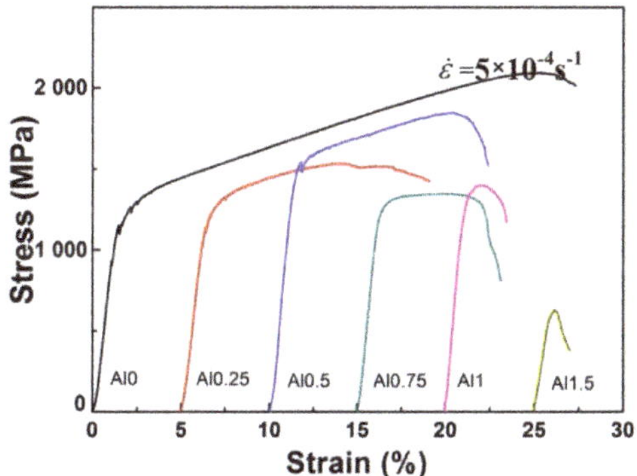

Figure 32. The engineering stress vs. engineering strain for the $Al_xNbTiMoV$ (x = 0, 0.2, 0.5, 0.75, 1, and 1.5) HEAs (reproduced from Reference [179]).

It is interesting to note that the presence of serrations in the sample which does not contain Al is in contrast with the results as reported by Yasuda et al., where serrations were observed in $Al_xCoCrFeNi$ HEA when x = 0 [191]. The serrations in the NbTiMoV alloy (i.e., x = 0) may be due to various reasons. For instance, the presence of interstitial impurities, such as C, may pin dislocations [295]. On the other hand, the observed serrations may be due to the other constituent elements pinning the dislocations, which supports the theory that any atom in the matrix can act as a solute in HEAs. Therefore, more work is required to elucidate the effect of impurities and additives, such as Al and C, on the serrated-flow behavior in HEAs.

3.7. $Al_5Cr_{12}Fe_{35}Mn_{28}Ni_{20}$ HEA

The serration behavior was examined in an $Al_5Cr_{12}Fe_{35}Mn_{28}Ni_{20}$ HEA that was tension tested at temperatures of 300 and 400 °C and a strain rate of 1×10^{-4} s^{-1} [167]. Here, the gauge section of the specimens was 3 mm in diameter and 12 mm in length. The stress vs. displacement and the corresponding CCDF curve are displayed in Figure 33a,b. It is apparent from the graph that the samples exhibited serrations at both temperatures. Moreover, the magnitudes of the serrations were found to increase as the sample deformed. From the results of the statistical analysis, as displayed in Figure 33b, the magnitude of the largest stress drop decreased with an increase in the temperature. Importantly, this result is consistent with the reported trend predicted by the simple mean-field model [222,252,256]. The authors hypothesized that the relatively smaller slip sizes at the higher temperature is attributed to the increased thermal-vibration energy of the pinning solute atoms. More specifically, the pinning effect of the solute atoms on the dislocations is reduced at higher temperatures, since the solute atoms have a greater tendency to "shake away" from their low-energy sites for pinning [167,187].

3.8. $Ag_{0.5}CoCrCuFeNi$ HEA

Laktionova et al. investigated the compression behavior of the $Ag_{0.5}CoCrCuFeNi$ HEA at temperatures ranging from 4.2 to 300 K (from −269 to 27 °C) [296]. Here, cylindrical samples with a diameter of 2 mm and length of 4 mm were undergoing compression testing at a strain rate of 4×10^{-4} s^{-1}. The loading of the sample was terminated when the strain reached values of ~20–30%. To perform the testing at temperatures below 77 K, samples were cooled with helium vapor, while the nitrogen vapor was used for temperatures ranging from 77 to 300 K. Figure 34 displays the stress–strain curves for the samples tested at the prescribed conditions. As can be observed, the samples exhibited the serrated flow at testing temperatures of 4.2 and 7.5 K, while serrations were not observed at any of the higher-temperature conditions. Furthermore, the strain at which the serrations commenced increased with the temperature. In particular, serrations began to appear at strains of 1.2% and 4.5% for the samples tested at 4.2 and 7.5 K, respectively. It was also reported that the stress-drop magnitude increased with an increase in the strain. For example, at 4.2 K, the magnitude increased from 20 MPa at a strain of ~2% to 67 MPa for a strain of roughly 23%.

3.9. CoCrFeMnNi HEA (Cantor Alloy)

Carroll et al. examined the serration behavior in multiple alloys, including CoCrFeMnNi HEA with CoCrFeNi HEA, CoFeNi medium-entropy alloy, CoNi low-entropy alloy (LEA), and pure Ni [153]. The above samples had configurational entropies ranging from 0R for the pure Ni to 1.61R for the CoCrFeMnNi HEA. Here, the specimens underwent tension tests with strain rates and test temperatures ranging from 1×10^{-5} to 1×10^{-2} s^{-1} and 300 to 700 °C, respectively. Figure 35 displays the stress–strain curves for the CoCrFeMnNi HEA tested at a strain rate of 1×10^{-4} s^{-1} for temperatures of 300–600 °C [181]. The observed serrated flow consisted of Type-A, Type-B, and Type-C serrations, where the serration type depended upon the temperature. Table 5 displays a summary of the serration type and the corresponding temperature for the tests performed at strain rates of 1×10^{-4} to 1×10^{-2} s^{-1} and 300 to 600 °C. For the sample tested at the highest strain rate, only Type-A serrations were observed. Furthermore, the sample that was tested at 1×10^{-4} s^{-1} exhibited serrations that were Types A and C at the highest and lowest temperatures, respectively. For the intermediate temperatures of 400 and 500 °C, Type-B serrations were observed. It should be noted that a similar trend was observed in $Al_{0.5}CoCrCuFeNi$ HEA [180], as discussed previously.

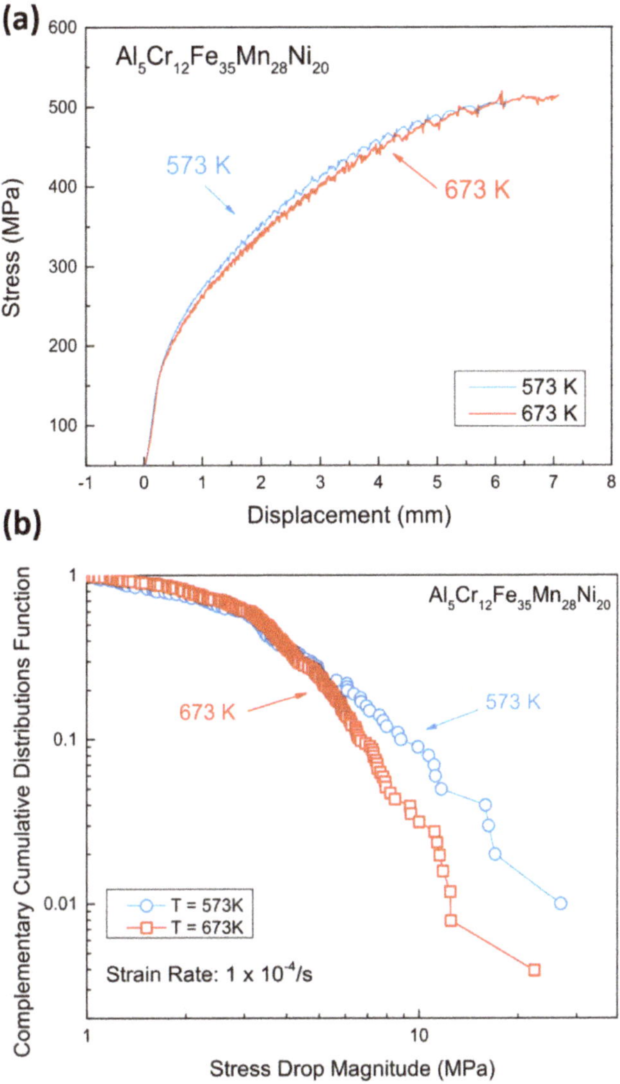

Figure 33. (**a**) Stress vs. displacement curves for the $Al_5Cr_{12}Fe_{35}Mn_{28}Ni_{20}$ HEA tension tested at temperatures of 573 and 673 K, at a strain rate of 1×10^{-4} s^{-1} and (**b**) the results of the CCDF analysis for the serration events for the $Al_5Cr_{12}Fe_{35}Mn_{28}Ni_{20}$ HEA in tension experiments under a constant strain rate of 1×10^{-4} s^{-1} (reproduced from Reference [167] with permission).

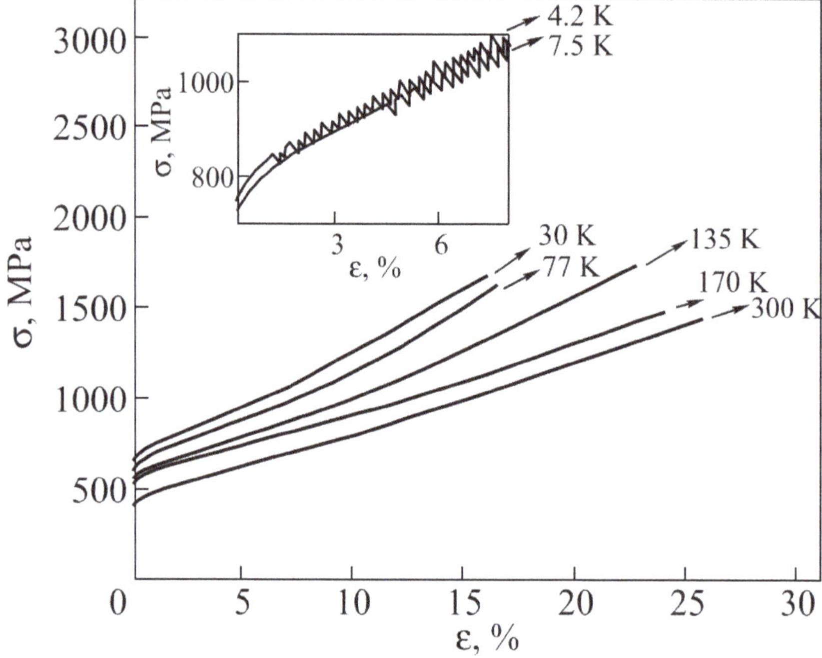

Figure 34. The stress–strain curves for the $Ag_{0.5}CoCrCuFeNi$ HEA that was undergoing tension testing at temperatures ranging from 4.2 to 300 K, at a strain rate of 4×10^{-4} s^{-1}. The inset displays a close-up of the serrated flow that occurred in the specimens tested at 4.2 and 7.5 K. (Reproduced from Reference [296] with permission).

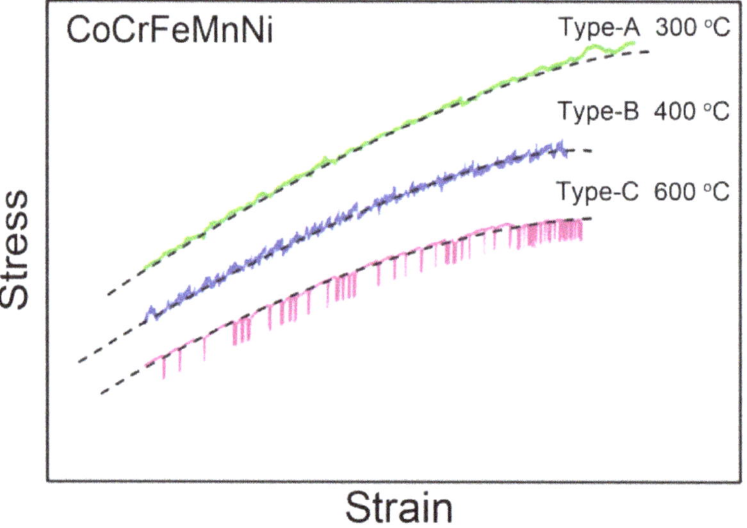

Figure 35. Stress vs. strain curves for the CoCrFeMnNi HEA tension tested at a strain rate of 1×10^{-4} s^{-1} and temperatures of 300–600 °C (reproduced from Reference [181]).

Table 5. Summary of the serration type exhibited by the CoCrFeMnNi HEA during tension at strain rates from 1×10^{-4} to 1×10^{-2} s^{-1} and temperatures of 300–600 °C (from Reference [153]).

Strain Rate (s^{-1})	Temperature (°C)	Serration Type
1×10^{-4}	300	A
	400	B
	500	B
	600	C
1×10^{-3}	300	A
	400	A
	500	B
	600	B
1×10^{-2}	400	A
	500	A
	600	A

It was found that, for a strain rate of 1×10^{-4} s^{-1}, the pure Ni and CoNi alloy did not exhibit serrations at any of the prescribed temperatures. Furthermore, the temperature range for which serrations were observed increased with an increasing chemical complexity of the alloy, such that the CoCrFeMnNi HEA was had the largest range of temperatures. From here, the authors suggested that the structure of the HEA prevents thermal vibrational from destroying the pinning effect, thereby allowing mobile solute atoms to pin the moving dislocations at lower temperatures.

Fu et al. examined the deformation behavior of a CoCrFeMnNi HEA that underwent homogenization, cold-rolling, and recrystallization [163]. The samples were homogenized at 1100 °C, for 24 h, in vacuum, and then subsequently cold-rolled, which led to a reduction in thickness of 40%. After cold-rolling, the sheets were recrystallized at 900 °C for 1 h. After the samples were fabricated, they underwent tension testing at temperatures ranging from RT to 800 °C and strain rates from 1.0×10^{-5} to 5.0×10^{-3} s^{-1}. The engineering stress vs. strain data for the samples tested at a strain rate of 3.0×10^{-4} s^{-1} and temperatures ranging from RT to 800 °C can be observed in Figure 36a,b. As can be seen in the figures, the serrations are well defined for temperatures ranging from 300 to 600 °C. Furthermore, the serration type was dependent on the test temperature (see Figure 36b). At 300 °C, Type-A serrations were observed in the graph, while Types A + B were seen in the sample tested at 400 °C. On the other hand, Type-B serrations occurred in the sample that was tested at 550 °C, while Type-C serrations were observed at 600 °C.

It was also found that the critical plastic strain for the onset of serrations was significantly affected by both the temperature and the applied strain rate. For example, at a strain rate of 3.0×10^{-4} s^{-1}, the critical strain decreased monotonically with respect to the temperature. In contrast, the critical strain increased with an increase in the strain rate. From the critical strain, it was determined that, for the above strain rate, there were two temperature regimes for the activation energy of the serrated flow. For the first region, which corresponded to test temperatures of 300–500 °C, the activation energy was 116 kJ/mol, and the solute pinning of dislocations was controlled by pipe diffusion. As for the second region, which occurred at temperatures ranging from 500 to 600 °C, the activation energy for serrations was 296 kJ/mol. In this region, the pinning-and-unpinning process was dominated by a cooperative lattice diffusion mechanism in which Ni was the rate-controlling constituent.

It should be noted that serrations were not observed for temperatures either below 300 °C or above 600 °C. For the lower-temperature conditions, the lack of observable serrations is most likely a consequence of diffusing solute atoms that are too slow to catch and pin mobile dislocations [153]. At higher temperatures, it is thought that the thermal vibration of atoms is large enough as to prohibit the effective locking of dislocations [155,297].

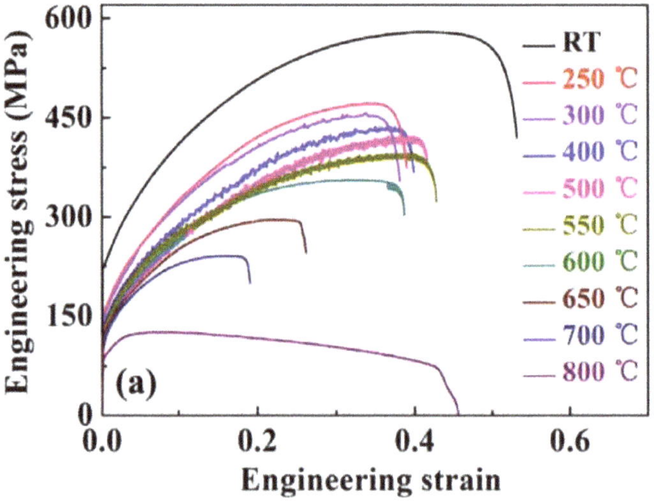

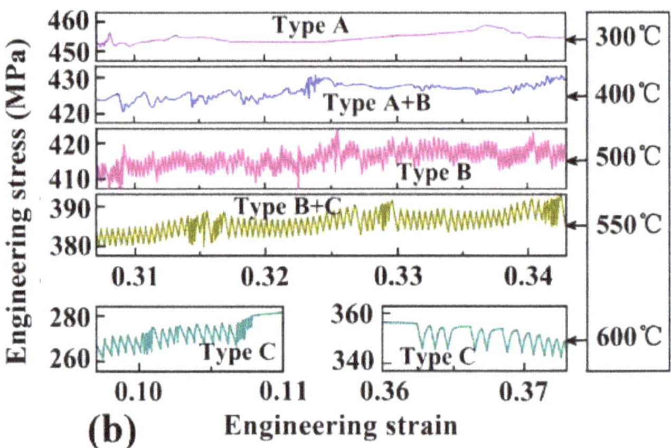

Figure 36. (**a**) Engineering stress–strain curves for the CoCrFeMnNi HEA performed at temperatures ranging from RT to 800 °C and a strain rate of 3×10^{-4} s^{-1} and (**b**) the partially enlarged segments for the curves that exhibited serrations (reproduced from Reference [163] with permission).

Figure 37a,b displays the engineering stress vs. strain curves for the samples tested at strain rates ranging of 1.0×10^{-5}–5.0×10^{-3} s^{-1} at a temperature of 500 °C. From the figures, it is apparent that the serration type is dependent on the strain rate. At the lowest strain rate, namely 1.0×10^{-5} s^{-1}, the sample exhibited Type-B + Type-C serrations, while Type-A serrations were observed at the highest strain rate. At intermediate strain rates, the stress-drop behavior exhibited Type-A + Type-B (1.0×10^{-5}) and Type-B (1.0×10^{-4}) serrated flow. Table 6 features the serration type for the temperature and strain-rate conditions of the experiments.

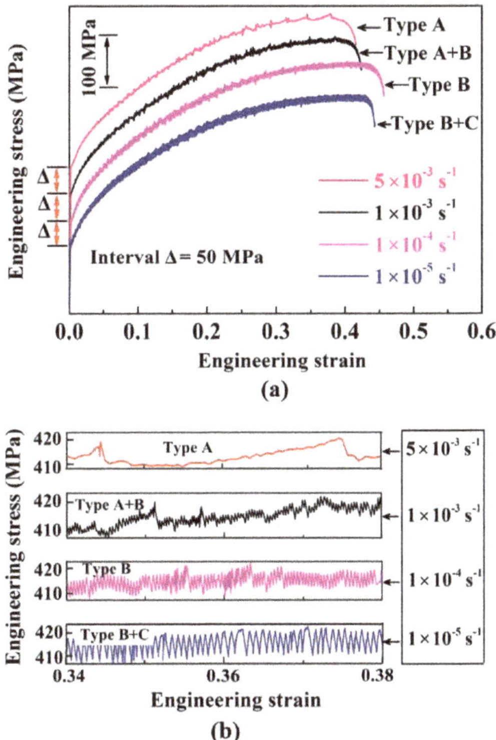

Figure 37. (**a**) The engineering stress vs. engineering strain curves for the sample tested at strain rates ranging from 1.0×10^{-5} to 5.0×10^{-3} s^{-1}, at a temperature of 500 °C. (**b**) The enlarged regions of (**a**) for engineering strain values ranging from 0.34 to 0.38 (reproduced from Reference [163] with permission).

Table 6. Summary of the serration type exhibited by the CoCrFeMnNi HEA during tension at strain rates of 1×10^{-5} s^{-1}–5×10^{-3} s^{-1} and temperatures of 300–600 °C (from Reference [163]).

Strain Rate (s^{-1})	Temperature (°C)	Serration Type
1×10^{-5}	500	B + C
1×10^{-4}		B
3×10^{-4}	300	A
	400	A + B
	500	B
	550	B + C
	600	C
1×10^{-3}	500	A + B
5×10^{-3}		A

In a later study, Fu et al. performed a similar study on the CoCrFeMnNi HEA, but this time they examined samples that had been cold rolled such that the sample thicknesses were reduced by 20%, 30%, and 40% [168]. Furthermore, the sheets were recrystallized at both 900 and 1000 °C (1 h) for each cold-rolling condition. XRD revealed that the alloy retained a simple FCC solid-solution phase after cold rolling and recrystallization. Samples undergoing quasi-static tensile tests were performed at temperatures ranging from RT to 800 °C. The fractured surfaces of the samples were examined by using SEM.

Figure 38a–c features the engineering stress vs. engineering strain for the HEA samples that were tested at temperatures ranging from RT to 800 °C and rolling conditions of 20%, 30%, and 40% reduced thicknesses. In the sample recrystallized at 900 °C for 1 h, the serrated flow was observed at the intermediate temperatures of 400 and 600 °C for all the rolling conditions. Interestingly, there was a discontinuous gap in the serrated flow for the samples tested at 600 °C. Similar to other studies involving the same alloy [153,163], the serrations appear to be of Types A and B at 400 °C, while they were characteristic of Type-C behavior at 600 °C. Moreover, for the samples tested at 600 °C, the discontinuous feature in the serrated flow was the most pronounced in the sample that had a thickness reduction of 40%. No serrations were observed in the samples tested at RT or 800 °C. The lack of observed serrations at 800 °C was attributed to the softening of the alloy at the higher temperature. As for the lack of observable serrations at RT, this trend was again due to the relatively low speed of the moving solutes, thus rendering them unable to reach and pin dislocations [163].

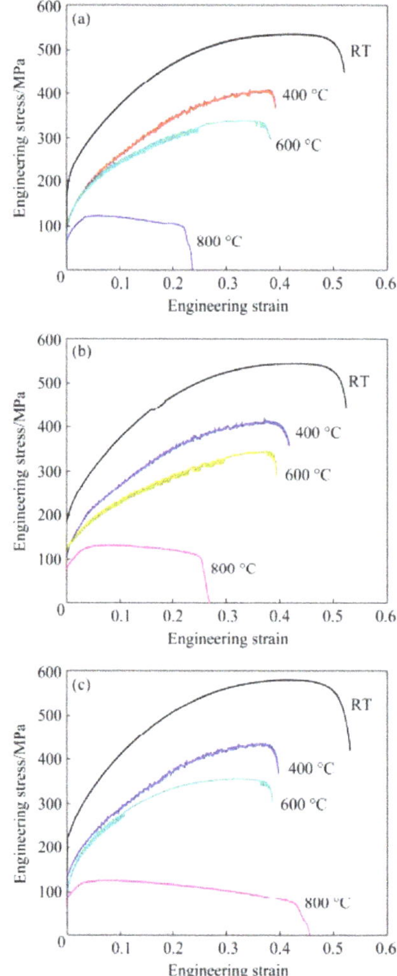

Figure 38. Graphs of the engineering stress–strain behavior for the recrystallized (900 °C, 1 h) and cold-rolled CoCrFeMnNi HEA with a rolling ratio of (**a**) 40%, (**b**) 30%, and (**c**) 20% (reproduced from Reference [168] with permission).

The SEM characterization revealed massive annular striation patterns located around the dimples on the fractured surface for the specimens tested at 400 and 600 °C. These striation patterns were attributed to a decrease in the ductility of the alloy. In contrast, no such striations were observed in the sample tested at 800 °C.

Wang et al. examined the deformation behavior of a spark-plasma-sintered CoCrFeMnNi HEA during high-strain-rate compression [166]. The results of the XRD characterization determined that the samples had a simple FCC structure. For the mechanical testing, samples were subjected to room-temperature impact tests, using a split-Hopkinson pressure bar system and strain rates ranging from 1×10^3 to 3×10^3 s^{-1}. During dynamic deformation, these strain rates amounted to loading rates that varied between 1200 and 2800 s^{-1}. Figure 39 displays the true stress vs. true strain behavior for the specimens, and as can be seen, there are observable serrations in the graph. Furthermore, these serrations appear comprise Type-A serrations. It was also determined that the serration behavior in the HEA is sensitive to changes in the applied-loading rate. Additionally, the general temperature of the sample approached values as high as 1300 K during testing. In this scenario, local hotspots in the matrix can weaken intergranular bonding, leading to the formation of cracks and microvoids in the severe-deformation zone.

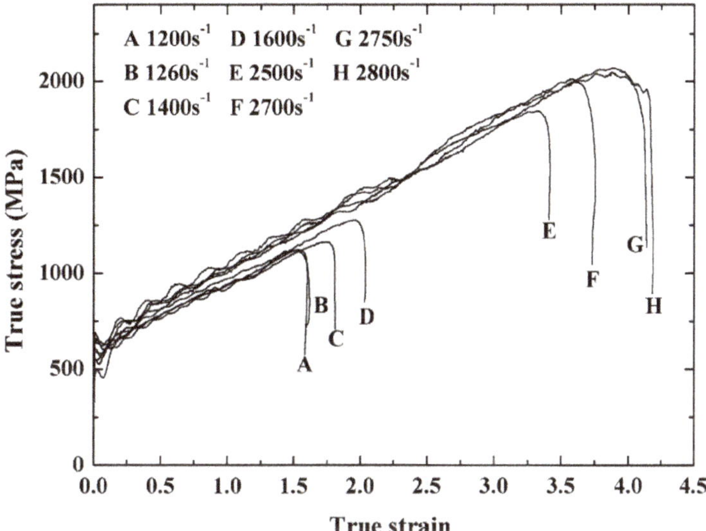

Figure 39. Compressive true stress–strain curves of the CoCrFeMnNi HEA at loading rates that varied from 1200 to 2800 s^{-1} (reproduced from Reference [166] with permission).

From the results, the authors hypothesized the following process involved in initializing the serrated flow. During the initial stages of deformation, dislocations form and then subsequently accumulate along the grain boundaries that become elongated during compression. As the deformation continues, the grains become more elongated, leading to a further decrease in the width of the grains. Meanwhile, the generated thermal hotspots weaken the intergranular bonding and ultimately lead to an eventual collapse of the grain boundaries. Consequently, numerous microvoids form along the grain boundaries, and shear bands are generated in the specimen, resulting in the observed serrated flow.

Before moving on, it is important to note that the split Hopkinson pressure bar test depends on the one-dimensional wave propagation in the bar [298]. The relatively smooth nature of the stress fluctuations, as can be seen in Figure 39, is indicative of poor pulse shaping during the test. This result suggests that the fluctuations may be due to machine noise rather than an actual physical metallurgical phenomenon.

Tirunilai et al. compared the deformation behavior of the CoCrFeMnNi HEA and pure Cu at cryogenic temperatures [299]. Here, tensile tests were performed at RT (295 K), 77, 8, and 4.2 K. For the experiments performed at 4.2 K, the samples were placed in a liquid He bath, whereas samples were placed in a vacuum cryostat when tested at 8 K. To examine the serration behavior in the HEA, tensile tests were performed at mean plastic strain rates of 6×10^{-5}, 3×10^{-4}, and 1×10^{-3} s^{-1}. A sample rate of 10 Hz was used for the data acquisition rate.

Figure 40a features a plot of the engineering stress–strain curve for the HEA tested at RT, 77 K, and 4.2 K and strain rate of 3×10^{-4} s^{-1}. As can be seen, there were pronounced serrated flows at 4.2 K whereas no apparent serrations were present at temperatures of 77 K and RT. Furthermore, deformation twinning was observed in the samples tested at 4.2 and 77 K. It was also found that the ductility in the sample deformed at 4.2 K remained relatively high.

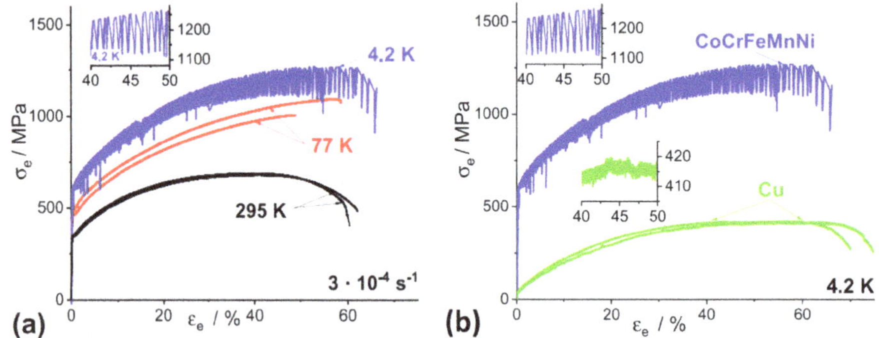

Figure 40. The stress–strain curves for the (**a**) CoCrFeMnNi HEA under tensile load at RT, 77, and 4.2 K. (**b**) CoCrFeMnNi HEA and pure Cu at 4.2 K (reproduced from Reference [299] with permission).

Interestingly, the authors reported that such serrated flow behavior was significantly less pronounced in pure Cu when tested at a strain rate of 3×10^{-4} s^{-1} at 4.2 K, as shown in Figure 40b. For instance, the serration amplitude was ~150 MPa in the HEA, whereas it was only 5 MPa in the Cu metal. The significantly more pronounced serrations observed in the HEA (as compared to the pure Cu) were attributed to the impact of the solid solution in the alloy. It was also suggested that the discrepancy in the serration behavior may be a result of the differences between the properties of the material systems, such as yield strength, ultimate tensile strength, work-hardening, thermal conductivity, and heat capacity.

Figure 41 features the stress vs. strain data for the samples that were tested at 8 K and strain rates ranging from 6×10^{-5} to 1×10^{-3} s^{-1}. As can be observed, the characteristics of the serrated flow were affected by the strain rate. For example, the time between stress-drops apparently decreased with an increase in the strain rate. The insets of the figure suggest that, for strain rates of 6×10^{-5} and 3×10^{-4} s^{-1}, the serrated flow consisted of Type-B serrations. On the other hand, the serrations were reminiscent of Type-D behavior at the highest strain rate.

Guo et al. examined the effects of carbon impurities on the deformation behavior of a CoCrFeMnNi HEA [192]. Here, the undoped and doped (0.93 atomic % C) samples were subjected to tension testing at RT and a strain rate of 1.6×10^{-3} s^{-1}. Figure 42a,b compares the true stress vs. strain of the carbon-doped and undoped HEA. As can be observed, serrations are clearly present in the deformation curve for the doped sample. This result is in contrast with the undoped specimen where serrations were absent. A magnified view of the serrated-flow behavior in the doped sample is displayed in Figure 42b. Here, the serrations exhibit a stair-step type pattern. The authors reported that the serrated flow exhibited Type-A serrations, although they could have instead been Type-D serrations (see Figure 10) [182].

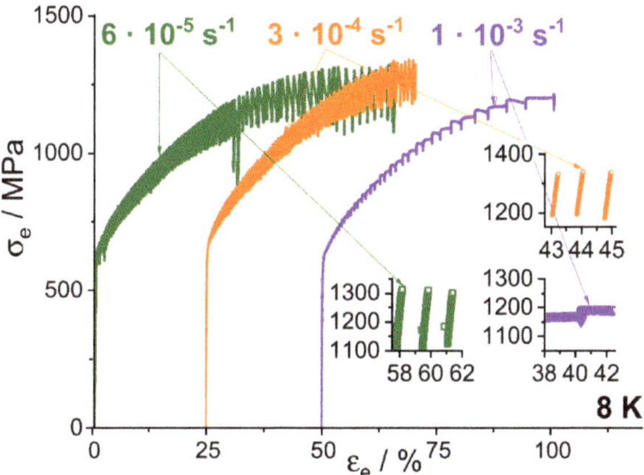

Figure 41. The engineering stress vs. strain curves for the CoCrFeMnNi HEA compressed at strain rates ranging from 6×10^{-5} to 1×10^{-3} s^{-1} at 8 K (reproduced from Reference [299] with permission).

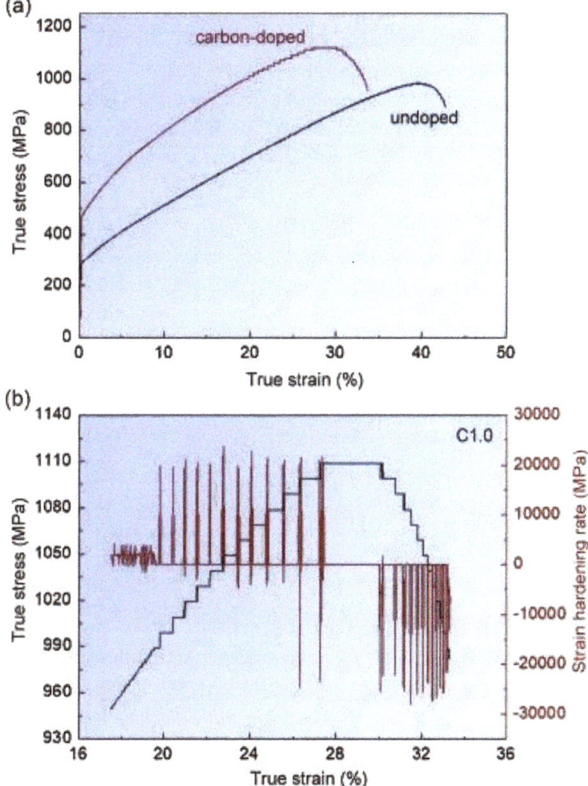

Figure 42. (**a**) True stress–true strain curves for CoCrFeMnNi HEA tested with and without carbon doping. The (**b**) enlarged image of the true stress–true strain curve and strain hardening rate–true strain curve in the strain range of 17.5–33.5% for the carbon-doped HEA. (Reproduced from Reference [192] with permission.)

The occurrence of the serrated flow in the HEA was attributed to the interplay among the carbon, SFs, and moving dislocations. The sharp increase in the stress during a serration event corresponds to the alteration of the short-range order by the dislocations that is accompanied by the migration of C from octahedral sites to tetrahedral sites and SFs. Since it is energetically unfavorable for the C to remain in the SF, the C will hop back into the octahedral site. Consequently, the SF energy is decreased, thus resulting in the accelerated plastic deformation that is characterized by the plateau region between successive serrations.

3.10. CoCuFeNiTi HEA

Samal et al. examined the serrated flow in an equiatomic CoCuFeNiTi HEA [162]. Here, cylindrical samples underwent isothermal hot-compression tests at temperatures ranging from 800 to 1000 °C and strain rates between 10^{-3} and 10^{-1} s^{-1}. Figure 43a,b presents the true stress vs. true strain for the samples compressed at strain rates of 10^{-1} and 10^{-3} s^{-1}, respectively. For both strain rates, the amplitude of the serrations increased with strain rate (at a given temperature) and a decrease in temperature (at a given strain rate). However, for the sample compressed at a strain rate of 10^{-3} s^{-1}, the serrations were only prominent for lower temperatures. The authors surmised that the decrease in the size of serrations at higher temperatures was a consequence of the increased thermal-vibration energy required to pin solute atoms. Such a hypothesis has also been discussed in a previous investigation [153]. Furthermore, they suggested that the increase in the serration amplitude at the lower strain rates was due to an increase in the time allotted for atoms to lock dislocations. Table 7 displays a summary of the serration types observed and their corresponding temperatures and strain rates. As can be seen, the serrations type was dependent on the experimental conditions. From the table, it is evident that, at the higher strain rate, the serration type evolved from Types A to A + B with temperature, whereas the serration type transitioned from Type-A + Type-B to Type-B at the lower strain rate.

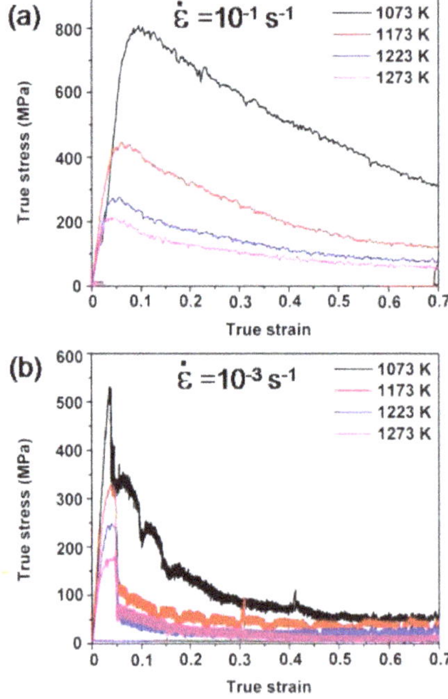

Figure 43. True stress–true strain plots at different temperatures and strain rates of (**a**) 10^{-1} and (**b**) 10^{-3} s^{-1} (reproduced from Reference [162] with permission).

Table 7. Summary of the serration type exhibited by the CoCuFeNiTi HEA during compression at strain rates of 1×10^{-3} s^{-1}–5×10^{-1} s^{-1} and temperatures of 800–1000 °C (from Reference [162]).

Strain Rate (s^{-1})	Temperature (°C)	Serration Type
1×10^{-3}	800	A + B
	900	A + B
	950	B
	1000	B
1×10^{-1}	800	A
	900	A + B
	950	A + B
	1000	A + B

3.11. CoCuFeMnNi HEA

Sonkusare et al. investigated the deformation behavior of a single-phase FCC CoCuFeMnNi HEA [300]. For their experiments, the HEA samples underwent RT compression tests at strain rates of 10^{-3} and 3×10^3 s^{-1}. To achieve the higher strain rate, a gas gun (6 kg/cm^2) was used to propel a 300 mm long striker bar into the sample. Figure 44 displays the true stress vs. true strain plot. From the graph it is apparent that the serrated flow occurred in the sample that was tested at the higher strain rate, while there were no observable serrations present in the deformation behavior of the sample compressed at a strain rate of 10^{-3} s^{-1}. Electron backscatter diffraction (EBSD) characterization revealed some interesting results. For instance, lenticular deformation twins were observed in the sample tested at the higher strain rate, suggesting that deformation twinning was likely the mechanism responsible for the serrated flow at the higher strain rate condition. Moreover, as compared to the lower strain-rate condition (10^{-3} s^{-1}), the microstructure for the sample tested at the higher strain rate consisted of relatively smaller-sized grains, a higher percentage of higher angle grain boundaries, and a smaller density of geometrically necessary dislocations.

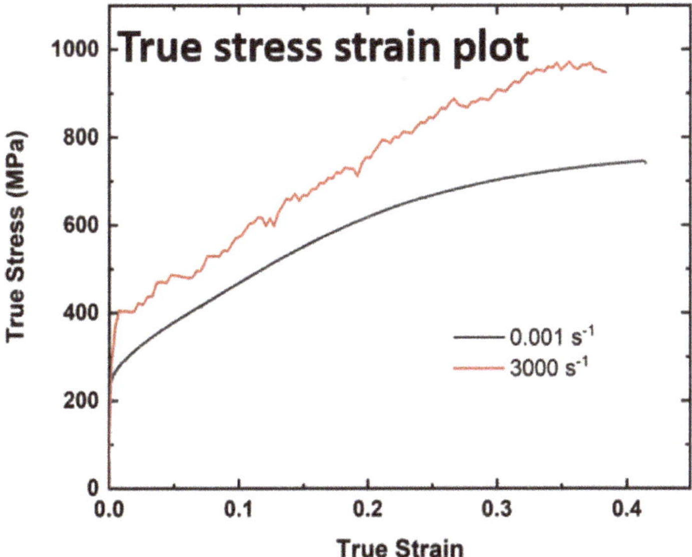

Figure 44. The true stress vs. true strain for the CoCuFeMnNi HEA compression tested at strain rates of 10^{-3} s^{-1} and 3×10^3 s^{-1} (reproduced from Reference [300] with permission).

3.12. CoCrFeNi HEA

The deformation behavior, during tension, was examined in a CoCrFeNi HEA for temperatures ranging from 4.2 to 293 K [159]. For the experiments, samples were tested at a strain rate of 10^{-3} s^{-1}. To cool the samples, liquid nitrogen (77 and 200 K), and liquid helium (4.2 K) were used. Figure 45a–d shows the results of the microstructural characterization, as determined by optical microscopy, XRD, and TEM. As can be observed from Figure 45a, the as-cast sample contained nearly equiaxed grains. Subsequent analysis determined that the grains consisted of a mean size of about 13 μm. The results of the XRD characterization, as displayed in Figure 45b, indicates that the as-cast sample consisted of an FCC structure. Figure 45c,d presents the TEM results, which reveals that both dislocations and {111} twins were present in the microstructure of the as-fabricated material.

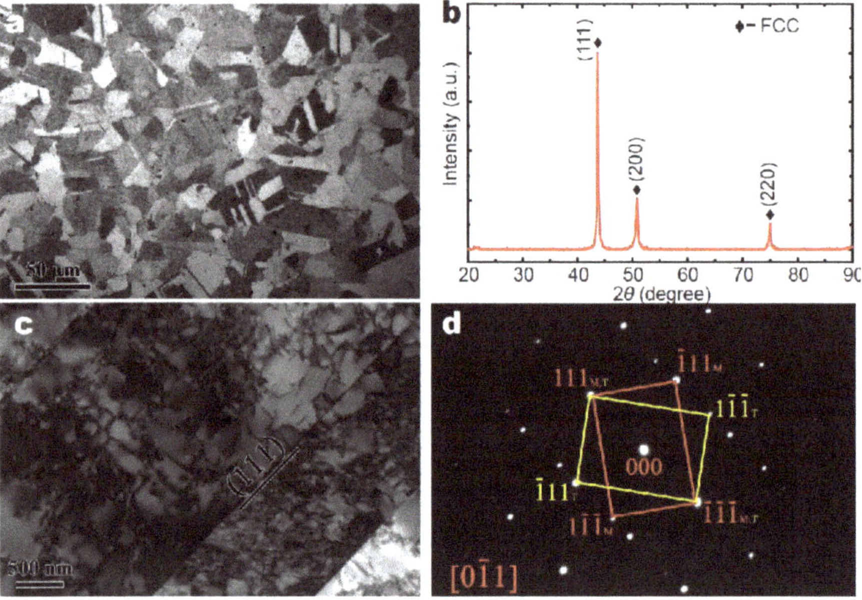

Figure 45. The results of the (**a**) optical microscopy imaging, (**b**) XRD characterization, (**c**) TEM image, and (**d**) selected area electron diffraction results for the CoCrFeNi HEA (reproduced from Reference [159] with permission).

Figure 46 displays the engineering stress vs. engineering strain for the samples tested at temperatures ranging from 4.2 to 293 K for a strain rate of 10^{-3} s^{-1}. It is evident from the figure that the serrated flow was only observed in the samples deformed at temperatures of 4.2 and 20 K. Upon analysis, the largest Lyapunov exponent for both conditions was found to be positive (0.05 for 4.2 K, and 0.001 for 20 K), indicating that the serration behavior was chaotic (unstable dynamics). Interestingly, TEM characterization revealed that the sample tested at 4.2 K underwent an FCC to HCP phase transformation during tension testing. Moreover, many high-density defects were present, such as nano-twinning, SFs, and the HCP stacking. From the above results, the authors concluded that the unstable behavior exhibited by the sample tested at 4.2 K was a consequence of the FCC to HCP transformation.

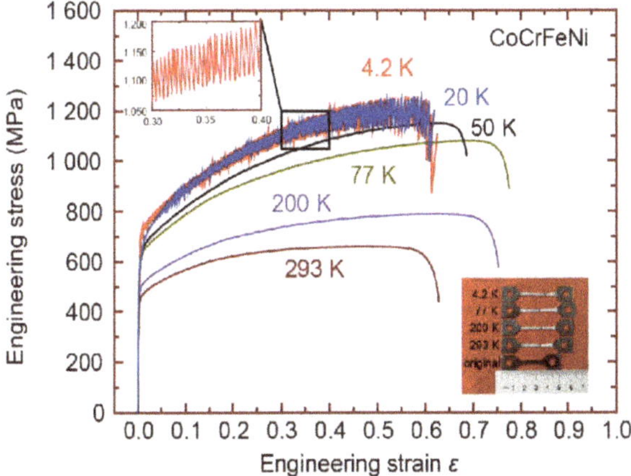

Figure 46. The engineering stress vs. engineering strain curves for the CoCrFeNi HEA tested at temperatures ranging from 4.2 to 293 K and a strain rate of 10^{-3} s^{-1}. The inset displays a magnification of the serrated flow behavior from the specimen tested at 4.2 K. The inset at the bottom-right portion of the figure displays a photograph of the dog-bone-shaped samples, before and after tensile tests. (Reproduced from [159] with permission.)

3.13. HfNbTaTiZr HEA

Chen et al. investigated the deformation behavior of a BCC HfNbTaTiZr HEA [301]. TEM and atomic probe tomography confirmed that the alloy did not contain any secondary phases. They performed tensile tests, under the displacement control, using a strain rate of 1×10^{-4} s^{-1}. During testing, samples were exposed to temperatures of 77, 298, 573, and 673 K. It was observed that, similar to conventional alloys, the yield and flow stress decrease, while the ductility increases with an increase in the temperature. Figure 47 presents the engineering stress–strain curve for the HfNbTaTiZr HEA that was tested at temperatures ranging from 77 to 673 K. As indicated by the graph, visible serrations were only exhibited by the sample that was tested at 673 K. However, it was determined that strain hardening was attributed to both forest hardening and DSA hardening in the specimens tested at 573 and 673 K. The authors surmised that there were two reasons for why DSA was likely a strengthening mechanism in the sample tested at 573 K (despite there being no visible serrations). Firstly, during deformation, the sample tested at the above temperature displayed similar hardening characteristics as the sample tested at 673 K. Secondly, serrations may not always be visible during the DSA [302].

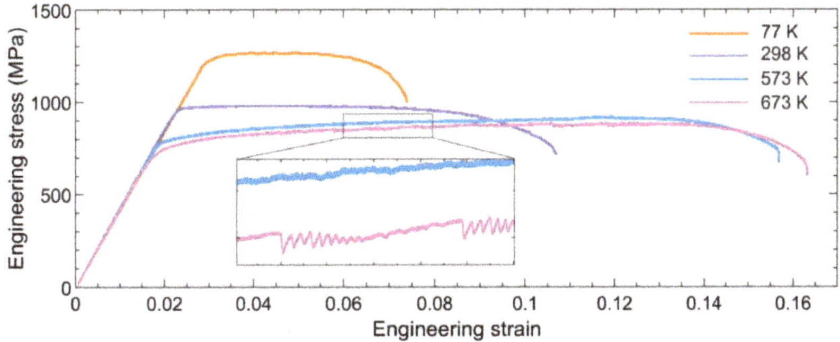

Figure 47. The engineering stress vs. engineering strain for the BCC HfNbTaTiZr HEA that was tested at temperatures ranging from 77 to 673 K, at a strain rate of 1×10^{-4} s^{-1} (reproduced from Reference [301] with permission).

4. Summary and Future Directions

Table 8 and Figure 48a–e both display the reported serration types for different kinds of HEAs. From the results, a few trends can be observed. In terms of the experimental conditions, serrations were reported for strain rates and temperatures ranging from 1×10^{-5} s^{-1} to 1×10^{-1} s^{-1} and RT to 1100 °C, respectively. The samples exhibited serration Type-A, Type-B, and Type-C serrations, as well as serration type combinations, such as Type-A + Type-B and Type-B + Type-C. On the other hand, none of the HEAs were reported to display Type-A + Type-C, Type-D, or Type-E serrations. The greatest variety of different strain rates were observed for the Type-A serrations. Of the serration types that were observed, Type-B serrations were the most common serration displayed by the HEAs.

Table 8. Reported serration type, as found from the literature, for Type-A, Type-B, Type-C, Type-A + Type-B, and Type-B + Type-C for different HEAs tested in a range of strain rates and temperatures. RT: room temperature.

Alloy	Test Type	Strain Rate (s^{-1})	Temperature (°C)	Serration Type	Source
CoCrFeMnNi	Tension	1×10^{-4}	300	A	[153]
			400	B	
			500	B	
			600	C	
		1×10^{-3}	300	A	
			400	A	
			500	B	
			600	B	
		1×10^{-2}	400	A	
			500	A	
			600	A	
CoCrFeMnNi	Tension	1×10^{-5}	500	B + C	[163]
		1×10^{-4}		B	
		3×10^{-4}	300	A	
			400	A + B	
			500	B	
			550	B + C	
			600	C	
		1×10^{-3}	500	A + B	
		5×10^{-3}		A	
CoCrFeMnNi (C ~0.9 at.%)	Tension	1.6×10^{-3}	RT	A	[192]
CoCuFeNiTi	Compression	1×10^{-3}	800	A + B	[162]
			900	A + B	
			950	B	
			1000	B	
		1×10^{-1}	800	A	
			900	A + B	
			950	A + B	
			1000	A + B	
$Al_{0.5}$CoCrCuFeNi	Compression	5×10^{-5}	400	A	[180]
			500	B	
			600	C	
		2×10^{-4}	400	A	
			500	B	
			600	C	
		2×10^{-3}	400	A	
			500	A	
			600	C	

Table 8. *Cont.*

Alloy	Test Type	Strain Rate (s^{-1})	Temperature (°C)	Serration Type	Source
$Al_{0.3}CoCrFeNi$	Compression	1×10^{-3}	400	C	[169]
			500	C	
			600	C	
			700	B + C	
			800	B	
$Al_{0.5}CoCrFeNi$	Tension	1×10^{-4}	200	A	[94]
			300	A + B	
			400	B + C	
		5×10^{-4}	300	A	
			400	A + B	
		1×10^{-3}	300	A	
			400	A + B	
			500	B + C	
AlCoCrFeNi	Compression	1×10^{-3}	1100	C	[303]
		1		B	
$Al_{0.4}CrMnFeCoNi$	Tension	3×10^{-4}	300	A + B	[304]
			400	B	
			500	B + C	
			600	C	
$Al_{0.5}CrMnFeCoNi$	Tension	3×10^{-4}	300	A + B	[304]
			400	B	
			500	B + C	
$Al_{0.6}CrMnFeCoNi$	Tension	3×10^{-4}	300	A + B	[304]
			400	B	
			500	C	

The data also reveal some important aspects about the serrated-flow phenomena in HEAs. For instance, out of all the alloys featured in Figure 48a–e and Table 8, the CoCrFeMnNi HEA was the only material to have exhibited all the reported serration types. Furthermore, the carbon-doped CoCrFeMnNi HEA was the only alloy reported to display serrations at RT. This effect of additives was also observed in [94], where Al was found to play a substantial role in the serration behavior of an $Al_{0.5}CoCrFeNi$ HEA. These findings, therefore, highlight the significance of additives to the serrated-flow process in HEAs.

Figure 49a,b displays a hypothetical schematic for the extent of serrations in HEAs as a function of the temperature and strain rate, as based on the reviewed literature. As can be seen in Figure 49a, there are two major temperature ranges where the serrated flow is observed. The first region, which corresponds to temperatures ranging from 4.2 to 77 K (cryogenic), represents the serrations that occur primarily due to twinning mechanisms. The findings from previous investigations suggest that the extent of the serrations may increase with a decrease in the temperature, which is due to an increase in the twinning mechanisms [166]. As for temperatures below 4.2 K, the authors are unaware of any investigations which have observed the serrated flow in this range, and this should therefore be the subject of future work. However, for temperatures between 77 K and RT serrations do not occur. The absence of serrations in this temperature regime is likely due to the absence of both twinning and dislocation pinning [153]. Serrated flow returns at temperatures between RT and 1100 and is primarily caused by the solute pinning of dislocations [152,180,192,303]. The extent of the serrations initially increases due to an increase in the migration speed of the solute atoms, which is accompanied by an acceleration in the rate of dislocation pinning [152]. However, after a certain temperature is exceeded, the enhanced thermal vibration of the solute atoms reduces their ability to pin dislocations, thereby decreasing the extent of serrations. Once the temperature rises above a certain value (1100 °C in the case of the reported HEAs), the thermal vibrations completely overcome the effect of solute atom pinning of dislocations, and therefore no serrated flow is observed [153].

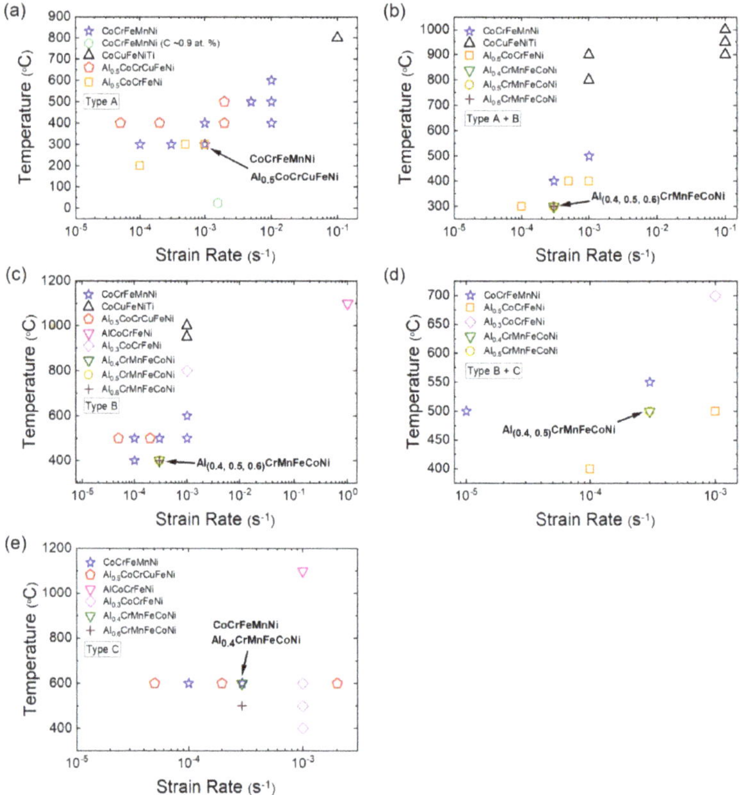

Figure 48. A graphical representation of the reported results that are listed in Table 8 for serration Types (a) A, (b) B, (c) C, (d) A + B, and (e) B + C.

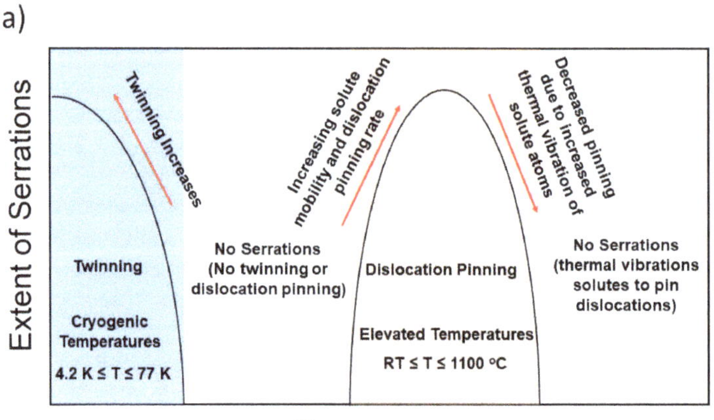

Figure 49. *Cont.*

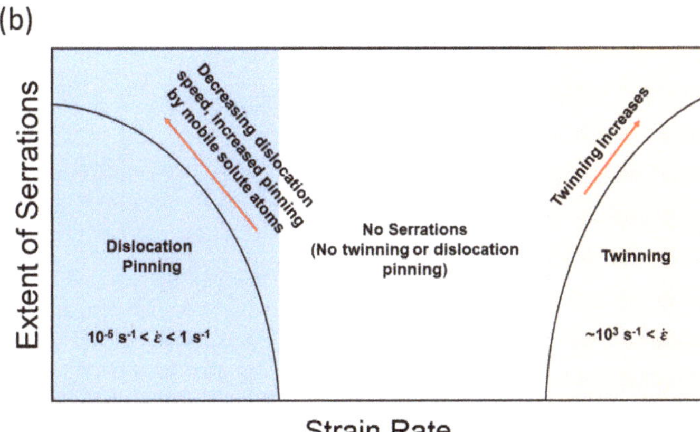

Figure 49. Basic schematics of the extent of serrations as a function of (**a**) temperature and (**b**) strain rate.

A similar graph depicting the extent of serrations as a function of the strain rate is presented in Figure 49b. For the lower strain rate regime (10^{-5}–1 s^{-1}), dislocation pinning is the primary mechanism underlying the serrated flow. For this regime, the extent of the serrations decreases with an increase in the strain rate, which is perhaps due to the increased propagation speed of dislocations [153]. Once the strain rate exceeds 1 s^{-1}, the dislocation velocity is such that no solute atmosphere can be formed [305] and no serrations occur. At strain rates greater than 10^3 s^{-1}, twinning appears to be the primary mechanism [166]. Based on the above study, the extent of the serrations appears to increase with strain rate due to the increase in the twinning. Between these two strain rate regions no, serrations have been reported to occur. The lack of observable serrations during plastic deformation is due to the absence of both twinning and dislocation pinning. It should be noted that, for strain rates less than 10^{-5} s^{-1} and above 10^3 s^{-1}, there is little available in the literature as to whether the serrated flow will occur in these ranges, and this should therefore be the subject of future investigations.

Based on the literature reviewed for the present work, the following important points can be summarized about the serrated-flow phenomenon in HEAs:

- At cryogenic temperatures (4.2–77 K, or from –269 to 196 °C), twinning is the primary mechanism of the serrated flow by hindering dislocation motion at twin boundaries.
- In most cases, serrations have been reported to occur at temperatures ranging from RT to 1100 °C and are a consequence of the pinning and unpinning of dislocations by solute atoms.
- The extent of the serrations in this temperature regime (RT to 1100 °C) initially increases with temperature due to the increased migration speed of pinning solutes until it reaches a maximum. After reaching the maximum, the extent of serrations decreases with temperature due to the increasing thermal vibration of solutes that reduces their ability to pin dislocations.
- At temperatures between the cryogenic regime and RT, serrated flows have not been observed. In this temperature range, solute atoms are not mobile enough to catch and pin moving dislocations, thus resulting in a lack of observable serrations during plastic deformation.
- Additive solutes, such as C and Al, play an important role in the serrated flow behavior in HEAs.
- The phase structure as well as the presence of nanoparticles (such as $L1_2$) in the matrix may play a role in the serration behavior.
- The serration types reported include Types A, B, C, A + B, and B + C.
- The serration type depends on temperature. For example, Type-A serrations generally occur at lower temperatures, as compared to Type-B and Type-C serrations.
- The serration type also depends on strain rate. Typically, Type-C serrations are observed at lower strain rates while Types A and B are seen at relatively higher strain rates.

- For strain rates less than 1 s^{-1}, the extent of serrations in HEAs increases with decreasing strain rate. This increase is due to the slowing down of dislocations, thus allowing solute atoms to more easily catch and pin dislocations.
- For strain rates greater than 10^3 s^{-1}, the extent of serrations increases with an increase strain rate due to the increase in twinning mechanisms.
- Multiple techniques, including complexity analysis, multifractal analysis, mean-field theory analysis, and chaos analysis, have been successfully employed to analyze and model the serrated flow behavior in HEAs.

Although much has been accomplished regarding the study of the serrated flow phenomena in HEAs, there is still considerable work to be done. Future work could investigate phenomena such as the effects of irradiation displacement damage on the deformation behavior of HEAs. Currently, there are not many investigations that have thoroughly examined how particle irradiation affects the serration behavior in HEAs. One example of such future studies could examine how neutron irradiation and thermal annealing influence the serrated flow in HEAs during either tension or compression. This study would be significant, since it would explore how irradiation-induced changes in the microstructure, such as the introduction of dislocation loops or precipitates into the matrix during particle bombardment, could affect the serrated-flow dynamics. Another important experiment could involve the micropillar compression or nanoindentation testing of samples irradiated by ions at a variety of temperatures and irradiation doses. Here, the avalanche statistics of the as-cast and irradiated specimens could be modeled and analyzed according to the MFT formalism for different compression rates or irradiation conditions. It is expected that such studies would provide the substantial insight into the effects of irradiation on the plastic-deformation behavior of HEAs.

Future investigations could also involve a more comprehensive study on the effects of additives, such as C and Al, as well as the presence of nanoparticles, on the serrated flow in HEAs. For these studies, the effects of additive concentration on the serration behavior in a wide array of HEAs would be investigated. Importantly, these studies would examine how the concentration of impurities would affect such factors as the temperature range or strain rate at which serrations would occur. Furthermore, perhaps it will be determined whether there exists a minimum required concentration for the serrated flow to occur. Machine-learning methods could also be employed to elucidate the relationship among the additive concentration, microstructure (such as dislocation dynamics), and serrated-flow behavior [306–310]. Consequently, a greater understanding of solute atom-dislocation pinning dynamics in HEAs would be achieved.

Additionally, more work is needed to better examine the twinning-induced serration phenomena in HEAs. These experiments could involve mechanically testing (compression and/or tension) HEA specimens at temperatures ranging from cryogenic to elevated temperatures and strain rates of greater than 10^3 s^{-1}. It would also be important to examine whether there is a minimum temperature or maximum strain rate at which the serrations would occur. For these experiments, the microstructure could be characterized by using techniques such as TEM and EBSD, while the serrated flow behavior could be analyzed and modeled by using different analytical techniques. The results of such experiments would provide fundamental insight on the relation between twinning and the serration behavior in HEAs.

Finally, there should be more studies that would model and analyze the serrated-flow behavior in HEAs (during tension and compression), using alternative methods, such as the complexity-analysis, multifractal-analysis, and the chaos-analysis techniques. Moreover, theoretical models covering a wide range of scales, including first principles, molecular dynamics, dislocation dynamics, finite-element methods, and crystal plasticity, could be used to quantify, predict, and simulate the effects of composition, temperature, strain rate, additive, irradiation, and environment on serrated flows of HEAs. The results of these types of analyses and modeling would be paired with advanced microstructural-characterization techniques, such as in situ TEM, XRD, and neutron diffraction. The combination of the microstructural and analytical/modeling results would likely yield unexpected and illuminating findings regarding the link between the microstructural behavior and the dynamics of plastic deformation during the serrated flow.

5. Conclusions

In this review, a clear summary of the serrated flow phenomenon, as it occurs in HEAs, was provided. Importantly, this review includes several key findings as revealed by the literature. For instance, additives, such as C and Al, can lead to the occurrence of the serrated flows in HEAs. In terms of the microstructure, serrations typically occur due to either twinning mechanisms (at cryogenic temperatures) or dislocation pinning by solute atoms at temperatures of RT and above. Interestingly, such behavior is not typically seen in conventional crystalline alloys. It has also been found that the serrated flow is significantly affected by experimental conditions, such as the strain rate and test temperature. In HEAs, several different serration types have been observed, including A, B, C, or combinations of serrations, such as A + B and B + C. It is interesting to note that the serration Type-A + Type-C, Type-D, and Type-E has not been reported to occur in this alloy system. Several different types of analytical methods, including the mean-field theory formalism and the complexity-analysis technique, have been successfully applied to examine the serrated flow in HEAs. The results of the analyses indicated that the serrated flow in HEAs consist of complex dynamical behavior. For example, serrations have exhibited chaotic fluctuations that were attributed to a phase transformation (FCC to HCP phases) at cryogenic temperatures. Lastly, further exploratory work is needed to gain a more fundamental perspective on the serrated-flow phenomenon in HEAs. Some areas in need of additional or more comprehensive research to analyze the serrated flow in HEAs have been identified as the analysis of irradiation effects, additives, such as C and Al, the presence of nanoparticles, twinning, and the use of alternate experimental methods, such as in situ TEM.

Author Contributions: Conceptualization, J.B. and S.C.; investigation, J.B., X.X., D.H., B.S., and C.L.; methodology, J.B., X.X., and C.F.; project administration, P.K.L.; supervision, J.B. and P.K.L.; validation, Y.S., A.M.C., and R.F.; resources, J.B.; writing—original draft, all authors; review and editing, C.F., E.I., F.G.M., P.D., R.J.S., and H.S. All authors have read and agreed to the published version of the manuscript.

Funding: S.C., C.L., R.F., X.X., and P.K.L. very much appreciate the supports of (1) the US National Science Foundation (DMR1611180 and 1809640) and (2) the US Army Research Office (W911NF-13-1-0438 and W911NF-19-2-0049), with J. Yang, G. Shiflet, D. Farkas, M. P. Bakas, D. M. Stepp, and S. Mathaudhu as program managers.

Conflicts of Interest: The authors declare no conflict of interest.

References

1. Yeh, J.W.; Chen, S.K.; Gan, J.Y.; Lin, S.J.; Chin, T.S.; Shun, T.T.; Tsau, C.H.; Chang, S.Y. Formation of simple crystal structures in Cu-Co-Ni-Cr-Al-Fe-Ti-V alloys with multiprincipal metallic elements. *Metall. Mater. Trans. A Phys. Metall. Mater. Sci.* **2004**, *35*, 2533–2536. [CrossRef]
2. Yeh, J.W.; Chen, S.K.; Lin, S.J.; Gan, J.Y.; Chin, T.S.; Shun, T.T.; Tsau, C.H.; Chang, S.Y. Nanostructured high-entropy alloys with multiple principal elements: Novel alloy design concepts and outcomes. *Adv. Eng. Mater.* **2004**, *6*, 299–303. [CrossRef]
3. Chen, T.K.; Shun, T.T.; Yeh, J.W.; Wong, M.S. Nanostructured nitride films of multi-element high-entropy alloys by reactive DC sputtering. *Surf. Coat. Technol.* **2004**, *188–189*, 193–200. [CrossRef]
4. Hsu, C.-Y.; Yeh, J.-W.; Chen, S.-K.; Shun, T.-T. Wear resistance and high-temperature compression strength of Fcc $CuCoNiCrAl_{0.5}Fe$ alloy with boron addition. *Metall. Mater. Trans. A* **2004**, *35*, 1465–1469. [CrossRef]
5. Huang, P.-K.; Yeh, J.-W.; Shun, T.-T.; Chen, S.-K. Multi-principal-element alloys with improved oxidation and wear resistance for thermal spray coating. *Adv. Eng. Mater.* **2004**, *6*, 74–78. [CrossRef]
6. Miracle, D.B. High-entropy alloys: A current evaluation of founding ideas and core effects and exploring "nonlinear alloys". *JOM* **2017**, *69*, 2130–2136. [CrossRef]
7. Hsu, C.Y.; Wang, W.R.; Tang, W.Y.; Chen, S.K.; Yeh, J.W. Microstructure and mechanical properties of new $AlCo_xCrFeMo_{0.5}Ni$ high-entropy alloys. *Adv. Eng. Mater.* **2010**, *12*, 44–49. [CrossRef]

8. Zhang, Y.; Liu, J.P.; Chen, S.Y.; Xie, X.; Liaw, P.K.; Dahmen, K.A.; Qiao, J.W.; Wang, Y.L. Serration and noise behaviors in materials. *Prog. Mater. Sci.* **2017**, *90*, 358–460. [CrossRef]
9. Zhang, Y.; Zuo, T.T.; Tang, Z.; Gao, M.C.; Dahmen, K.A.; Liaw, P.K.; Lua, Z.P. Microstructures and properties of high-entropy alloys. *Prog. Mater. Sci.* **2014**, *61*, 1–93. [CrossRef]
10. Cantor, B.; Chang, I.T.H.; Knight, P.; Vincent, A.J.B. Microstructural development in equiatomic multicomponent alloys. *Mater. Sci. Eng. A Struct. Mater. Prop. Microstruct. Process.* **2004**, *375*, 213–218. [CrossRef]
11. Gao, M.C.; Carney, C.S.; Doğan, Ö.N.; Jablonksi, P.D.; Hawk, J.A.; Alman, D.E. Design of refractory high-entropy alloys. *JOM* **2015**, *67*, 2653–2669. [CrossRef]
12. Haas, S.; Mosbacher, M.; Senkov, O.N.; Feuerbacher, M.; Freudenberger, J.; Gezgin, S.; Völkl, R.; Glatzel, U. Entropy determination of single-phase high entropy alloys with different crystal structures over a wide temperature range. *Entropy* **2018**, *20*, 654. [CrossRef]
13. Yeh, J.-W. Alloy design strategies and future trends in high-entropy alloys. *JOM* **2013**, *65*, 1759–1771. [CrossRef]
14. Eißmann, N.; Klöden, B.; Weißgärber, T.; Kieback, B. High-entropy alloy CoCrFeMnNi produced by powder metallurgy. *Powder Metall.* **2017**, *60*, 184–197. [CrossRef]
15. Ruiz-Yi, B.; Bunn, J.K.; Stasak, D.; Mehta, A.; Besser, M.; Kramer, M.J.; Takeuchi, I.; Hattrick-Simpers, J. The different roles of entropy and solubility in high entropy alloy stability. *ACS Comb. Sci.* **2016**, *18*, 596–603. [CrossRef]
16. Cantor, B. Multicomponent and high entropy alloys. *Entropy* **2014**, *16*, 4749–4768. [CrossRef]
17. Lee, C.; Song, G.; Gao, M.C.; Feng, R.; Chen, P.; Brechtl, J.; Chen, Y.; An, K.; Guo, W.; Poplawsky, J.D.; et al. Lattice distortion in a strong and ductile refractory high-entropy alloy. *Acta Mater.* **2018**, *160*, 158–172. [CrossRef]
18. Miracle, D.B.; Senkov, O.N. A critical review of high entropy alloys and related concepts. *Acta Mater.* **2017**, *122*, 448–511. [CrossRef]
19. Tsai, K.Y.; Tsai, M.H.; Yeh, J.W. Sluggish diffusion in Co-Cr-Fe-Mn-Ni high-entropy alloys. *Acta Mater.* **2013**, *61*, 4887–4897. [CrossRef]
20. Tsai, M.-H.; Wang, C.-W.; Tsai, C.-W.; Shen, W.-J.; Yeh, J.-W.; Gan, J.-Y.; Wu, W.-W. Thermal stability and performance of NbSiTaTiZr high-entropy alloy barrier for copper metallization. *J. Electrochem. Soc.* **2011**, *158*, H1161. [CrossRef]
21. Hsu, C.-Y.; Juan, C.-C.; Wang, W.-R.; Sheu, T.-S.; Yeh, J.-W.; Chen, S.-K. On the superior hot hardness and softening resistance of $AlCoCr_xFeMo_{0.5}Ni$ high-entropy alloys. *Mater. Sci. Eng. A* **2011**, *528*, 3581–3588. [CrossRef]
22. Paul, A. Comments on "Sluggish diffusion in Co-Cr-Fe-Mn-Ni high-entropy alloys" by KY Tsai, MH Tsai and JW Yeh, Acta Materialia 61 (2013) 4887–4897. *Scr. Mater.* **2017**, *135*, 153–157. [CrossRef]
23. Tsai, K.Y.; Tsai, M.H.; Yeh, J.W. Reply to comments on "Sluggish diffusion in Co-Cr-Fe-Mn-Ni high-entropy alloys" by KY Tsai, MH Tsai and JW Yeh, Acta Materialia 61 (2013) 4887–4897. *Scr. Mater.* **2017**, *135*, 158–159. [CrossRef]
24. Beke, D.L.; Erdélyi, G. On the diffusion in high-entropy alloys. *Mater. Lett.* **2016**, *164*, 111–113. [CrossRef]
25. Zhang, C.; Zhang, F.; Jin, K.; Bei, H.B.; Chen, S.L.; Cao, W.S.; Zhu, J.; Lv, D.C. Understanding of the elemental diffusion behavior in concentrated solid solution alloys. *J. Phase Equilibria Diffus.* **2017**, *38*, 434–444. [CrossRef]
26. Pickering, E.J.; Jones, N.G. High-entropy alloys: A critical assessment of their founding principles and future prospects. *Int. Mater. Rev.* **2016**, *61*, 183–202. [CrossRef]
27. Murty, B.S.; Yeh, J.W.; Ranganathan, S.; Bhattacharjee, P.P. *High-Entropy Alloys*; Elsevier Science: Amsterdam, The Netherlands, 2019.
28. Yeh, J.-W. Physical metallurgy of high-entropy alloys. *JOM* **2015**, *67*, 2254–2261. [CrossRef]
29. Zhang, Y.; Zuo, T.; Cheng, Y.; Liaw, P.K. High-entropy alloys with high saturation magnetization, electrical resistivity and malleability. *Sci. Rep.* **2013**, *3*, 1455. [CrossRef]
30. Chou, Y.L.; Wang, Y.C.; Yeh, J.W.; Shih, H.C. The effect of molybdenum on the corrosion behaviour of the high-entropy alloys $Co_{1.5}CrFeNi_{1.5}Ti_{0.5}Mo_x$ in aqueous environments. *Corros. Sci.* **2010**, *52*, 1026–1034. [CrossRef]
31. Kao, Y.F.; Lee, T.D.; Chen, S.K.; Chang, Y.S. Electrochemical passive properties of $Al_xCoCrFeNi$ (x = 0, 0.25, 0.50, 1.00) alloys in sulfuric acids. *Corros. Sci.* **2017**, *52*, 1026–1034. [CrossRef]

32. Shi, Y.; Yang, B.; Xie, X.; Brechtl, J.; Dahmen, K.A.; Liaw, P.K. Corrosion of $Al_xCoCrFeNi$ high-entropy alloys: Al-content and potential scan-rate dependent pitting behavior. *Corros. Sci.* **2017**, *119*, 33–45. [CrossRef]
33. Tang, Z.; Huang, L.; He, W.; Liaw, P.K. Alloying and processing effects on the aqueous corrosion behavior of high-entropy alloys. *Entropy* **2014**, *16*, 895–911. [CrossRef]
34. Nair, R.B.; Arora, H.S.; Mukherjee, S.; Singh, S.; Singh, H.; Grewel, H.S. Exceptionally high cavitation erosion and corrosion resistance of a high entropy alloy. *Ultrason. Sonochem.* **2018**, *41*, 252–260. [CrossRef]
35. Shi, Y.Z.; Yang, B.; Liaw, P.K. Corrosion-resistant high-entropy alloys: A review. *Metals* **2017**, *7*, 43. [CrossRef]
36. Hemphill, M.A.; Yuan, T.; Wang, G.Y.; Yeh, J.W.; Tsai, C.W.; Chuang, A.; Liaw, P.K. Fatigue behavior of $Al_{0.5}CoCrCuFeNi$ high entropy alloys. *Acta Mater.* **2012**, *60*, 5723–5734. [CrossRef]
37. Seifi, M.; Li, D.Y.; Yong, Z.; Liaw, P.K.; Lewandowski, J.J. Fracture toughness and fatigue crack growth behavior of as-cast high-entropy alloys. *JOM* **2015**, *67*, 2288–2295. [CrossRef]
38. Thurston, K.V.S.; Gludovatz, B.; Hohenwarter, A.; Laplanche, G.; George, E.P.; Ritchie, R.O. Effect of temperature on the fatigue-crack growth behavior of the high-entropy alloy CrMnFeCoNi. *Intermetallics* **2017**, *88*, 65–72. [CrossRef]
39. Tang, Z.; Yuan, T.; Tsai, C.W.; Yeh, J.W.; Lundin, C.D.; Liaw, P.K. Fatigue behavior of a wrought $Al_{0.5}CoCrCuFeNi$ two-phase high-entropy alloy. *Acta Mater.* **2015**, *99*, 247–258. [CrossRef]
40. Chen, P.Y.; Lee, C.; Wang, S.Y.; Seifi, M.; Lewandowski, J.J.; Dahmen, K.A.; Jia, H.L.; Xie, X.; Chen, B.L.; Yeh, J.W.; et al. Fatigue behavior of high-entropy alloys: A review. *Sci. China Technol. Sci.* **2018**, *61*, 168–178. [CrossRef]
41. Liu, K.M.; Komarasamy, M.; Gwalani, B.; Shukla, S.; Mishra, R.S. Fatigue behavior of ultrafine grained triplex $Al_{0.3}CoCrFeNi$ high entropy alloy. *Scr. Mater.* **2019**, *158*, 116–120. [CrossRef]
42. Gao, X.; Lu, Y.; Zhang, B.; Liang, N.; Wu, G.; Sha, G.; Liu, J.; Zhao, Y. Microstructural origins of high strength and high ductility in an $AlCoCrFeNi_{2.1}$ eutectic high-entropy alloy. *Acta Mater.* **2017**, *141*, 59–66. [CrossRef]
43. Koch, C.C. Nanocrystalline high-entropy alloys. *J. Mater. Res.* **2017**, *32*, 3435–3444. [CrossRef]
44. Lilensten, L.; Couzinie, J.-P.; Perriere, L.; Hocini, A.; Keller, C.; Dirras, G.; Guillot, I. Study of a BCC multi-principal element alloy: Tensile and simple shear properties and underlying deformation mechanisms. *Acta Mater.* **2018**, *142*, 131–141. [CrossRef]
45. Cai, B.; Liu, B.; Kabra, S.; Wang, Y.; Yan, K.; Lee, P.K.; Liu, Y. Deformation mechanisms of Mo alloyed FeCoCrNi high entropy alloy: In situ neutron diffraction. *Acta Mater.* **2017**, *127*, 471–480. [CrossRef]
46. Gludovatz, B.; Hohenwarter, A.; Catoor, D.; Chang, E.H.; George, E.P.; Ritchie, R.O. A fracture-resistant high-entropy alloy for cryogenic applications. *Science* **2014**, *345*, 1153–1158. [CrossRef]
47. George, E.P.; Raabe, D.; Ritchie, R.O. High-entropy alloys. *Nat. Rev. Mater.* **2019**, *4*, 515–534. [CrossRef]
48. Li, Z.; Zhao, S.; Diao, H.; Liaw, P.K.; Meyers, M.A. High-velocity deformation of $Al_{0.3}CoCrFeNi$ high-entropy alloy: Remarkable resistance to shear failure. *Sci. Rep.* **2017**, *7*, 42742. [CrossRef]
49. Egami, T.; Guo, W.; Rack, P.D.; Nagase, T. Irradiation resistance of multicomponent alloys. *Metall. Mater. Trans. A* **2014**, *45A*, 180–183. [CrossRef]
50. Xia, S.Q.; Yang, X.; Yang, T.F.; Liu, S.; Zhang, Y. Irradiation resistance in $Al_xCoCrFeNi$ high entropy alloys. *JOM* **2015**, *67*, 2340–2344. [CrossRef]
51. Xia, S.; Gao, M.C.; Yang, T.; Liaw, P.K.; Zhang, Y. Phase stability and microstructures of high entropy alloys ion irradiated to high doses. *J. Nucl. Mater.* **2016**, *480*, 100–108. [CrossRef]
52. Kumar, N.A.P.K.; Li, C.; Leonard, K.J.; Bei, H.; Zinkle, S.J. Microstructural stability and mechanical behavior of FeNiMnCr high entropy alloy under ion irradiation. *Acta Mater.* **2016**, *113*, 230–244. [CrossRef]
53. Yang, T.; Xia, S.; Liu, S.; Wang, C.; Liu, S.; Fang, Y.; Zhang, Y.; Xue, J.; Yan, S.; Wang, Y. Precipitation behavior of $Al_xCoCrFeNi$ high entropy alloys under ion irradiation. *Sci. Rep.* **2016**, *6*, 32146. [CrossRef] [PubMed]
54. Xia, S.-Q.; Wang, Z.; Yang, T.-F.; Zhang, Y. Irradiation behavior in high entropy alloys. *J. Iron Steel Res.* **2015**, *22*, 879–884. [CrossRef]
55. He, M.-R.; Wang, S.; Jin, K.; Bei, H.; Yasuda, K.; Matsumura, S.; Higashida, K.; Robertson, I.M. Enhanced damage resistance and novel defect structure of CrFeCoNi under in situ electron irradiation. *Scr. Mater.* **2016**, *125*, 5–9. [CrossRef]
56. Yang, T.; Xia, S.; Guo, W.; Hu, R.; Poplawsky, J.D.; Sha, G.; Fang, Y.; Yan, Z.; Wang, C.; Li, C.; et al. Effects of temperature on the irradiation responses of $Al_{0.1}CoCrFeNi$ high entropy alloy. *Scr. Mater.* **2018**, *144*, 31–35. [CrossRef]

57. Zhang, Y.; Stocks, G.M.; Jin, K.; Lu, C.; Bei, H.; Sales, B.C.; Wang, L.; Béland, L.K.; Stoller, R.E.; Samolyuk, G.D.; et al. Influence of chemical disorder on energy dissipation and defect evolution in concentrated solid solution alloys. *Nat. Commun.* **2015**, *6*, 8736. [CrossRef]
58. Yang, T.; Li, C.; Zinkle, S.J.; Zhao, S.; Bei, H.; Zhang, Y. Irradiation responses and defect behavior of single-phase concentrated solid solution alloys. *J. Mater. Res.* **2018**, *33*, 3077–3091. [CrossRef]
59. Ching, W.-Y.; San, S.; Brechtl, J.; Sakidja, R.; Zhang, M.; Liaw, P.K. Fundamental electronic structure and multiatomic bonding in 13 biocompatible high-entropy alloys. *NPJ Comput. Mater.* **2020**, *6*, 45. [CrossRef]
60. Todai, M.; Nagase, T.; Hori, T.; Matsugaki, A.; Sekita, A.; Nakano, T. Novel TiNbTaZrMo high-entropy alloys for metallic biomaterials. *Scr. Mater.* **2017**, *129*, 65–68. [CrossRef]
61. Yang, W.; Liu, Y.; Pang, S.; Liaw, P.K.; Zhang, T. Bio-corrosion behavior and in vitro biocompatibility of equimolar TiZrHfNbTa high-entropy alloy. *Intermetallics* **2020**, *124*, 106845. [CrossRef]
62. Jo, Y.H.; Jung, S.; Choi, W.M.; Sohn, S.S.; Kim, H.S.; Lee, B.J.; Kim, N.J.; Lee, S. Cryogenic strength improvement by utilizing room-temperature deformation twinning in a partially recrystallized VCrMnFeCoNi high-entropy alloy. *Nat. Commun.* **2017**, *8*, 15719. [CrossRef] [PubMed]
63. Li, D.; Li, C.X.; Feng, T.; Zhang, Y.D.; Sha, G.; Lewandowski, J.J.; Liaw, P.K.; Zhang, Y. High-entropy $Al_{0.3}CoCrFeNi$ alloy fibers with high tensile strength and ductility at ambient and cryogenic temperatures. *Acta Mater.* **2017**, *123*, 285–294. [CrossRef]
64. Qiao, J.W.; Ma, S.G.; Huang, E.W.; Chuang, C.P.; Liaw, P.K.; Zhang, Y. Microstructural characteristics and mechanical behaviors of AlCoCrFeNi high-entropy alloys at ambient and cryogenic temperatures. In *Nano-Scale and Amourphous Materials*; Wang, R.M., Wu, Y., Wu, X.F., Eds.; Trans Tech Publications Ltd.: Durnten-Zurich, Switzerland, 2011; Volume 688, pp. 419–425.
65. Lyu, Z.; Fan, X.; Lee, C.; Wang, S.-Y.; Feng, R.; Liaw, P.K. Fundamental understanding of mechanical behavior of high-entropy alloys at low temperatures: A review. *J. Mater. Res.* **2018**, *33*, 2998–3010. [CrossRef]
66. Chuang, M.-H.; Tsai, M.H.; Wang, W.R.; Lin, S.J.; Yeh, J.W. Microstructure and wear behavior of $Al_xCo_{1.5}CrFeNi_{1.5}Ti_y$ high-entropy alloys. *Acta Mater.* **2011**, *59*, 6308–6317. [CrossRef]
67. Chuang, M.-H.; Tsai, M.-H.; Tsai, C.-W.; Yang, N.-H.; Chang, S.-Y.; Yeh, J.-W.; Chen, S.-K.; Lin, S.-J. Intrinsic surface hardening and precipitation kinetics of $Al_{0.3}CrFe_{1.5}MnNi_{0.5}$ multi-component alloy. *J. Alloys Compd.* **2013**, *551*, 12–18. [CrossRef]
68. Diao, H.Y.; Feng, R.; Dahmen, K.A.; Liaw, P.K. Fundamental deformation behavior in high-entropy alloys: An overview. *Curr. Opin. Solid State Mater. Sci.* **2017**, *21*, 252–266. [CrossRef]
69. Stevenson, A.; Waite, M. *Concise Oxford English Dictionary: Luxury Edition*; OUP Oxford: New York, NY, USA, 2011.
70. Moyer, J.K.; Bemis, W.E. Shark teeth as edged weapons: Serrated teeth of three species of selachians. *Zoology* **2017**, *120*, 101–109. [CrossRef]
71. Stefanita, C.G. *From Bulk to Nano: The Many Sides of Magnetism*; Springer: Berlin/Heidelberg, Germany, 2008.
72. Mascarenas, D.; Lockhart, M.; Lienert, T. Barkhausen noise as an intrinsic fingerprint for ferromagnetic components. *Smart Mater. Struct.* **2019**, *28*, 17. [CrossRef]
73. Bohn, F.; Durin, G.; Correa, M.A.; Machado, N.R.; Della Pace, R.D.; Chesman, C.; Sommer, R.L. Playing with universality classes of Barkhausen avalanches. *Sci. Rep.* **2018**, *8*, 12. [CrossRef]
74. Travesset, A.; White, R.A.; Dahmen, K.A. Crackling noise, power spectra, and disorder-induced critical scaling. *Phys. Rev. B* **2002**, *66*, 11. [CrossRef]
75. Varotsos, P.A.; Sarlis, N.V.; Skordas, E.S.; Lazaridou-Varotsos, M.S. MW9 tohoku earthquake in 2011 in Japan: Precursors uncovered by natural time analysis. *Earthq. Sci.* **2017**, *30*, 183–191. [CrossRef]
76. Uhl, J.T.; Pathak, S.; Schorlemmer, D.; Liu, X.; Swindeman, R.; Brinkman, B.A.W.; LeBlanc, M.; Tsekenis, G.; Friedman, N.; Behringer, R.; et al. Universal quake statistics: From compressed nanocrystals to earthquakes. *Sci. Rep.* **2015**, *5*, 16493. [CrossRef] [PubMed]
77. Sethna, J.P.; Dahmen, K.A.; Myers, C.R. Crackling noise. *Nature* **2001**, *410*, 242–250. [CrossRef]
78. Kurata, M.; Li, X.; Fujita, K.; Yamaguchi, M. Piezoelectric dynamic strain monitoring for detecting local seismic damage in steel buildings. *Smart Mater. Struct.* **2013**, *22*, 115002. [CrossRef]
79. Da Silva, A.C.; Maganini, N.D.; de Almeida, E.F. Multifractal analysis of Bitcoin market. *Phys. A Stat. Mech. Its Appl.* **2018**, *512*, 954–967. [CrossRef]
80. Ferreira, P. Assessing the relationship between dependence and volume in stock markets: A dynamic analysis. *Phys. A Stat. Mech. Its Appl.* **2019**, *516*, 90–97. [CrossRef]

81. Xing, Y.N.; Wang, J. Statistical volatility duration and complexity of financial dynamics on Sierpinski gasket lattice percolation. *Phys. A Stat. Mech. Its Appl.* **2019**, *513*, 234–247. [CrossRef]
82. Bashir, U.; Zebende, G.F.; Yu, Y.G.; Hussain, M.; Ali, A.; Abbas, G. Differential market reactions to pre and post Brexit referendum. *Phys. A Stat. Mech. Its Appl.* **2019**, *515*, 151–158. [CrossRef]
83. Begusic, S.; Kostanjcar, Z.; Stanley, H.E.; Podobnik, B. Scaling properties of extreme price fluctuations in Bitcoin markets. *Phys. A Stat. Mech. Its Appl.* **2018**, *510*, 400–406. [CrossRef]
84. Thiagarajan, T. Interpreting electrical signals from the brain. *Acta Phys. Pol. B* **2018**, *49*, 2095–2125. [CrossRef]
85. Scarpetta, S.; Apicella, I.; Minati, L.; de Candia, A. Hysteresis, neural avalanches, and critical behavior near a first-order transition of a spiking neural network. *Phys. Rev. E* **2018**, *97*, 13. [CrossRef] [PubMed]
86. Martinello, M.; Hidalgo, J.; Maritan, A.; Di Santo, S.; Plenz, D.; Munoz, M.A. Neutral theory and scale-free neural dynamics. *Phys. Rev. X* **2017**, *7*, 11. [CrossRef]
87. Cocchi, L.; Gollo, L.L.; Zalesky, A.; Breakspear, M. Criticality in the brain: A synthesis of neurobiology, models and cognition. *Prog. Neurobiol.* **2017**, *158*, 132–152. [CrossRef] [PubMed]
88. Karimipanah, Y.; Ma, Z.Y.; Miller, J.E.K.; Yuste, R.; Wessel, R. Neocortical activity is stimulus- and scale-invariant. *PLoS ONE* **2017**, *12*, e0177396. [CrossRef]
89. Bellay, T.; Klaus, A.; Seshadri, S.; Plenz, D. Irregular spiking of pyramidal neurons organizes as scale-invariant neuronal avalanches in the awake state. *Elife* **2015**, *4*, 25. [CrossRef]
90. Steriade, M.; McCormick, D.A.; Sejnowski, T.J. Thalamocortical oscillations in the sleeping and aroused brain. *Science* **1993**, *262*, 679–685. [CrossRef]
91. Destexhe, A.; Contreras, D. The fine structure of slow-wave sleep oscillations: From single neurons to large networks. In *Sleep and Anesthesia: Neural Correlates in Theory and Experiment*; Hutt, A., Ed.; Springer: New York, NY, USA, 2011; pp. 69–105. [CrossRef]
92. Destexhe, A. *Nonlinear Dynamics of the Rhythmical Activity of the Brain (Aspects Non Linéaires de l'Activité Rythmique du Cerveau)*; Université Libre de Bruxelles: Brussels, Belgium, 1992.
93. Neuhauser, H. Collective microshear processes and plastic instabilities in crystalline and amorphous structures. *Int. J. Plast.* **1993**, *9*, 421–435. [CrossRef]
94. Niu, S.Z.; Kou, H.C.; Zhang, Y.; Wang, J.; Li, J.S. The characteristics of serration in $Al_{0.5}CoCrFeNi$ high entropy alloy. *Mater. Sci. Eng. A Struct. Mater. Prop. Microstruct. Process.* **2017**, *702*, 96–103. [CrossRef]
95. Abbadi, M.; Hahner, P.; Zeghloul, A. On the characteristics of Portevin-Le Chatelier bands in aluminum alloy 5182 under stress-controlled and strain-controlled tensile testing. *Mater. Sci. Eng. A Struct. Mater. Prop. Microstruct. Process.* **2002**, *337*, 194–201. [CrossRef]
96. Reed, J.M.; Walter, M.E. Observations of serration characteristics and acoustic emission during serrated flow of an Al–Mg alloy. *Mater. Sci. Eng. A* **2003**, *359*, 1–10. [CrossRef]
97. Golovin, Y.I.; Ivolgin, V.I.; Khonik, V.A.; Kitagawa, K.; Tyurin, A.I. Serrated plastic flow during nanoindentation of a bulk metallic glass. *Scr. Mater.* **2001**, *45*, 947–952. [CrossRef]
98. François, D.; Pineau, A.; Zaoui, A. *Mechanical Behaviour of Materials: Volume 1: Micro- and Macroscopic Constitutive Behaviour*; Springer: Dordrecht, The Netherlands, 2012.
99. Pustovalov, V.V. Serrated deformation of metals and alloys at low temperatures (Review). *Low Temp. Phys.* **2008**, *34*, 683–723. [CrossRef]
100. Gindin, I.; Khotkevich, V.; Starodubov, Y.A. Ductile characteristics of aluminum at low temperatures. *Fiz. Met. Metalloved* **1959**, *7–10*, 794.
101. Blewitt, T.H.; Coltman, R.R.; Redman, J.K. Low-temperature deformation of copper single crystals. *J. Appl. Phys.* **1957**, *28*, 651–660. [CrossRef]
102. Lebedev, V.P.; Krylovskiy, V.S.; Lebedev, S.V. Low-temperature jump-like strains on different scales in the normal state of Pb-(4–49) at. % In alloys. *Low Temp. Phys.* **2012**, *38*, 248–254. [CrossRef]
103. Kuzmenko, I.; Lubenets, S.; Pustovalov, V.; Fomenko, L. Vliyanie sverkhprovodyashchego perekhoda na skolzhenie i dvoynikovanie indiya i ego splavov [Effect of the superconducting transition on slip and twinning in indium and its alloys]. *Sov. J. Low Temp. Phys.* **1983**, *9*, 450–453.
104. Sarkar, A.; Barat, P.; Mukherjee, P. Multiscale entropy analysis of the Portevin-Le Chatelier effect in an Al-2.5%Mg alloy. *Fractals* **2010**, *18*, 319–325. [CrossRef]
105. Chatterjee, A.; Sarkar, A.; Barat, P.; Mukherjee, P.; Gayathri, N. Character of the deformation bands in the (A + B) regime of the Portevin-Le Chatelier effect in Al-2.5%Mg alloy. *Mater. Sci. Eng. A* **2009**, *508*, 156–160. [CrossRef]

106. Valdes-Tabernero, M.A.; Sancho-Cadenas, R.; Sabirov, I.; Murashkin, M.Y.; Ovid'ko, I.A.; Galvez, F. Effect of SPD processing on mechanical behavior and dynamic strain aging of an Al-Mg alloy in various deformation modes and wide strain rate range. *Mater. Sci. Eng. A* **2017**, *696*, 348–359. [CrossRef]
107. Chibane, N.; Ait-Amokhtar, H.; Fressengeas, C. On the strain rate dependence of the critical strain for plastic instabilities in Al-Mg alloys. *Scr. Mater.* **2017**, *130*, 252–255. [CrossRef]
108. Yuzbekova, D.; Mogucheva, A.; Zhemchuzhnikova, D.; Lebedkina, T.; Lebyodkin, M.; Kaibyshev, R. Effect of microstructure on continuous propagation of the Portevin-Le Chatelier deformation bands. *Int. J. Plast.* **2017**, *96*, 210–226. [CrossRef]
109. Jobba, M.; Mishra, R.K.; Niewczas, M. Flow stress and work-hardening behaviour of Al-Mg binary alloys. *Int. J. Plast.* **2015**, *65*, 43–60. [CrossRef]
110. Shibkov, A.A.; Gasanov, M.F.; Zheltov, M.A.; Zolotov, A.E.; Ivolgin, V.I. Intermittent plasticity associated with the spatio-temporal dynamics of deformation bands during creep tests in an AlMg polycrystal. *Int. J. Plast.* **2016**, *86*, 37–55. [CrossRef]
111. Chihab, K. On the apparent strain rate sensitivity of Portevin—Le Chatelier effect. *Ann. Chim. Sci. Des. Mater.* **2004**, *29*, 15–23. [CrossRef]
112. Park, D.; Morris, J.G. The tensile deformation-behavior of AA-3004 aluminum-alloy. *Metall. Mater. Trans. A Phys. Metall. Mater. Sci.* **1994**, *25*, 357–364. [CrossRef]
113. Tian, B.; Paris, O.; Prem, M.; Pink, E.; Fratzl, P. Serrated flow and related microstructures in an Al-8.4 at.% Li alloy. *J. Mater. Sci.* **2002**, *37*, 1355–1361. [CrossRef]
114. Tian, B.H. Comparing characteristics of serrations in Al-Li and Al-Mg alloys. *Mater. Sci. Eng. A Struct. Mater. Prop. Microstruct. Process.* **2003**, *360*, 330–338. [CrossRef]
115. Thomas, G.; Srinivasan, N.K. Effect of quenching temperature on nature of serrations in an aluminum-alloy. *Scr. Metall.* **1974**, *8*, 1163–1166. [CrossRef]
116. Ananthakrishna, G.; Fressengeas, C.; Grosbras, M.; Vergnol, J.; Engelke, C.; Plessing, J.; Neuhäuser, H.; Bouchaud, E.; Planès, J.; Kubin, L.P. On the existence of chaos in jerky flow. *Scr. Metall. Mater.* **1995**, *32*, 1731–1737. [CrossRef]
117. Rowcliffe, A.F.; Zinkle, S.J.; Hoelzer, D.T. Effect of strain rate on the tensile properties of unirradiated and irradiated V-4Cr-4Ti. *J. Nucl. Mater.* **2000**, *283*, 508–512. [CrossRef]
118. Koyama, M.; Fukumoto, K.; Matsui, H. Effects of purity on high temperature mechanical properties of vanadium alloys. *J. Nucl. Mater.* **2004**, *329*, 442–446. [CrossRef]
119. Sarkar, A.; Maloy, S.A.; Murty, K.L. Investigation of Portevin-LeChatelier effect in HT-9 steel. *Mater. Sci. Eng. A* **2015**, *631*, 120–125. [CrossRef]
120. Field, D.M.; Aken, D.C.V. Dynamic strain aging phenomena and tensile response of medium-Mn TRIP steel. *Metall. Mater. Trans. A* **2018**, *49*, 1152–1166. [CrossRef]
121. Lan, P.; Zhang, J.Q. Serrated flow and dynamic strain aging in Fe-Mn-C TWIP steel. *Metall. Mater. Trans. A Phys. Metall. Mater. Sci.* **2018**, *49*, 147–161. [CrossRef]
122. Zavattieri, P.D.; Savic, V.; Hector, L.G.; Fekete, J.R.; Tong, W.; Xuan, Y. Spatio-temporal characteristics of the Portevin-Le Chatelier effect in austenitic steel with twinning induced plasticity. *Int. J. Plast.* **2009**, *25*, 2298–2330. [CrossRef]
123. Madivala, M.; Schwedt, A.; Wong, S.L.; Roters, F.; Prahl, U.; Bleck, W. Temperature dependent strain hardening and fracture behavior of TWIP steel. *Int. J. Plast.* **2018**, *104*, 80–103. [CrossRef]
124. Yang, F.; Luo, H.W.; Pu, E.X.; Zhang, S.L.; Dong, H. On the characteristics of Portevin-Le Chatelier bands in cold-rolled 7Mn steel showing transformation induced plasticity. *Int. J. Plast.* **2018**, *103*, 188–202. [CrossRef]
125. Alomari, A.S.; Kumar, N.; Murty, K.L. Enhanced ductility in dynamic strain aging regime in a Fe-25Ni-20Cr austenitic stainless steel. *Mater. Sci. Eng. A Struct. Mater. Prop. Microstruct. Process.* **2018**, *729*, 157–160. [CrossRef]
126. Hong, Y.Y.; Li, S.L.; Li, H.J.; Li, J.; Sun, G.G.; Wang, Y.D. Development of intergranular residual stress and its implication to mechanical behaviors at elevated temperatures in AL6XN austenitic stainless steel. *Metall. Mater. Trans. A Phys. Metall. Mater. Sci.* **2018**, *49*, 3237–3246. [CrossRef]
127. Li, Q.S.; Shen, Y.Z.; Han, P.C. Serrated flow behavior of Aisi 316l austenitic stainless steel for nuclear reactors. In *3rd Annual International Workshop on Materials Science and Engineering*; IOP Publishing Ltd.: Bristol, UK, 2017; Volume 250.

128. Ferrero, C.; Monforte, R.; Marinari, C.; Martino, E. Correlation between serration effect and temperature. *Cryogenics* **1994**, *34*, 473–476. [CrossRef]
129. Kim, D.W.; Ryu, W.S.; Hong, J.H.; Choi, S.K. Effect of nitrogen on the dynamic strain ageing behaviour of type 316L stainless steel. *J. Mater. Sci.* **1998**, *33*, 675–679. [CrossRef]
130. Koyama, M.; Sawaguchi, T.; Tsuzaki, K. Overview of dynamic strain aging and associated phenomena in Fe–Mn–C austenitic steels. *ISIJ Int.* **2018**, *58*, 1383–1395. [CrossRef]
131. Brechtl, J.; Chen, B.; Xie, X.; Ren, Y.; Venable, J.D.; Liaw, P.K.; Zinkle, S.J. Entropy modeling on serrated flows in carburized steels. *Mater. Sci. Eng. A* **2019**, *753*, 135–145. [CrossRef]
132. Denisov, D.V.; Lőrincz, K.A.; Wright, W.J.; Hufnagel, T.C.; Nawano, A.; Gu, X.; Uhl, J.T.; Dahmen, K.A.; Schall, P. Universal slip dynamics in metallic glasses and granular matter–linking frictional weakening with inertial effects. *Sci. Rep.* **2017**, *7*, 43376. [CrossRef] [PubMed]
133. Antonaglia, J.; Xie, X.; Schwarz, G.; Wraith, M.; Qiao, J.; Zhang, Y.; Liaw, P.K.; Uhl, J.T.; Dahmen, K.A. Tuned critical avalanche scaling in bulk metallic glasses. *Sci. Rep.* **2014**, *4*, 4382. [CrossRef]
134. Antonaglia, J.; Wright, W.J.; Gu, X.; Byer, R.R.; Hufnagel, T.C.; LeBlanc, M.; Uhl, J.T.; Dahmen, K.A. Bulk metallic glasses deform via slip avalanches. *Phys. Rev. Lett.* **2014**, *112*, 1–5. [CrossRef] [PubMed]
135. Li, J.J.; Wang, Z.; Qiao, J.W. Power-law scaling between mean stress drops and strain rates in bulk metallic glasses. *Mater. Des.* **2016**, *99*, 427–432. [CrossRef]
136. Torre, F.H.D.; Klaumünzer, D.; Maaß, R.; Löffler, J.F. Stick–slip behavior of serrated flow during inhomogeneous deformation of bulk metallic glasses. *Acta Mater.* **2010**, *58*, 3742–3750. [CrossRef]
137. Maaß, R.; Klaumünzer, D.; Löffler, J.F. Propagation dynamics of individual shear bands during inhomogeneous flow in a Zr-based bulk metallic glass. *Acta Mater.* **2011**, *59*, 3205–3213. [CrossRef]
138. Shi, B.; Luan, S.; Jin, P. Crossover from free propagation to cooperative motions of shear bands and its effect on serrated flow in metallic glass. *J. Non-Cryst. Solids* **2018**, *482*, 126–131. [CrossRef]
139. Jiang, W.H.; Fan, G.J.; Liu, F.X.; Wang, G.Y.; Choo, H.; Liaw, P.K. Spatiotemporally inhomogeneous plastic flow of a bulk-metallic glass. *Int. J. Plast.* **2008**, *24*, 1–16. [CrossRef]
140. Xie, X.; Lo, Y.-C.; Tong, Y.; Qiao, J.; Wang, G.; Ogata, S.; Qi, H.; Dahmen, K.A.; Gao, Y.; Liaw, P.K. Origin of serrated flow in bulk metallic glasses. *J. Mech. Phys. Solids* **2019**, *124*, 634–642. [CrossRef]
141. Schuh, C.A.; Nieh, T.G. A nanoindentation study of serrated flow in bulk metallic glasses. *Acta Mater.* **2003**, *51*, 87–99. [CrossRef]
142. Wei, B.C.; Zhang, T.H.; Li, W.H.; Sun, Y.F.; Yu, Y.; Wang, Y.R. Serrated plastic flow during nanoindentation in Nd-based bulk metallic glasses. *Intermetallics* **2004**, *12*, 1239–1243. [CrossRef]
143. Liu, L.; Chan, K.C. Plastic deformation of Zr-based bulk metallic glasses under nanoindentation. *Mater. Lett.* **2005**, *59*, 3090–3094. [CrossRef]
144. Li, W.H.; Wei, B.C.; Zhang, T.H.; Xing, D.M.; Zhang, L.C.; Wang, Y.R. Study of serrated flow and plastic deformation in metallic glasses through instrumented indentation. *Intermetallics* **2007**, *15*, 706–710. [CrossRef]
145. Jiang, W.H.; Jiang, F.; Liu, F.X.; Choo, H.; Liaw, P.K. Temperature dependence of serrated flows in compression in a bulk-metallic glass. *Appl. Phys. Lett.* **2006**, *89*. [CrossRef]
146. Georgarakis, K.; Aljerf, M.; Li, Y.; LeMoulec, A.; Charlot, F.; Yavari, A.R.; Chornokhvostenko, K.; Tabachnikova, E.; Evangelakis, G.A.; Miracle, D.B.; et al. Shear band melting and serrated flow in metallic glasses. *Appl. Phys. Lett.* **2008**, *93*, 3. [CrossRef]
147. Tang, C.G.; Li, Y.; Zeng, K.Y. Effect of residual shear bands on serrated flow in a metallic glass. *Mater. Lett.* **2005**, *59*, 3325–3329. [CrossRef]
148. Yu, G.S.; Lin, J.G.; Li, W.; Lin, Z.W. Structural relaxation and serrated flow due to annealing treatments in Zr-based metallic glasses. *J. Alloys Compd.* **2010**, *489*, 558–561. [CrossRef]
149. Kuo, C.N.; Chen, H.M.; Du, X.H.; Huang, J.C. Flow serrations and fracture morphologies of Cu-based bulk metallic glasses in energy release perspective. *Intermetallics* **2010**, *18*, 1648–1652. [CrossRef]
150. Brechtl, J.; Wang, Z.; Xie, X.; Qiao, J.-W.; Liaw, P.K. Relation between the defect interactions and the serration dynamics in a zr-based bulk metallic glass. *Appl. Sci.* **2020**, *10*, 3892. [CrossRef]
151. Brechtl, J.; Xie, X.; Wang, Z.; Qiao, J.; Liaw, P.K. Complexity analysis of serrated flows in a bulk metallic glass under constrained and unconstrained conditions. *Mater. Sci. Eng. A* **2020**, *771*, 138585. [CrossRef]
152. Chen, S.; Xie, X.; Chen, B.L.; Qiao, J.W.; Zhang, Y.; Ren, Y.; Dahmen, K.A.; Liaw, P.K. Effects of temperature on serrated flows of $Al_{0.5}$CoCrCuFeNi high-entropy alloy. *JOM* **2015**, *67*, 2314–2320. [CrossRef]

153. Carroll, R.; Lee, C.; Tsai, C.-W.; Yeh, J.-W.; Antonaglia, J.; Brinkman, B.A.W.; LeBlanc, M.; Xie, X.; Chen, S.; Liaw, P.K.; et al. Experiments and model for serration statistics in low-entropy, medium-entropy, and high-entropy alloys. *Sci. Rep.* **2015**, *5*, 16997. [CrossRef] [PubMed]
154. Chen, S.; Yu, L.; Ren, J.; Xie, X.; Li, X.; Xu, Y.; Zhao, G.; Li, P.; Yang, F.; Ren, Y.; et al. Self-similar random process and chaotic behavior in serrated flow of high entropy alloys. *Sci. Rep.* **2016**, *6*, 29798. [CrossRef]
155. Chen, S.; Xie, X.; Li, W.; Feng, R.; Chen, B.; Qiao, J.; Ren, Y.; Zhang, Y.; Dahmen, K.A.; Liaw, P.K. Temperature effects on the serrated behavior of an $Al_{0.5}$CoCrCuFeNi high-entropy alloy. *Mater. Chem. Phys.* **2018**, *210*, 20–28. [CrossRef]
156. Komarasamy, M.; Kumar, N.; Mishra, R.S.; Liaw, P.K. Anomalies in the deformation mechanism and kinetics of coarse-grained high entropy alloy. *Mater. Sci. Eng. A* **2016**, *654*, 256–263. [CrossRef]
157. Chen, S.; Li, W.; Xie, X.; Brechtl, J.; Chen, B.; Li, P.; Zhao, G.; Yang, F.; Qiao, J.; Dahmen, K.A.; et al. Nanoscale serration and creep characteristics of $Al_{0.5}$CoCrCuFeNi high-entropy alloys. *J. Alloys Compd.* **2018**, *752*, 464–475. [CrossRef]
158. Tong, C.J.; Chen, M.R.; Chen, S.K.; Yeh, J.W.; Shun, T.T.; Lin, S.J.; Chang, S.Y. Mechanical performance of the Al_xCoCrCuFeNi high-entropy alloy system with multiprincipal elements. *Metall. Mater. Trans. A Phys. Metall. Mater. Sci.* **2005**, *36A*, 1263–1271. [CrossRef]
159. Liu, J.; Guo, X.; Lin, Q.; Zhanbing, H.; An, X.; Li, L.; Liaw, P.K.; Liao, X.; Yu, L.; Lin, J.; et al. Excellent ductility and serration feature of metastable CoCrFeNi high-entropy alloy at extremely low temperatures. *Sci. China Mater.* **2019**, in press. [CrossRef]
160. Hu, Y.; Shu, L.; Yang, Q.; Guo, W.; Liaw, P.K.; Dahmen, K.A.; Zuo, J.M. Dislocation avalanche mechanism in slowly compressed high entropy alloy nanopillars. *Commun. Phys.* **2018**, *1*, 8. [CrossRef]
161. Jiao, Z.M.; Chu, M.Y.; Yang, H.J.; Wang, Z.H.; Qiao, J.W. Nanoindentation characterised plastic deformation of a $Al_{0.5}$CoCrFeNi high entropy alloy. *Mater. Sci. Technol.* **2015**, *31*, 1244–1249. [CrossRef]
162. Samal, S.; Rahul, M.R.; Kottada, R.S.; Phanikumar, G. Hot deformation behaviour and processing map of Co-Cu-Fe-Ni-Ti eutectic high entropy alloy. *Mater. Sci. Eng. A Struct. Mater. Prop. Microstruct. Process.* **2016**, *664*, 227–235. [CrossRef]
163. Fu, J.X.; Cao, C.M.; Tong, W.; Hao, Y.X.; Peng, L.M. The tensile properties and serrated flow behavior of a thermomechanically treated CoCrFeNiMn high-entropy alloy. *Mater. Sci. Eng. A Struct. Mater. Prop. Microstruct. Process.* **2017**, *690*, 418–426. [CrossRef]
164. Komarasamy, M.; Alagarsamy, K.; Mishra, R.S. Serration behavior and negative strain rate sensitivity of $Al_{0.1}$CoCrFeNi high entropy alloy. *Intermetallics* **2017**, *84*, 20–24. [CrossRef]
165. Guo, X.X.; Xie, X.; Ren, J.L.; Laktionova, M.; Tabachnikova, E.; Yu, L.P.; Cheung, W.-S.; Dahmen, K.A.; Liaw, P.K. Plastic dynamics of the $Al_{0.5}$CoCrCuFeNi high entropy alloy at cryogenic temperatures: Jerky flow, stair-like fluctuation, scaling behavior, and non-chaotic state. *Appl. Phys. Lett.* **2017**, *111*. [CrossRef]
166. Wang, B.F.; Fu, A.; Huang, X.X.; Liu, B.; Liu, Y.; Li, Z.Z.; Zan, X. Mechanical properties and microstructure of the CoCrFeMnNi high entropy alloy under high strain rate compression. *J. Mater. Eng. Perform.* **2016**, *25*, 2985–2992. [CrossRef]
167. Antonaglia, J.; Xie, X.; Tang, Z.; Tsai, C.-W.; Qiao, J.W.; Zhang, Y.; Laktionova, M.O.; Tabachnikova, E.D.; Yeh, J.W.; Senkov, O.N.; et al. Temperature effects on deformation and serration behavior of high-entropy alloys (HEAs). *JOM* **2014**, *66*, 2002–2008. [CrossRef]
168. Fu, J.X.; Cao, C.M.; Tong, W.; Peng, L.M. Effect of thermomechanical processing on microstructure and mechanical properties of CoCrFeNiMn high entropy alloy. *Trans. Nonferrous Met. Soc. China* **2018**, *28*, 931–938. [CrossRef]
169. Zhang, Y.; Li, J.S.; Wang, J.; Wang, W.Y.; Kou, H.C.; Beaugnon, E. Temperature dependent deformation mechanisms of Al0.3CoCrFeNi high-entropy alloy, starting from serrated flow behavior. *J. Alloys Compd.* **2018**, *757*, 39–43. [CrossRef]
170. Wang, B.F.; Yao, X.R.; Wang, C.; Zhang, X.Y.; Huang, X.X. Mechanical properties and microstructure of a nicrfecomn high-entropy alloy deformed at high strain rates. *Entropy* **2018**, *20*, 892. [CrossRef]
171. Zhang, H.T.; Siu, K.W.; Liao, W.B.; Wang, Q.; Yang, Y.; Lu, Y. In situ mechanical characterization of CoCrCuFeNi high-entropy alloy micro/nano-pillars for their size-dependent mechanical behavior. *Mater. Res. Express* **2016**, *3*, 8. [CrossRef]

172. Zhang, L.J.; Yu, P.F.; Cheng, H.; Zhang, H.; Diao, H.Y.; Shi, Y.Z.; Chen, B.L.; Chen, P.Y.; Feng, R.; Bai, J.; et al. Nanoindentation creep behavior of an Al0.3CoCrFeNi high-entropy alloy. *Metall. Mater. Trans. A Phys. Metall. Mater. Sci.* **2016**, *47*, 5871–5875. [CrossRef]
173. Ge, S.F.; Fu, H.M.; Zhang, L.; Mao, H.H.; Li, H.; Wang, A.M.; Li, W.R.; Zhang, H.F. Effects of Al addition on the microstructures and properties of MoNbTaTiV refractory high entropy alloy. *Mater. Sci. Eng. A Struct. Mater. Prop. Microstruct. Process.* **2020**, *784*, 9. [CrossRef]
174. Chen, W.Y.; Liu, X.; Chen, Y.R.; Yeh, J.W.; Tseng, K.K.; Natesan, K. Irradiation effects in high entropy alloys and 316H stainless steel at 300 degrees C. *J. Nucl. Mater.* **2018**, *510*, 421–430. [CrossRef]
175. Kang, M.; Lim, K.R.; Won, J.W.; Na, Y.S. Effect of Co content on the mechanical properties of A2 and B2 phases in $AlCo_xCrFeNi$ high-entropy alloys. *J. Alloys Compd.* **2018**, *769*, 808–812. [CrossRef]
176. Chen, M.; Liu, Y.; Li, Y.X.; Chen, X. Microstructure and mechanical properties of $AlTiFeNiCuCr_x$ high-entropy alloy with multi-principal elements. *Acta Metall. Sin.* **2007**, *43*, 1020–1024.
177. Yu, L.P.; Chen, S.Y.; Ren, J.L.; Ren, Y.; Yang, F.Q.; Dahmen, K.A.; Liaw, P.K. Plasticity performance of $Al_{0.5}CoCrCuFeNi$ high-entropy alloys under nanoindentation. *J. Iron Steel Res. Int.* **2017**, *24*, 390–396. [CrossRef]
178. Zhang, B.; Liaw, P.K.; Brechtl, J.; Ren, J.; Guo, X.; Zhang, Y. Effects of Cu and Zn on microstructures and mechanical behavior of the medium-entropy aluminum alloy. *J. Alloy. Compd.* **2020**, *820*, 153092. [CrossRef]
179. Chen, S.Y.; Yang, X.; Dahmen, K.A.; Liaw, P.K.; Zhang, Y. Microstructures and crackling noise of $Al_xNbTiMoV$ high entropy alloys. *Entropy* **2014**, *16*, 870–884. [CrossRef]
180. Brechtl, J.; Chen, S.Y.; Xie, X.; Ren, Y.; Qiao, J.W.; Liaw, P.K.; Zinkle, S.J. Towards a greater understanding of serrated flows in an Al-containing high-entropy-based alloy. *Int. J. Plast.* **2019**, *115*, 71–92. [CrossRef]
181. Tsai, C.-W.; Lee, C.; Lin, P.-T.; Xie, X.; Chen, S.; Carroll, R.; LeBlanc, M.; Brinkman, B.A.W.; Liaw, P.K.; Dahmen, K.A.; et al. Portevin-Le Chatelier mechanism in face-centered-cubic metallic alloys from low to high entropy. *Int. J. Plast.* **2019**, *122*, 212–224. [CrossRef]
182. Rodriguez, P. Serrated plastic flow. *Bull. Mater. Sci.* **1984**, *6*, 653–663. [CrossRef]
183. Weertman, J. Theory of infinitesimal dislocations distributed on a plane applied to discontinuous yield phenomena. *Can. J. Phys.* **1967**, *45*, 797–807. [CrossRef]
184. Pöhl, F. Pop-in behavior and elastic-to-plastic transition of polycrystalline pure iron during sharp nanoindentation. *Sci. Rep.* **2019**, *9*, 15350. [CrossRef]
185. Samuel, K.G.; Rodriguez, P. Age-Softening and Yield Points in Beta Brass. *Trans. Indian Inst. Metals.* **1980**, *33*, 285–295.
186. Soboyejo, W. *Mechanical Properties of Engineered Materials*; CRC Press: Boca Raton, FL, USA, 2002.
187. Cottrell, A.H.; Bilby, B.A. Dislocation theory of yielding and strain ageing of iron. *Proc. Phys. Soc. Sect. A* **1949**, *62*, 49–62. [CrossRef]
188. Herman, H. *Treatise on Materials Science and Technology*; Elsevier Science: London, UK, 2017; Volume 4.
189. Agrawal, B.K. *Introduction to Engineering Materials*; Tata McGraw-Hill: New Delhi, India, 1988.
190. Kozłowska, A.; Grzegorczyk, B.; Morawiec, M.; Grajcar, A. Explanation of the PLC effect in advanced high-strength medium-mn steels. a review. *Materials* **2019**, *12*, 4175. [CrossRef]
191. Yasuda, H.Y.; Shigeno, K.; Nagase, T. Dynamic strain aging of $Al_{0.3}CoCrFeNi$ high entropy alloy single crystals. *Scr. Mater.* **2015**, *108*, 80–83. [CrossRef]
192. Guo, L.; Gu, J.; Gong, X.; Li, K.; Ni, S.; Liu, Y.; Song, M. Short-range ordering induced serrated flow in a carbon contained FeCoCrNiMn high entropy alloy. *Micron* **2019**, *126*, 102739. [CrossRef] [PubMed]
193. Tabachnikova, E.D.; Podolskiy, A.V.; Laktionova, M.O.; Bereznaia, N.A.; Tikhonovsky, M.A.; Tortika, A.S. Mechanical properties of the $CoCrFeNiMnV_x$ high entropy alloys in temperature range 4.2–300 K. *J. Alloys Compd.* **2017**, *698*, 501–509. [CrossRef]
194. Salishchev, G.A.; Tikhonovsky, M.A.; Shaysultanov, D.G.; Stepanov, N.D.; Kuznetsov, A.V.; Kolodiy, I.V.; Tortika, A.S.; Senkov, O.N. Effect of Mn and V on structure and mechanical properties of high-entropy alloys based on CoCrFeNi system. *J. Alloys Compd.* **2014**, *591*, 11–21. [CrossRef]
195. Bazlov, A.I.; Churyumov, A.Y.; Louzguine-Luzgin, D.V. Investigation of the structure and properties of the Fe-Ni-Co-Cu-V multiprincipal element alloys. *Metall. Mater. Trans. A* **2018**, *49*, 5646–5652. [CrossRef]
196. Churyumov, A.Y.; Pozdniakov, A.V.; Bazlov, A.I.; Mao, H.; Polkin, V.I.; Louzguine-Luzgin, D.V. Effect of Nb addition on microstructure and thermal and mechanical properties of Fe-Co-Ni-Cu-Cr multiprincipal-element (high-entropy) alloys in As-Cast and heat-treated state. *JOM* **2019**, *71*, 3481–3489. [CrossRef]

197. Liu, L.H.; Liu, Z.Y.; Huan, Y.; Wu, X.Y.; Lou, Y.; Huang, X.S.; He, L.J.; Li, P.J.; Zhang, L.C. Effect of structural heterogeneity on serrated flow behavior of Zr-based metallic glass. *J. Alloys Compd.* **2018**, *766*, 908–917. [CrossRef]
198. Liu, Z.Y.; Yang, Y.; Liu, C.T. Yielding and shear banding of metallic glasses. *Acta Mater.* **2013**, *61*, 5928–5936. [CrossRef]
199. Suryanarayana, C.; Inoue, A. *Bulk Metallic Glasses*, 2nd ed.; CRC Press: New York, NY, USA, 2017.
200. *Bulk Metallic Glasses: An Overview*; Miller, M., Liaw, P.K., Eds.; Springer: New York, NY, USA, 2008.
201. Maaß, R.; Löffler, J.F. Shear-band dynamics in metallic glasses. *Adv. Funct. Mater.* **2015**, *25*, 2353–2368. [CrossRef]
202. Zhao, L.Z.; Xue, R.J.; Zhu, Z.G.; Lu, Z.; Axinte, E.; Wang, W.H.; Bai, H.Y. Evaluation of flow units and free volumes in metallic glasses. *J. Appl. Phys.* **2014**, *116*, 103516. [CrossRef]
203. Alrasheedi, N.H.; Yousfi, M.A.; Hajlaoui, K.; Mahfoudh, B.J.; Tourki, Z.; Yavari, A.R. On the modelling of the transient flow behavior of metallic glasses: Analogy with portevin-Le chatelier effect. *Metals* **2016**, *6*, 48. [CrossRef]
204. Pink, E.; Grinberg, A. Serrated flow in a ferritic stainless steel. *Mater. Sci. Eng.* **1981**, *51*, 1–8. [CrossRef]
205. Cai, Y.L.; Tian, C.G.; Fu, S.H.; Han, G.M.; Cui, C.Y.; Zhang, Q.C. Influence of gamma′ precipitates on Portevin-Le chatelier effect of Ni-based superalloys. *Mater. Sci. Eng. A Struct. Mater. Prop. Microstruct. Process.* **2015**, *638*, 314–321. [CrossRef]
206. Chandravathi, K.S.; Laha, K.; Parameswaran, P.; Mathew, M.D. Effect of microstructure on the critical strain to onset of serrated flow in modified 9Cr–1Mo steel. *Int. J. Press. Vessel. Pip.* **2012**, *89*, 162–169. [CrossRef]
207. Bayramin, B.; Şimşir, C.; Efe, M. Dynamic strain aging in DP steels at forming relevant strain rates and temperatures. *Mater. Sci. Eng. A* **2017**, *704*, 164–172. [CrossRef]
208. Calcagnotto, M.; Ponge, D.; Demir, E.; Raabe, D. Orientation gradients and geometrically necessary dislocations in ultrafine grained dual-phase steels studied by 2D and 3D EBSD. *Mater. Sci. Eng. A* **2010**, *527*, 2738–2746. [CrossRef]
209. Sarosiek, A.M.; Owen, W.S. The work hardening of dual-phase steels at small plastic strains. *Mater. Sci. Eng.* **1984**, *66*, 13–34. [CrossRef]
210. Gopinath, K.; Gogia, A.K.; Kamat, S.V.; Ramamurty, U. Dynamic strain ageing in Ni-base superalloy 720Li. *Acta Mater.* **2009**, *57*, 1243–1253. [CrossRef]
211. Sakthivel, T.; Laha, K.; Nandagopal, M.; Chandravathi, K.S.; Parameswaran, P.; Panneer Selvi, S.; Mathew, M.D.; Mannan, S.K. Effect of temperature and strain rate on serrated flow behaviour of Hastelloy X. *Mater. Sci. Eng. A* **2012**, *534*, 580–587. [CrossRef]
212. Robinson, J.M.; Shaw, M.P. Microstructural and mechanical influences on dynamic strain aging phenomena. *Int. Mater. Rev.* **1994**, *39*, 113–122. [CrossRef]
213. Choudhary, B.K. Influence of strain rate and temperature on serrated flow in 9Cr–1Mo ferritic steel. *Mater. Sci. Eng. A* **2013**, *564*, 303–309. [CrossRef]
214. Worthington, P.J.; Brindley, B.J. Serrated yielding in substitutional alloys. *Philos. Mag. A J. Theor. Exp. Appl. Phys.* **1969**, *19*, 1175–1178. [CrossRef]
215. Fu, S.; Zhang, Q.; Hu, Q.; Gong, M.; Cao, P.; Liu, H. The influence of temperature on the PLC effect in Al-Mg alloy. *Sci. China Technol. Sci.* **2011**, *54*, 1389. [CrossRef]
216. Bharathi, M.S.; Lebyodkin, M.; Ananthakrishna, G.; Fressengeas, C.; Kubin, L.P. Multifractal burst in the spatiotemporal dynamics of jerky flow. *Phys. Rev. Lett.* **2001**, *87*, 4. [CrossRef]
217. Fu, S.-H.; Cai, Y.-L.; Yang, S.-L.; Zhang, Q.-C.; Wu, X.-P. The mechanism of critical strain of serrated yielding in strain rate domain. *Chin. Phys. Lett.* **2016**, *33*, 026201.
218. Yuan, B.; Li, J.-J.; Qiao, J.-W. Statistical analysis on strain-rate effects during serrations in a Zr-based bulk metallic glass. *J. Iron Steel Res. Int.* **2017**, *24*, 455–461. [CrossRef]
219. Brechtl, J.; Wang, H.; Kumar, N.A.P.K.; Yang, T.; Lin, Y.R.; Bei, H.; Neuefeind, J.; Dmowski, W.; Zinkle, S.J. Investigation of the thermal and neutron irradiation response of BAM-11 bulk metallic glass. *J. Nucl. Mater.* **2019**, *526*, 151771. [CrossRef]
220. Li, W.; Bei, H.; Tong, Y.; Dmowski, W.; Gao, Y.F. Structural heterogeneity induced plasticity in bulk metallic glasses: From well-relaxed fragile glass to metal-like behavior. *Appl. Phys. Lett.* **2013**, *103*, 171910. [CrossRef]
221. Lebedkina, T.A.; Lebyodkin, M.A. Effect of deformation geometry on the intermittent plastic flow associated with the Portevin–Le Chatelier effect. *Acta Mater.* **2008**, *56*, 5567–5574. [CrossRef]

222. Friedman, N.; Jennings, A.T.; Tsekenis, G.; Kim, J.-Y.; Tao, M.; Uhl, J.T.; Greer, J.R.; Dahmen, K.A. Statistics of dislocation slip avalanches in nanosized single crystals show tuned critical behavior predicted by a simple mean field model. *Phys. Rev. Lett.* **2012**, *109*, 1–5. [CrossRef]
223. Lebedkina, T.A.; Lebyodkin, M.A.; Lamark, T.T.; Janeček, M.; Estrin, Y. Effect of equal channel angular pressing on the Portevin–Le Chatelier effect in an Al_3Mg alloy. *Mater. Sci. Eng. A* **2014**, *615*, 7–13. [CrossRef]
224. Lebyodkin, M.A.; Lebedkina, T.A. Multifractality and randomness in the unstable plastic flow near the lower strain-rate boundary of instability. *Phys. Rev. E* **2008**, *77*, 8. [CrossRef] [PubMed]
225. Sarkar, A.; Chatterjee, A.; Barat, P.; Mukherjee, P. Comparative study of the Portevin-Le Chatelier effect in interstitial and substitutional alloy. *Mater. Sci. Eng. A Struct. Mater. Prop. Microstruct. Process.* **2007**, *459*, 361–365. [CrossRef]
226. Costa, M.; Goldberger, A.L.; Peng, C.K. Multiscale entropy analysis of complex physiologic time series. *Phys. Rev. Lett.* **2002**, *89*, 068102. [CrossRef] [PubMed]
227. Costa, M.; Goldberger, A.L.; Peng, C.K. Multiscale entropy analysis of biological signals. *Phys. Rev. E* **2005**, *71*, 021906. [CrossRef]
228. Xia, J.A.; Shang, P.J. Multiscale entropy analysis of financial time series. *Fluct. Noise Lett.* **2012**, *11*, 12. [CrossRef]
229. Costa, M.D.; Henriques, T.; Munshi, M.N.; Segal, A.R.; Goldberger, A.L. Dynamical glucometry: Use of multiscale entropy analysis in diabetes. *Chaos* **2014**, *24*, 5. [CrossRef]
230. Pincus, S.M. Approximate entropy as a measure of system complexity. *Proc. Natl. Acad. Sci. USA* **1991**, *88*, 2297–2301. [CrossRef]
231. Weinstein, M.; Hermalin, A.I.; Stoto, M.A. *Population Health and Aging: Strengthening the Dialogue between Epidemiology and Demography*; New York Academy of Sciences: New York, NY, USA, 2001.
232. Yentes, J.M.; Hunt, N.; Schmid, K.K.; Kaipust, J.P.; McGrath, D.; Stergiou, N. The appropriate use of approximate entropy and sample entropy with short data sets. *Ann. Biomed. Eng.* **2013**, *41*, 349–365. [CrossRef]
233. Volos, C.; Jafari, S.; Kengne, J.; Munoz-Pacheco, J.M.; Rajagopal, K. *Nonlinear Dynamics and Entropy of Complex Systems with Hidden and Self-Excited Attractors*; MDPI AG: Basel, Switzerland, 2019.
234. Wu, S.-D.; Wu, C.-W.; Lin, S.-G.; Wang, C.-C.; Lee, K.-Y. Time series analysis using composite multiscale entropy. *Entropy* **2013**, *15*, 1069–1084. [CrossRef]
235. Wu, S.-D.; Wu, C.-W.; Lin, S.-G.; Lee, K.-Y.; Peng, C.-K. Analysis of complex time series using refined composite multiscale entropy. *Phys. Lett. A* **2014**, *378*, 1369–1374. [CrossRef]
236. Brechtl, J.; Xie, X.; Liaw, P.K.; Zinkle, S.J. Complexity modeling and analysis of chaos and other fluctuating phenomena. *Chaos Solitons Fractals* **2018**, *116*, 166–175. [CrossRef]
237. Blons, E.; Arsac, L.M.; Gilfriche, P.; Deschodt-Arsac, V. Multiscale entropy of cardiac and postural control reflects a flexible adaptation to a cognitive task. *Entropy* **2019**, *21*, 1024. [CrossRef]
238. Iliopoulos, A.C.; Nikolaidis, N.S.; Aifantis, E.C. Analysis of serrations and shear bands fractality in UFGs. *J. Mech. Behav. Mater.* **2015**, *24*, 1–9. [CrossRef]
239. Heister, T.; Rebholz, L.G. *Scientific Computing: For. Scientists and Engineers*; De Gruyter: Berlin, Germany, 2015.
240. Costa, M.D.; Goldberger, A.L. Generalized multiscale entropy analysis: Application to quantifying the complex volatility of human heartbeat time series. *Entropy* **2015**, *17*, 1197–1203. [CrossRef] [PubMed]
241. Ihlen, E. Introduction to multifractal detrended fluctuation analysis in matlab. *Front. Physiol.* **2012**, *3*. [CrossRef] [PubMed]
242. Lebyodkin, M.A.; Estrin, Y. Multifractal analysis of the Portevin-Le Chatelier effect: General approach and application to AlMg and $AlMg/Al_2O_3$ alloys. *Acta Mater.* **2005**, *53*, 3403–3413. [CrossRef]
243. Lebyodkin, M.A.; Kobelev, N.P.; Bougherira, Y.; Entemeyer, D.; Fressengeas, C.; Lebedkina, T.A.; Shashkov, I.V. On the similarity of plastic flow processes during smooth and jerky flow in dilute alloys. *Acta Mater.* **2012**, *60*, 844–850. [CrossRef]
244. Lebyodkin, M.A.; Lebedkina, T.A.; Jacques, A. *Multifractal Analysis of Unstable Plastic Flow*; Nova Science Publishers, Inc.: New York, NY, USA, 2009.
245. Lebyodkin, M.A.; Lebedkina, T.A. Multifractal analysis of evolving noise associated with unstable plastic flow. *Phys. Rev. E* **2006**, *73*, 8. [CrossRef]
246. Ren, J.L.; Chen, C.; Liu, Z.Y.; Li, R.; Wang, G. Plastic dynamics transition between chaotic and self-organized critical states in a glassy metal via a multifractal intermediate. *Phys. Rev. B* **2012**, *86*, 134303. [CrossRef]

247. Aifantis, E.C. Chapter one—Internal Length Gradient (ILG) material mechanics across scales and disciplines. In *Advances in Applied Mechanics*; Bordas, S.P.A., Balint, D.S., Eds.; Elsevier: Amsterdam, The Netherlands, 2016; Volume 49, pp. 1–110.
248. Salat, H.; Murcio, R.; Arcaute, E. Multifractal methodology. *Physica A* **2017**, *473*, 467–487. [CrossRef]
249. Dahmen, K.A.; Uhl, J.T.; Wright, W.J. Why the crackling deformations of single crystals, metallic glasses, rock, granular materials, and the earth's crust are so surprisingly similar. *Front. Phys.* **2019**, *7*. [CrossRef]
250. Long, A.A.; Denisov, D.V.; Schall, P.; Hufnagel, T.C.; Gu, X.; Wright, W.J.; Dahmen, K.A. From critical behavior to catastrophic runaways: Comparing sheared granular materials with bulk metallic glasses. *Granul. Matter* **2019**, *21*, 99. [CrossRef]
251. Denisov, D.V.; Lörincz, K.A.; Uhl, J.T.; Dahmen, K.A.; Schall, P. Universality of slip avalanches in flowing granular matter. *Nat. Commun.* **2016**, *7*, 10641. [CrossRef]
252. Dahmen, K.A.; Ben-Zion, Y.; Uhl, J.T. A simple analytic theory for the statistics of avalanches in sheared granular materials. *Nat. Phys.* **2011**, *7*, 554–557. [CrossRef]
253. Salje, E.K.H.; Saxena, A.; Planes, A. *Avalanches in Functional Materials and Geophysics*; Springer International Publishing: Berlin/Heidelberg, Germany, 2016.
254. Egami, T.; Ojha, M.; Khorgolkhuu, O.; Nicholson, D.M.; Stocks, G.M. Local electronic effects and irradiation resistance in high-entropy alloys. *JOM* **2015**, *67*, 2345–2349. [CrossRef]
255. Falk, M.L.; Langer, J.S. Deformation and failure of amorphous, solidlike materials. *Annu. Rev. Condens. Matter Phys.* **2011**, *2*, 353–373. [CrossRef]
256. Dahmen, K.A.; Ben-Zion, Y.; Uhl, J.T. Micromechanical model for deformation in solids with universal predictions for stress-strain curves and slip avalanches. *Phys. Rev. Lett.* **2009**, *102*, 175501. [CrossRef]
257. Bian, X.L.; Wang, G.; Chan, K.C.; Ren, J.L.; Gao, Y.L.; Zhai, Q.J. Shear avalanches in metallic glasses under nanoindentation: Deformation units and rate dependent strain burst cut-off. *Appl. Phys. Lett.* **2013**, *103*, 101907. [CrossRef]
258. Toker, D.; Sommer, F.T.; D'Esposito, M. A simple method for detecting chaos in nature. *Commun. Biol.* **2020**, *3*, 11. [CrossRef]
259. Xu, T.; Scaffidi, T.; Cao, X. Does scrambling equal chaos? *Phys. Rev. Lett.* **2020**, *124*, 140602. [CrossRef]
260. Brechtl, J.; Xie, X.; Liaw, P.K. Investigation of chaos and memory effects in the Bonhoeffer-van der Pol oscillator with a non-ideal capacitor. *Commun. Nonlinear Sci. Numer. Simul.* **2019**, *73*, 195–216. [CrossRef]
261. Sekikawa, M.; Shimizu, K.; Inaba, N.; Kita, H.; Endo, T.; Fujimoto, K.; Yoshinaga, T.; Aihara, K. Sudden change from chaos to oscillation death in the Bonhoeffer-van der Pol oscillator under weak periodic perturbation. *Phys. Rev. E* **2011**, *84*, 8. [CrossRef] [PubMed]
262. Kousaka, T.; Ogura, Y.; Shimizu, K.; Asahara, H.; Inaba, N. Analysis of mixed-mode oscillation-incrementing bifurcations generated in a nonautonomous constrained Bonhoeffer-van der Pol oscillator. *Phys. D Nonlinear Phenom.* **2017**, *353*, 48–57. [CrossRef]
263. Krishnamurthy, V. Predictability of weather and climate. *Earth Space Sci.* **2019**, *6*, 1043–1056. [CrossRef] [PubMed]
264. Selvam, A. *Chaotic Climate Dynamics*; Luniver Press: London, UK, 2007.
265. Buizza, R. Chaos and weather prediction a review of recent advances in numerical weather prediction Ensemble forecasting and adaptive observation targeting. *Nuovo Cim. C* **2001**, *24*, 273–301.
266. Korn, H.; Faure, P. Is there chaos in the brain? II. Experimental evidence and related models. *C. R. Biol.* **2003**, *326*, 787–840. [CrossRef]
267. Destexhe, A. Oscillations, complex spatiotemporal behavior, and information transport in networks of excitatory and inhibitory neurons. *Phys. Rev. E* **1994**, *50*, 1594–1606. [CrossRef]
268. Tavares, B.S.; de Paula Vidigal, G.; Garner, D.M.; Raimundo, R.D.; de Abreu, L.C.; Valenti, V.E. Effects of guided breath exercise on complex behaviour of heart rate dynamics. *Clin. Physiol. Funct. Imaging* **2017**, *37*, 622–629. [CrossRef]
269. Kaizoji, T. Intermittent chaos in a model of financial markets with heterogeneous agents. *Chaos Solitons Fractals* **2004**, *20*, 323–327. [CrossRef]
270. Tiwari, A.K.; Gupta, R. Chaos in G7 stock markets using over one century of data: A note. *Res. Int. Bus. Financ.* **2019**, *47*, 304–310. [CrossRef]
271. Tsionas, M.G.; Michaelides, P.G. Neglected chaos in international stock markets: Bayesian analysis of the joint return–volatility dynamical system. *Phys. A Stat. Mech. Its Appl.* **2017**, *482*, 95–107. [CrossRef]

272. Nepomuceno, E.G.; Perc, M. Computational chaos in complex networks. *J. Complex Netw.* **2019**, *8*. [CrossRef]
273. Bharathi, M.S.; Lebyodkin, M.; Ananthakrishna, G.; Fressengeas, C.; Kubin, L.P. The hidden order behind jerky flow. *Acta Mater.* **2002**, *50*, 2813–2824. [CrossRef]
274. Panday, P.; Samanta, S.; Pal, N.; Chattopadhyay, J. Delay induced multiple stability switch and chaos in a predator–prey model with fear effect. *Math. Comput. Simul.* **2020**, *172*, 134–158. [CrossRef]
275. Ruxton, G.D. Chaos in a three-species food chain with a lower bound on the bottom population. *Ecology* **1996**, *77*, 317–319. [CrossRef]
276. Telesh, I.V.; Schubert, H.; Joehnk, K.D.; Heerkloss, R.; Schumann, R.; Feike, M.; Schoor, A.; Skarlato, S.O. Chaos theory discloses triggers and drivers of plankton dynamics in stable environment. *Sci. Rep.* **2019**, *9*, 20351. [CrossRef]
277. Ananthakrishna, G.; Valsakumar, M.C. Chaotic flow in a model for repeated yielding. *Phys. Lett. A* **1983**, *95*, 69–71. [CrossRef]
278. Lorenz, E.N. Deterministic nonperiodic flow. *J. Atmos. Sci.* **1963**, *20*, 130–141. [CrossRef]
279. Barreira, L. *Lyapunov Exponents*; Springer: Berlin/Heidelberg, Germany, 2017.
280. Wolf, A.; Swift, J.B.; Swinney, H.L.; Vastano, J.A. Determining lyapunov exponents from a time series. *Phys. D Nonlinear Phenom.* **1985**, *16*, 285–317. [CrossRef]
281. Oliver, W.C.; Pharr, G.M. Nanoindentation in materials research: Past, present, and future. *MRS Bull.* **2010**, *35*, 897–907. [CrossRef]
282. Schuh, C.A. Nanoindentation studies of materials. *Mater. Today* **2006**, *9*, 32–40. [CrossRef]
283. Qiao, J.W.; Zhang, Y.; Liaw, P.K. Serrated flow kinetics in a Zr-based bulk metallic glass. *Intermetallics* **2010**, *18*, 2057–2064. [CrossRef]
284. Chan, P.Y.; Tsekenis, G.; Dantzig, J.; Dahmen, K.A.; Goldenfeld, N. Plasticity and dislocation dynamics in a phase field crystal model. *Phys. Rev. Lett.* **2010**, *105*, 015502. [CrossRef] [PubMed]
285. Xia, S.Q.; Zhang, Y. Deformation mechanisms of $Al_{0.1}CoCrFeNi$ high entropy alloy at ambient and cryogenic temperatures. *Mater. Sci. Eng. A Struct. Mater. Prop. Microstruct. Process.* **2018**, *733*, 408–413. [CrossRef]
286. Gerold, V.; Karnthaler, H.P. On the origin of planar slip in f.c.c. alloys. *Acta Metall.* **1989**, *37*, 2177–2183. [CrossRef]
287. Qiang, J.; Tsuchiya, K.; Diao, H.Y.; Liaw, P.K. Vanishing of room-temperature slip avalanches in a face-centered-cubic high-entropy alloy by ultrafine grain formation. *Scr. Mater.* **2018**, *155*, 99–103. [CrossRef]
288. Zhang, L.; Ohmura, T. Plasticity initiation and evolution during nanoindentation of an iron–3% silicon crystal. *Phys. Rev. Lett.* **2014**, *112*, 145504. [CrossRef]
289. Wang, L.; Bei, H.; Li, T.L.; Gao, Y.F.; George, E.P.; Nieh, T.G. Determining the activation energies and slip systems for dislocation nucleation in body-centered cubic Mo and face-centered cubic Ni single crystals. *Scr. Mater.* **2011**, *65*, 179–182. [CrossRef]
290. Daoud, H.M.; Manzoni, A.M.; Wanderka, N.; Glatzel, U. High-temperature tensile strength of $Al_{10}Co_{25}Cr_8Fe_{15}Ni_{36}Ti_6$ compositionally complex alloy (high-entropy alloy). *JOM* **2015**, *67*, 2271–2277. [CrossRef]
291. Lu, Y.; Dong, Y.; Guo, S.; Jiang, L.; Kang, H.; Wang, T.; Wen, B.; Wang, Z.; Jie, J.; Cao, Z.; et al. A promising new class of high-temperature alloys: Eutectic high-entropy alloys. *Sci. Rep.* **2014**, *4*, 6200. [CrossRef]
292. Basu, I.; Ocelik, V.; De Hosson, J.T.M. Size dependent plasticity and damage response in multiphase body centered cubic high entropy alloys. *Acta Mater.* **2018**, *150*, 104–116. [CrossRef]
293. Fischer-Cripps, A.C. *Nanoindentation*; Springer: New York, NY, USA, 2013.
294. Cottrell, A.H. LXXXVI. A note on the Portevin-Le chatelier effect. *Lond. Edinb. Dublin Philos. Mag. J. Sci.* **1953**, *44*, 829–832. [CrossRef]
295. Yang, Y.; Ma, L.; Gan, G.-Y.; Wang, W.; Tang, B.-Y. Investigation of thermodynamic properties of high entropy (TaNbHfTiZr)C and (TaNbHfTiZr)N. *J. Alloy. Compd.* **2019**, *788*, 1076–1083. [CrossRef]
296. Laktionova, M.A.; Tabchnikova, E.D.; Tang, Z.; Liaw, P.K. Mechanical properties of the high-entropy alloy $Ag_{0.5}CoCrCuFeNi$ at temperatures of 4.2–300 K. *Low Temp. Phys.* **2013**, *39*, 630–632. [CrossRef]
297. Mohamed, F.A.; Murty, K.L.; Langdon, T.G. The portevin-le chatelier effect in Cu_3Au. *Acta Metall.* **1974**, *22*, 325–332. [CrossRef]
298. Sridharan, S. *Delamination Behaviour of Composites*; Elsevier Science: Cambridge, UK, 2008.

299. Tirunilai, A.S.; Sas, J.; Weiss, K.-P.; Chen, H.; Szabó, D.V.; Schlabach, S.; Haas, S.; Geissler, D.; Freudenberger, J.; Heilmaier, M.; et al. Peculiarities of deformation of CoCrFeMnNi at cryogenic temperatures. *J. Mater. Res.* **2018**, *33*, 3287–3300. [CrossRef]
300. Sonkusare, R.; Jain, R.; Biswas, K.; Parameswaran, V.; Gurao, N.P. High strain rate compression behaviour of single phase CoCuFeMnNi high entropy alloy. *J. Alloy. Compd.* **2020**, *823*, 153763. [CrossRef]
301. Chen, S.Y.; Wang, L.; Li, W.D.; Tong, Y.; Tseng, K.K.; Tsai, C.W.; Yeh, J.W.; Ren, Y.; Guo, W.; Poplawsky, J.D.; et al. Peierls barrier characteristic and anomalous strain hardening provoked by dynamic-strain-aging strengthening in a body-centered-cubic high-entropy alloy. *Mater. Res. Lett.* **2019**, *7*, 475–481. [CrossRef]
302. Hähner, P. On the physics of the Portevin-Le Châtelier effect part 1: The statistics of dynamic strain ageing. *Mater. Sci. Eng. A* **1996**, *207*, 208–215. [CrossRef]
303. Tian, Q.; Zhang, G.; Yin, K.; Wang, L.; Wang, W.; Cheng, W.; Wang, Y.; Huang, J.C. High temperature deformation mechanism and microstructural evolution of relatively lightweight AlCoCrFeNi high entropy alloy. *Intermetallics* **2020**, *119*, 106707. [CrossRef]
304. Xu, J.; Cao, C.-m.; Gu, P.; Peng, L.-m. Microstructures, tensile properties and serrated flow of AlxCrMnFeCoNi high entropy alloys. *Trans. Nonferrous Met. Soc. China* **2020**, *30*, 746–755. [CrossRef]
305. Suzuki, T.; Takeuchi, S.; Yoshinaga, H. *Dislocation Dynamics and Plasticity*; Springer: Berlin/Heidelberg, Germany, 2013.
306. Isayev, O.; Tropsha, A.; Curtarolo, S. *Materials Informatics: Methods, Tools, and Applications*; Wiley: Berlin, Germany, 2019.
307. Pagan, D.C.; Phan, T.Q.; Weaver, J.S.; Benson, A.R.; Beaudoin, A.J. Unsupervised learning of dislocation motion. *Acta Mater.* **2019**, *181*, 510–518. [CrossRef]
308. Shen, C.; Wang, C.; Wei, X.; Li, Y.; van der Zwaag, S.; Xu, W. Physical metallurgy-guided machine learning and artificial intelligent design of ultrahigh-strength stainless steel. *Acta Mater.* **2019**, *179*, 201–214. [CrossRef]
309. Morand, L.; Helm, D. A mixture of experts approach to handle ambiguities in parameter identification problems in material modeling. *Comput. Mater. Sci.* **2019**, *167*, 85–91. [CrossRef]
310. Salmenjoki, H.; Alava, M.J.; Laurson, L. Machine learning plastic deformation of crystals. *Nat. Commun.* **2018**, *9*, 5307. [CrossRef] [PubMed]

Article

Activation Volume and Energy for Dislocation Nucleation in Multi-Principal Element Alloys

Sanghita Mridha, Maryam Sadeghilaridjani and Sundeep Mukherjee *

Department of Materials Science and Engineering, University of North Texas, Denton, TX 76203, USA; sanghitamridha@my.unt.edu (S.M.); Maryam.Sadeghilaridjani@unt.edu (M.S.)

* Correspondence: sundeep.mukherjee@unt.edu; Tel.: +1-940-565-4170; Fax: +1-940-565-2944

Received: 25 January 2019; Accepted: 21 February 2019; Published: 23 February 2019

Abstract: Incipient plasticity in multi-principal element alloys, CoCrNi, CoCrFeMnNi, and $Al_{0.1}$CoCrFeNi was evaluated by nano-indentation and compared with pure Ni. The tests were performed at a loading rate of 70 μN/s in the temperature range of 298 K to 473 K. The activation energy and activation volume were determined using a statistical approach of analyzing the "pop-in" load marking incipient plasticity. The CoCrFeMnNi and $Al_{0.1}$CoCrFeNi multi-principal element alloys showed two times higher activation volume and energy compared to CoCrNi and pure Ni, suggesting complex cooperative motion of atoms for deformation in the five component systems. The small calculated values of activation energy and activation volume indicate heterogeneous dislocation nucleation at point defects like vacancy and hot-spot.

Keywords: multi-principal element alloys; dislocation nucleation; activation volume; activation energy; nano-indentation; high/medium entropy alloys

---

1. Introduction

Traditional alloy design employs the addition of alloying elements to a principal constituent to enhance its properties via multi-phase complex microstructure. Multi-principal element alloys (MPEAs) represent a new generation of material system consisting of several elements in equimolar or near equimolar proportions. Despite the complex chemistry, high configurational entropy may favor formation of a simple solid solution instead of complex intermetallic compounds in these alloys, which are also referred to as high entropy alloys (HEAs) [1]. Certain range in atomic size difference, electronegativity difference, and mixing entropy/enthalpy of the constituent elements result in multiple concentrated phases in the microstructure [2–5].

The mechanical properties of MPEAs have been widely studied in recent years over a range of temperatures for potential applications in nuclear and aerospace industries. These include tensile/compressive properties, fatigue behavior, and fracture toughness [6–12]. It has been hypothesized that MPEAs have a distorted lattice structure which makes dislocation movement difficult and non-traditional in these alloy systems [1,6–8,13–18]. To understand dislocation nucleation and the activation process, nano-indentation technique has been widely used. The displacement burst (pop-in) in the nano-indentation load versus contact depth curve is attributed to dislocation nucleation. The "pop-in" represents the transition from elastic to plastic deformation in materials and typically happens at a small indentation depth of <100 nm [19–26]. The stress associated with the pop-in is approximately equal to the theoretical strength of the material and happens in small indented volume with a very low probability of having pre-existing dislocations. Therefore, many researchers have attributed the pop-in event to dislocation nucleation [21,25,27–33], and others have proposed vacancy assisted mechanisms [34]. It has been demonstrated that the pop-in during nano-indentation is time and temperature dependent, a kinetically limiting process that requires activation energy. It occurs in thermally and mechanically favorable positions, which makes it a probabilistic event, i.e., the load at which the pop-in takes place is distributed over a range rather than occurring at a fixed value each time, and this distribution is temperature and time dependent.

Nano-indentation technique has been extensively used to study the local scale elastic-plastic deformation behavior of metals/alloys and to determine the critical shear stress required for dislocation nucleation [19–26]. However, there are few reports and limited understanding of the effect of intrinsic length-scale and submicron plasticity in MPEAs. Wang et al. [35] and Zhu et al. [36] have investigated the activation volume and energy for the onset of plastic deformation in two body centered cubic (BCC) HEAs (of compositions NbTaTiZr and NbMoTaTiZr) and a face centered cubic (FCC) HEA (of composition CoCrFeMnNi) by nano-indentation. They reported a larger activation volume for HEAs compared to conventional metals, attributing it to complex nucleation mechanism. Still, there is limited understanding of the dependence of dislocation nucleation on alloy composition and microstructure, and there are multiple discrepancies in the published studies [2,36].

Here, incipient plasticity and dislocation nucleation were studied using nano-indentation for three different multi-principal element alloys, CoCrNi, CoCrFeMnNi, and $Al_{0.1}CoCrFeNi$. All the three alloys showed a single-phase FCC microstructure without any secondary phases. The three-component CoCrNi alloy showed higher strength than five-component CoCrFeMnNi [7,37] and was chosen to study the effect of the number of constituents. A comparison was made with a pure metal, Ni, to understand solid solution strengthening effects. In addition, $Al_{0.1}CoCrFeNi$ alloy was chosen because it is a model FCC system for comparison with the results of CoCrFeMnNi. The present study paves the way for a fundamental understanding of small-scale deformation mechanisms in multi-principal element alloys as a function of alloy chemistry and microstructure.

2. Experimental

Alloys with nominal composition of CoCrNi, CoCrFeMnNi, and $Al_{0.1}CoCrFeNi$ (in mole fractions) were prepared by arc melting the constituent pure elements (purity > 99.99%) in Ti-gettered Ar atmosphere. As-cast alloys were rolled up to a 70% reduction in thickness. After rolling, they were annealed at 1173 K for 20 h to obtain equiaxed dislocation-free grains and to remove residual stress from processing. The annealed samples were polished with SiC abrasive paper followed by microfiber cloth with 1 μm diamond suspension. The polished samples were then put in vibromet with 0.04 μm colloidal silica suspension to produce a surface finish equivalent in texture to an electro-polished sample. Crystal structure and phase identification were done using Rigaku III Ultima X-ray diffractometer (XRD Rigaku Corporation, Tokyo, Japan) with Cu-Kα radiation with wavelength of 1.54 Å. Scanning electron microscopy (SEM) was done using FEI Quanta ESEM (FEI Company, Hillsboro, OR, USA) to analyze the grain size and microstructure of the alloys.

Nano-indentation (Bruker, Minneapolis, MN, USA) was performed using a Berkovich diamond probe. The pop-in behavior of alloys was evaluated with a maximum load of 350 μN at 298 K, 373 K, and 473 K. One hundred indentations were made at each temperature with 10 μm distance between two indents to avoid overlap of their plastic zones. The indentations were done in load control mode with a constant loading rate of 70 μN/s. Elevated temperature indentations were performed using XSol600 heating stage with bottom and top heating to avoid any thermal gradients. To prevent oxidation of the sample, a continuous flow of Ar + 5% H_2 gas mixture was maintained with a flow rate of 1 lit/min.

3. Results and Discussion

Figure 1a–c show the XRD patterns for the three alloys studied, which confirmed that they formed a simple FCC solid solution and were free of any secondary phases as reported previously [38–40]. The backscattered SEM images of the alloys are shown in Figure 2a–c. All three alloys showed equiaxed microstructure with an average grain size of ~30–60 μm. They also showed a significant number of annealing twins, which indicates low stacking fault energy in these alloys.

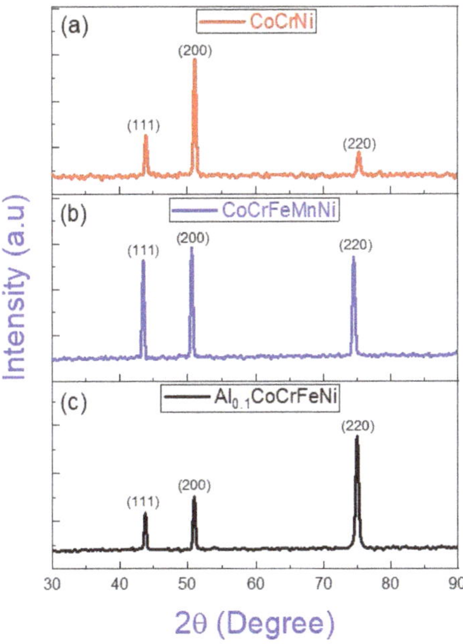

Figure 1. X-ray diffraction (XRD) patterns of multi-principal (**a**) CoCrNi, (**b**) CoCrFeMnNi, and (**c**) $Al_{0.1}CoCrFeNi$ alloys.

Figure 2. Backscattered scanning electron microscopy (SEM) images of (**a**) CoCrNi, (**b**) CoCrFeMnNi, and (**c**) $Al_{0.1}CoCrFeNi$ show equiaxed grains after annealing with grain size ~30–60 μm.

Deformation during nano-indentation is initially elastic and follows the Hertzian equation [34].

$$P = 4/3E_r R^{\frac{1}{2}} h^{\frac{3}{2}} \tag{1}$$

where, P is the load applied by the indenter, R is radius of the tip, and h is the indentation depth, E_r is the reduced modulus of the indenter-sample combination and is defined by: $\frac{1}{E_r} = \frac{1-v^2}{E} + \frac{1-v_i^2}{E_i}$. E, v, E_i = 1141 GPa, and v_i = 0.07 are the Young's modulus and Poisson's ratio of the sample and diamond indenter, respectively. Figure 3a shows the typical load-displacement curve obtained from nano-indentation along with the Hertzian fit. The reduced modulus of the alloys was obtained from the fitting coefficient of the elastic part and values are summarized in Table 1 and are similar to earlier reports [41,42]. The initial part of the load-displacement curve followed Hertzian contact theory and subsequently deviated due to plastic deformation or elastic-to-plastic transition [34]. The load and displacement at the deviation point was determined for each indent. Figure 3b shows the temperature dependence of the pop-in load for the CoCrFeMnNi alloy. The load at first pop-in was found to decrease with increasing temperature; other alloys (CoCrNi and $Al_{0.1}$CoCrFeNi) and pure Ni showed a similar trend. This thermal softening has also been reported in several previous studies [34,36].

Table 1. Reduced modulus, activation volume, and activation energy of present face-centered cubic (FCC) alloys.

Alloys	E_r (GPa)	Activation Volume (Å^3)	Activation Energy (eV)
Ni	195 ± 7	4.12	0.23 ± 0.008
CoCrNi	195 ± 6	4.69	0.27 ± 0.01
CoCrFeMnNi	180 ± 4	9.01	0.50 ± 0.007
$Al_{0.1}$CoCrFeNi	190 ± 5	8.04	0.47 ± 0.009

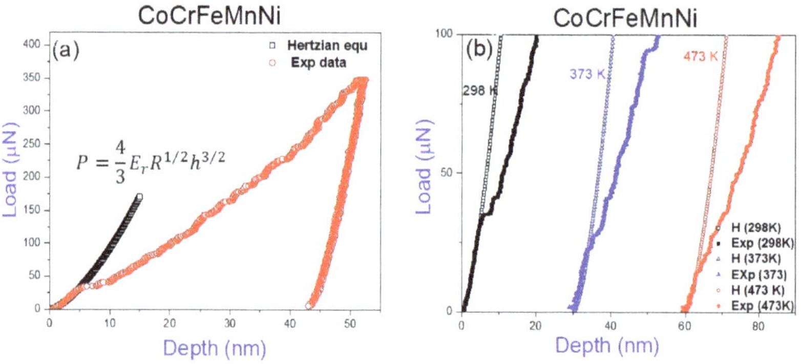

Figure 3. (**a**) Representative load-displacement curve of CoCrFeMnNi alloy at room temperature along with the Hertzian; (**b**) load-displacement curve of CoCrFeMnNi alloy at different temperatures along with the Hertzian fitting for the elastic section.

Figure 4 shows the statistical distribution of the pop-in load at room temperature (298 K) and 473 K for Ni, CoCrNi, CoCrFeMnNi, and $Al_{0.1}$CoCrFeNi. The experimentally obtained data at 373 K followed a similar trend and has not been included in the figure for clarity. It is also evident from Figure 4 that the pop-in load decreased with increasing temperature. The temperature dependence of displacement burst in multi-principal alloys indicates a stress-biased thermally activated mechanism similar to pure metals [34]. The shear stress beneath the indenter at pop-in is evaluated using Hertzian contact theory as [34]:

$$p_m = \left(\frac{6PE_r^2}{\pi^3 R^2}\right)^{1/3} \tag{2}$$

$$\tau_m = 0.31 p_m \tag{3}$$

where, p_m is the mean contact pressure and τ_m is the maximum shear stress at pop-in, which was in the range of $\sim\frac{\mu}{16}$–$\frac{\mu}{10}$ (where μ is the shear modulus) at room temperature for all alloys. As the ideal shear strength of a metal also lies in a similar range ($\frac{\mu}{30}$–$\frac{\mu}{5}$) [36], it may be concluded that the pop-in event involves dislocation nucleation. Using the probabilistic distribution of pop-in stress, an analytical solution has been proposed to evaluate the activation volume and activation energy associated with the displacement burst [34]. Nucleation of dislocation involves an activation barrier which may be overcome by mechanical work done by the indenter or by the combined effect of thermal and mechanical work. The nucleation rate per unit volume, $\dot{n}$ is represented by [34]:

$$\dot{n} = \eta \exp\left(-\frac{\varepsilon - \sigma V}{kT}\right) \tag{4}$$

where η is a pre-exponential frequency factor, ϵ is the activation energy barrier for the process, and σV represents the stress bias. In the present condition, the stress bias is the maximum shear stress beneath the indenter. The activation volume of the process may be expressed as [34]:

$$V = \frac{\pi}{0.47}\left(\frac{3R}{4E_R}\right)^{2/3} kT\alpha \tag{5}$$

where, α is a time-independent parameter and is obtained from the slope of $\ln[-\ln(1-F)]$ versus $P^{1/3}$ plot (as shown in Figure 5), and F is the cumulative probability of the event (as shown in Figure 4). Figure 5 shows the $\ln[-\ln(1-F)]$ versus $P^{1/3}$ for Ni and the three MPEAs at room temperature and 473 K. Using averaged α value obtained from the slope of the curves and Equation (5), the activation volume was calculated. Activation volume for the three alloys and Ni are included in Table 1.

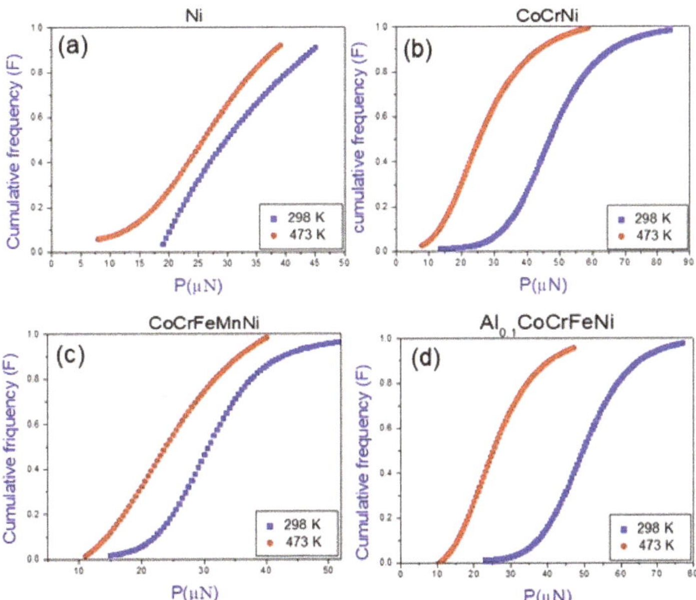

Figure 4. Cumulative fraction of experiment loads at the pop-in events plotted at 298 K and 473 K for (**a**) Ni, (**b**) CoCrNi, (**c**) CoCrFeMnNi, and (**d**) $Al_{0.1}$CoCrFeNi alloys.

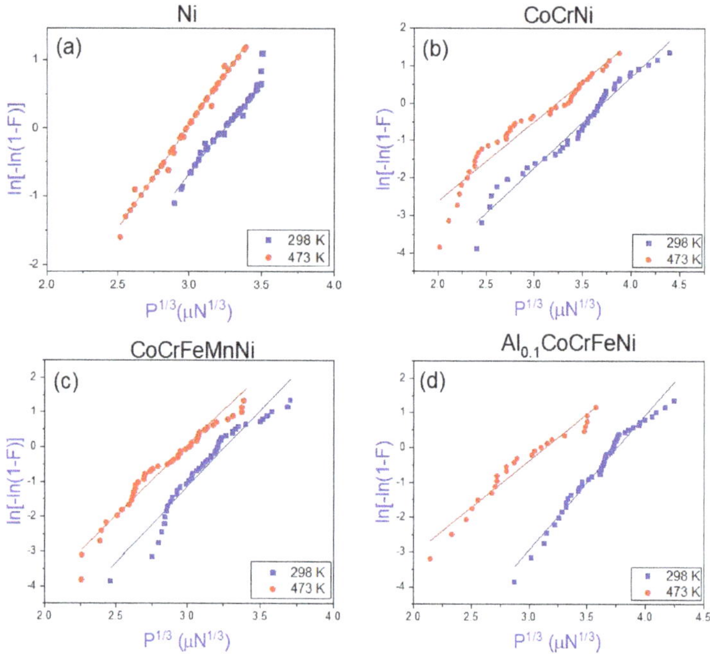

Figure 5. Plot of $\ln[-\ln(1-F)]$ vs. $P^{1/3}$ at 298 K and 473 K to determine the activation volume from experimental data analysis for (**a**) Ni, (**b**) CoCrNi, (**c**) CoCrFeMnNi, and (**d**) $Al_{0.1}$CoCrFeNi.

Equation (4) indicates that ϵ also plays an important role in the nucleation process. The activation enthalpy or ϵ may be evaluated from the pop-in load and activation volume using the equation [34]:

$$P^{1/3} = \gamma kT + \frac{\pi}{0.47}\left(\frac{3R}{4E_R}\right)^{2/3}\frac{\varepsilon}{V} \tag{6}$$

Figure 6 shows the $P^{1/3}$ vs. temperature (T) curves for pure Ni and the MPEAs at three different cumulative frequencies (F). $P^{1/3}$ vs. T shows a linear relation with slope γk and the intercept at 0 K in y-axis representing the second part of Equation (6). With the obtained $P^{1/3}$ at 0 K and the activation volume, ϵ was evaluated. This represents the energy required for nucleation at 0 K, i.e., all the work done is mechanical work, and there is no thermal work. The activation energy for pure Ni and the MPEAs are summarized in Table 1.

The average activation volume for pop-in in Ni was determined to be 4.12 Å^3, which falls within a similar range for FCC-Pt [34] and hexagonal close packed (HCP)-Mg [43]. The activation volume and activation energy evaluated for MPEAs using first-order shear bias statistical model were in the range of 4 to 9 Å^3 and 0.23 to 0.50 eV, respectively. These values are much lower than that required for homogenous dislocation nucleation which is higher than 1 eV. The discrepancy in activation enthalpy between the experimentally measured value and that expected for homogenous nucleation has been explained by the presence of subcritical loops and thermal fluctuations, while other studies have ruled out these possibilities [44–46]. Vacancy migration during nano-indentation has been proposed as another possible mechanism [2]. But the self-diffusion rate is very slow at room temperature, and vacancy migration energy lies in the range of 1–2 eV. The obtained activation enthalpy for the three MPEAs is much lower than the vacancy migration energy. In addition, due to lower diffusivity in equi-atomic multi-component alloys than pure metals, vacancy migration is more difficult in these alloys. Several simulation studies have shown that surface ledges may act as potential sites for heterogeneous dislocation nucleation by reducing the activation energy [47,48]. Another proposed mechanism for heterogeneous dislocation nucleation involves hot-spots. At high temperatures, thermal vibrations generate asymmetry which create hot-spot defects below the surface (in the bulk) and act as potential sites for dislocation nucleation, whereas at 0 K the dislocation nucleates at surface defects [45,49]. In the current study, it was observed that the displacement at pop-in decreased with increasing temperature. The average pop-in depth for CoCrFeMnNi alloy corresponding to 298 K, 373 K, and 473 K was 8 nm, 5 nm, and 3 nm, respectively. A similar trend was observed for the other two alloys, CoCrNi and $Al_{0.1}$CoCrFeNi. The decreasing trend of pop-in depth with increasing temperature supports the hot-spot mechanism because with increasing temperature the fraction of hot-spot defects increases. This leads to lowering of the depth and load for pop-in. Vacancy clusters and impurity atoms have also been proposed as potential sites for dislocation nucleation [34]. Since the alloys were air cooled after annealing, there is a possibility of having vacancy concentration higher than the equilibrium concentration.

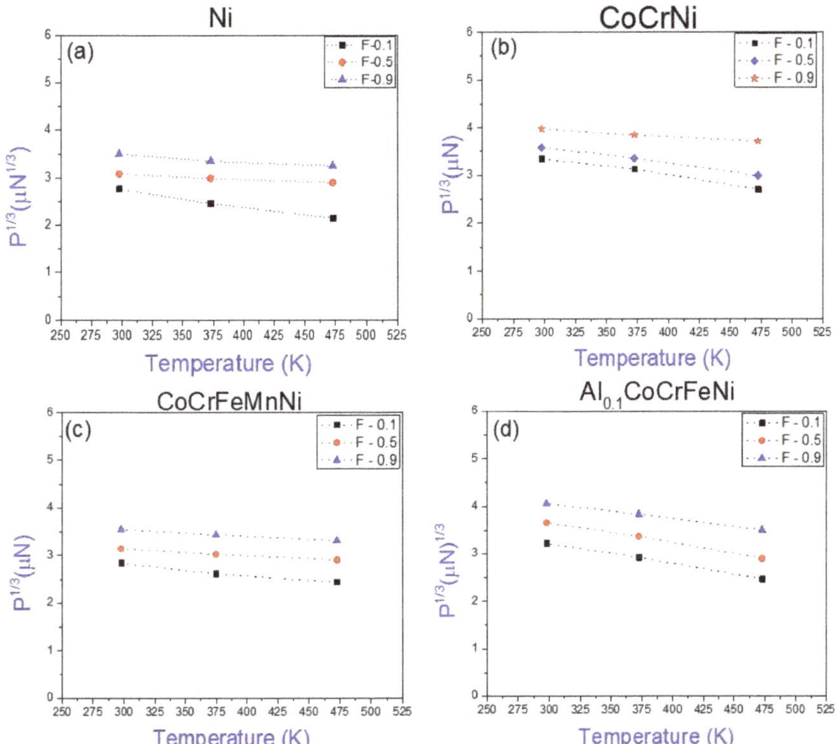

Figure 6. Plot of $P^{1/3}$ vs. T to evaluate the activation enthalpy of the first burst for (**a**) Ni, (**b**) CoCrNi, (**c**) CoCrFeMnNi, and (**d**) $Al_{0.1}$CoCrFeNi alloys.

The activation volume and energy of incipient plasticity for CoCrFeMnNi and $Al_{0.1}$CoCrFeNi were roughly two times that of pure Ni. This indicates that in the case of MPEAs, the cooperative motion of several atoms is necessary for deformation in comparison with one to one atomic motion in pure metals. The activation volume and energy of the three-component CoCrNi alloy was marginally higher than pure Ni (14% higher) but significantly lower compared to the five-component alloys (~50% lower). This suggests that the dislocation mechanism in this medium entropy alloy (CoCrNi) may be similar to pure metals. The activation volume for dislocation nucleation in BCC TiZNbTa and TiZrNbTaMo HEAs were determined to be ~30 Å^3 [35] which is more than that of the current studied MPEAs. It should also be noted that the obtained activation volume and activation energy for CoCrFeMnNi in the current work is different compared to that earlier reported by Zhu et al. for the same composition [36]. This may be attributed to the sample preparation conditions and the resulting surface state.

4. Conclusions

Shallow depth nano-indentation was performed for multi-principal five-component $Al_{0.1}$CoCrFeNi and CoCrFeMnNi alloys and three-component CoCrNi alloy to elucidate the mechanism associated with the onset of plasticity in these alloys. Nano-indentation was performed at three different temperatures of 298 K, 373 K, and 473 K. The experimentally obtained data were analyzed within the framework of first-order shear bias statistical model, and the activation volume and activation enthalpy of the process were calculated. The following can be concluded from the obtained experimental and analytical results:

1. The pop-in load marking incipient plasticity was found to decrease with the increase of temperature for each of the three alloys indicating a thermally activated event.
2. At all the temperatures, the pop-in shear stress was not constant but rather occurred over a range of loads. At room temperature, the values of the pop-in shear stress ranged over $\frac{\mu}{16}$–$\frac{\mu}{10}$ for three alloys.
3. Using a statistical approach, the evaluated activation volume and enthalpy for the five-component multi-principal alloys were around 10 Å^3 and 0.5 eV, respectively. The small values of activation volume and energy suggest heterogeneous dislocation nucleation.
4. The CoCrFeMnNi and $Al_{0.1}$CoCrFeNi multi-principal alloys showed two times higher activation volume and energy compared to CoCrNi medium entropy alloy and pure Ni suggesting complex cooperative motion of atoms for deformation in these systems.
5. Vacancy migration was ruled out as a possible nucleation mechanism because of the small values of activation volume and energy.
6. Surface defects like ledges, vacancy clusters, and hot-spots created by asymmetry generated due to thermal vibration could be potential sites for heterogeneous dislocation nucleation in these alloys during nano-indentation.

Author Contributions: Conceptualization, S.M. (Sundeep Mukherjee); methodology, S.M. (Sanghita Mridha) and S.M. (Sundeep Mukherjee); validation, S.M. (Sanghita Mridha) and M.S.; formal analysis, M.S.; investigation, S.M. (Sanghita Mridha) and M.S.; resources, S.M. (Sundeep Mukherjee); data curation, S.M. (Sanghita Mridha) and M.S.; writing—original draft preparation, S.M. (Sanghita Mridha) and M.S.; writing—review and editing, M.S. and S.M. (Sundeep Mukherjee); visualization, M.S.; supervision, S.M. (Sundeep Mukherjee); project administration, S.M. (Sundeep Mukherjee).

Funding: This research received no external funding.

Conflicts of Interest: The authors declare no conflict of interest.

References

1. Yeh, J.W.; Chen, S.K.; Lin, S.J.; Gan, J.Y.; Chin, T.S.; Shun, T.T.; Tsai, C.H.; Chang, S.Y. Nanostructured high-entropy alloys with multiple principal elements: Novel alloy design concepts and outcomes. *Adv. Eng. Mater.* **2004**, *6*, 299–303. [CrossRef]
2. Wu, D.; Jang, J.S.C.; Nieh, T.G. Elastic and plastic deformations in a high entropy alloy investigated using a nanoindentation method. *Intermetallics* **2016**, *68*, 118–127. [CrossRef]
3. Miracle, D.B.; Senkov, O.N. A critical review of high entropy alloys and related concepts. *Acta Mater.* **2017**, *122*, 448–511. [CrossRef]
4. Guo, S.; Ng, C.; Lu, J.; Liu, C.T. Effect of valence electron concentration on stability of fcc or bcc phase in high entropy alloys. *J. Appl. Phys.* **2011**, *109*, 103505. [CrossRef]

5. Zhang, Y.; Zhou, Y.J.; Lin, J.P.; Chen, G.L.; Liaw, P.K. Solid-solution phase formation rules for multi-component alloys. *Adv. Eng. Mater.* **2008**, *10*, 534–538. [CrossRef]
6. Zhang, Y.; Zuo, T.T.; Tang, Z.; Gao, M.C.; Dahmen, K.A.; Liaw, P.K.; Lu, Z.P. Microstructures and properties of high-entropy alloys. *Prog. Mater. Sci.* **2014**, *61*, 1–93. [CrossRef]
7. Gludovatz, B.; Hohenwarter, A.; Thurston, K.V.; Bei, H.; Wu, Z.; George, E.P.; Ritchie, R.O. Exceptional damage-tolerance of a medium-entropy alloy CrCoNi at cryogenic temperatures. *Nat. Commun.* **2016**, *7*, 1–8. [CrossRef] [PubMed]
8. Gali, A.; George, E.P. Tensile properties of high-and medium-entropy alloys. *Intermetallics* **2013**, *39*, 74–78. [CrossRef]
9. Schneider, M.; Werner, F.; Langenkämper, D.; Reinhart, C.; Laplanche, G. Effect of Temperature and Texture on Hall–Petch Strengthening by Grain and Annealing Twin Boundaries in the MnFeNi Medium-Entropy Alloy. *Metals* **2019**, *9*, 84. [CrossRef]
10. Klimova, M.; Stepanov, N.; Shaysultanov, D.; Chernichenko, R.; Yurchenko, N.; Sanin, V.; Zherebtsov, S. Microstructure and Mechanical Properties Evolution of the Al, C-Containing CoCrFeNiMn-Type High-Entropy Alloy during Cold Rolling. *Materials* **2018**, *11*, 53. [CrossRef] [PubMed]
11. Schuh, B.; Pippan, R.; Hohenwarter, A. Tailoring bimodal grain size structures in nanocrystalline compositionally complex alloys to improve ductility. *Mater. Sci. Eng. A* **2019**, *748*, 379–385. [CrossRef]
12. Gwalani, B.; Gorsse, S.; Choudhuri, D.; Styles, M.; Zheng, Y.; Mishra, R.S.; Banerjee, R. Modifying transformation pathways in high entropy alloys or complex concentrated alloys via thermo-mechanical processing. *Acta Mater.* **2018**, *153*, 169–185. [CrossRef]
13. Gao, M.C. Progress in high-entropy alloys. *JOM* **2015**, *67*, 2251–2253. [CrossRef]
14. Kozak, R.; Sologubenko, A.; Steurer, W. Single-phase high-entropy alloys–an overview. *Z. Kristallogr.-Cryst. Mater.* **2015**, *230*, 55–68. [CrossRef]
15. Tsai, M.H.; Yeh, J.W. High-entropy alloys: A critical review. *Mater. Res. Lett.* **2014**, *2*, 107–123. [CrossRef]
16. Yeh, J.W.; Chen, Y.L.; Lin, S.J.; Chen, S.K. High-entropy alloys—A new era of exploitation. *Mater. Sci. Forum* **2007**, *560*, 1–9. [CrossRef]
17. Komarasamy, M.; Kumar, N.; Tang, Z.; Mishra, R.; Liaw, P. Effect of microstructure on the deformation mechanism of friction stir-processed $Al_{0.1}$CoCrFeNi high entropy alloy. *Mater. Res. Lett.* **2015**, *3*, 30–34. [CrossRef]
18. Mishra, R.S.; Kumar, N.; Komarasamy, M. Lattice strain framework for plastic deformation in complex concentrated alloys including high entropy alloys. *Mater. Sci. Technol.* **2015**, *31*, 1259–1263. [CrossRef]
19. Schuh, C.A.; Lund, A.C. Application of nucleation theory to the rate dependence of incipient plasticity during nanoindentation. *J. Mater. Res.* **2004**, *19*, 2152–2158. [CrossRef]
20. Chiu, Y.L.; Ngan, A.H.W. A TEM investigation on indentation plastic zones in Ni_3Al (Cr, B) single crystals. *Acta Mater.* **2002**, *50*, 2677–2691. [CrossRef]
21. Kiely, J.D.; Houston, J.E. Nanomechanical properties of Au (111), (001), and (110) surfaces. *Phys. Rev. B* **1998**, *57*, 12588–12594. [CrossRef]
22. Tymiak, N.I.; Daugela, A.; Wyrobek, T.J.; Warren, O.L. Acoustic emission monitoring of the earliest stages of contact-induced plasticity in sapphire. *Acta Mater.* **2004**, *52*, 553–563. [CrossRef]
23. Lorenz, D.; Zeckzer, A.; Hilpert, U.; Grau, P.; Johansen, H.; Leipner, H.S. Pop-in effect as homogeneous nucleation of dislocations during nanoindentation. *Phys. Rev. B* **2003**, *67*, 172101. [CrossRef]
24. Syed Asif, S.A.; Pethica, J.B. Nanoindentation creep of single-crystal tungsten and gallium arsenide. *Philos. Mag. A* **1997**, *76*, 1105–1118. [CrossRef]
25. Corcoran, S.G.; Colton, R.J.; Lilleodden, E.T.; Gerberich, W.W. Anomalous plastic deformation at surfaces: Nanoindentation of gold single crystals. *Phys. Rev. B* **1997**, *55*, R16057. [CrossRef]
26. Bahr, D.F.; Kramer, D.E.; Gerberich, W.W. Non-linear deformation mechanisms during nanoindentation. *Acta Mater.* **1998**, *46*, 3605–3617. [CrossRef]
27. Chiu, Y.L.; Ngan, A.H.W. Time-dependent characteristics of incipient plasticity in nanoindentation of a Ni_3Al single crystal. *Acta Mater.* **2002**, *50*, 1599–1611. [CrossRef]
28. Wang, W.; Jiang, C.B.; Lu, K. Deformation behavior of Ni_3Al single crystals during nanoindentation. *Acta Mater.* **2003**, *51*, 6169–6180. [CrossRef]
29. Page, T.F.; Oliver, W.C.; McHargue, C.J. The deformation behavior of ceramic crystals subjected to very low load (nano) indentations. *J. Mater. Res.* **1992**, *7*, 450–473. [CrossRef]

30. Gouldstone, A.; Koh, H.; Zeng, K.Y.; Giannakopoulos, A.E.; Suresh, S. Discrete and continuous deformation during nanoindentation of thin films. *Acta Mater.* **2000**, *48*, 2277–2295. [CrossRef]
31. Suresh, S.; Nieh, T.G.; Choi, B.W. Nano-indentation of copper thin films on silicon substrates. *Scr. Mater.* **1999**, *41*, 951–957. [CrossRef]
32. Van Vliet, K.J.; Li, J.; Zhu, T.; Yip, S.; Suresh, S. Quantifying the early stages of plasticity through nanoscale experiments and simulations. *Phys. Rev. B* **2003**, *67*, 104105. [CrossRef]
33. Gannepalli, A.; Mallapragada, S.K. Atomistic studies of defect nucleation during nanoindentation of Au (001). *Phys. Rev. B* **2002**, *66*, 104103. [CrossRef]
34. Mason, J.K.; Lund, A.C.; Schuh, C.A. Determining the activation energy and volume for the onset of plasticity during nanoindentation. *Phys. Rev. B* **2006**, *73*, 054102. [CrossRef]
35. Wang, S.P.; Xu, J. Incipient plasticity and activation volume of dislocation nucleation for TiZrNbTaMo high-entropy alloys characterized by nanoindentation. *J. Mater. Sci. Technol.* **2019**, *35*, 812–816. [CrossRef]
36. Zhu, C.; Lu, Z.P.; Nieh, T.G. Incipient plasticity and dislocation nucleation of FeCoCrNiMn high-entropy alloy. *Acta Mater.* **2013**, *61*, 2993–3001. [CrossRef]
37. Cao, F.; Munroe, P.; Zhou, Z.; Xie, Z. Medium entropy alloy CoCrNi coatings: Enhancing hardness and damage-tolerance through a nanotwinned structuring. *Surf. Coat. Technol.* **2018**, *335*, 257–264. [CrossRef]
38. Ahmad, A.S.; Su, Y.; Liu, S.Y.; Stahl, K.; Wu, Y.D.; Hui, X.D.; Ruett, U.; Gutowski, O.; Glazyrin, K.; Liermann, H.P.; et al. Structural stability of high entropy alloys under pressure and temperature. *J. Appl. Phys.* **2017**, *121*, 235901. [CrossRef]
39. Sathiyamoorthi, P.; Bae, J.W.; Asghari-Rad, P.; Park, J.M.; Kim, J.G.; Kim, H.S. Effect of annealing on microstructure and tensile behavior of CoCrNi medium entropy alloy processed by high-pressure torsion. *Entropy* **2018**, *20*, 849. [CrossRef]
40. Yang, T.; Tang, Z.; Xie, X.; Carroll, R.; Wang, G.; Wang, Y.; Dahmen, K.; Liaw, P.; Yanwen, Z. Deformation mechanisms of $Al_{0.1}$CoCrFeNi at elevated temperature. *Mater. Sci. Eng. A* **2017**, *684*, 552–558. [CrossRef]
41. Laplanche, G.; Gadaud, P.; Horst, O.; Otto, F.; Eggeler, G.; George, E.P. Temperature dependencies of the elastic moduli and thermal expansion coefficient of an equiatomic, single-phase CoCrFeMnNi high-entropy alloy. *J. Alloys Compd.* **2015**, *623*, 348–353. [CrossRef]
42. Haglund, A.; Koehler, M.; Catoor, D.; George, E.P.; Keppens, V. Polycrystalline elastic moduli of a high-entropy alloy at cryogenic temperatures. *Intermetallics* **2015**, *58*, 62–64. [CrossRef]
43. Somekawa, H.; Schuh, C.A. Effect of solid solution elements on nanoindentation hardness, rate dependence, and incipient plasticity in fine grained magnesium alloys. *Acta Mater.* **2011**, *59*, 7554–7563. [CrossRef]
44. Sun, Y.Q.; Hazzledine, P.M.; Hirsch, P.B. Cooperative nucleation of shear dislocation loops. *Phys. Rev. Lett.* **2002**, *88*, 065503. [CrossRef] [PubMed]
45. Zuo, L.; Ngan, A.H.W.; Zheng, G.P. Size dependence of incipient dislocation plasticity in Ni_3Al. *Phys. Rev. Lett.* **2005**, *94*, 095501. [CrossRef] [PubMed]
46. Wo, P.C.; Zuo, L.; Ngan, A.H.W. Time-dependent incipient plasticity in Ni_3Al as observed in nanoindentation. *J. Mater. Res.* **2005**, *20*, 489–495. [CrossRef]
47. Salehinia, I.; Bahr, D.F. The impact of a variety of point defects on the inception of plastic deformation in dislocation-free metals. *Scr. Mater.* **2012**, *66*, 339–342.
48. Salehinia, I.; Perez, V.; Bahr, D. Effect of vacancies on incipient plasticity during contact loading. *Philos. Mag.* **2012**, *92*, 550–570.
49. Wagner, R.J.; Ma, L.; Tavazza, F.; Levine, L.E. Dislocation nucleation during nanoindentation of aluminum. *J. Appl. Phys.* **2008**, *104*, 114311. [CrossRef]

Article

High-Temperature Nano-Indentation Creep Behavior of Multi-Principal Element Alloys under Static and Dynamic Loads

Maryam Sadeghilaridjani and Sundeep Mukherjee *

Department of Materials Science and Engineering, University of North Texas, Denton, TX 76203, USA; Maryam.Sadeghilaridjani@unt.edu

* Correspondence: sundeep.mukherjee@unt.edu; Tel.: +1-940-565-4170; Fax: 1-940-565-2944

Received: 14 January 2020; Accepted: 9 February 2020; Published: 13 February 2020

Abstract: Creep is a serious concern reducing the efficiency and service life of components in various structural applications. Multi-principal element alloys are attractive as a new generation of structural materials due to their desirable elevated temperature mechanical properties. Here, time-dependent plastic deformation behavior of two multi-principal element alloys, CoCrNi and CoCrFeMnNi, was investigated using nano-indentation technique over the temperature range of 298 K to 573 K under static and dynamic loads with applied load up to 1000 mN. The stress exponent was determined to be in the range of 15 to 135 indicating dislocation creep as the dominant mechanism. The activation volume was ~$25b^3$ for both CoCrNi and CoCrFeMnNi alloys, which is in the range indicating dislocation glide. The stress exponent increased with increasing indentation depth due to higher density and entanglement of dislocations, and decreased with increasing temperature owing to thermally activated dislocations. The results for the two multi-principal element alloys were compared with pure Ni. CoCrNi showed the smallest creep displacement and the highest activation energy among the three systems studied indicating its superior creep resistance.

Keywords: multi-principal element alloys; creep; nano-indentation; stress exponent; activation volume; activation energy

1. Introduction

Material degradation related to creep is a serious concern reducing the efficiency and service life of components in different structural applications. Therefore, there is strong demand for materials with inherently high creep resistance. Multi-principal element alloys (MPEAs) have attracted much attention as a new generation of structural materials with excellent mechanical properties including creep resistance [1–3]. MPEAs are composed of several elements in equiatomic or near-equiatomic proportion and despite complex chemistry, they usually form a simple single- or multi-phase solid solution [1–3]. The presence of different sized atoms in the unit cell leads to high degree of lattice strain in MPEAs [1] and results in high hardness and strength, good fracture toughness at cryogenic and elevated temperatures, high fatigue resistance, good wear, erosion and corrosion resistance [3–7]. Local lattice distortion, nano-clustering, and inhomogeneity at microstructural length scale in MPEAs [8] may significantly affect their mechanical properties including creep resistance. Therefore, probing the small-scale deformation behavior and local creep processes in MPEAs is critical in establishing their application worthiness. Nano-indentation technique has been widely used to characterize the small-scale deformation behavior of materials [9–15]. However, there are limited reports on the creep behavior of MPEAs using nano-indentation, and majority of the studies are at room temperature with a maximum load of 100 mN [16–22]. There are no reports on the deformation behavior of MPEAs as a function of temperature comparing static and dynamic loads. This is important in wide ranging applications where components are subject to both static and cyclic loads such as aerospace, nuclear, automotive, and oil and gas industries [12].

Here, we report the nano-indentation creep behavior of two MPEAs, namely CoCrNi and CoCrFeMnNi. These were chosen as model alloys with excellent mechanical properties [23] and a single-phase face centered cubic (FCC) microstructure [10,24]. Nano-indentation creep was studied under static and dynamic loads as a function of temperature and peak load. Stress exponent, activation volume, and activation energy of the alloys were evaluated. Nickel (Ni) was used as the reference metal with FCC crystal structure for comparison with the two MPEAs.

2. Experimental

Alloys with nominal compositions of CoCrNi and CoCrFeMnNi in equimolar ratios were prepared by arc-melting high purity elements (>99.9%) in a Ti-gettered argon atmosphere. To ensure chemical homogeneity, the ingots were flipped and remelted several times. As-cast alloys and Ni were then rolled up to 70% reduction in thickness followed by annealing at 1173 K for 20 h to get equiaxed grains with low dislocation density. The annealed samples were polished with silicon carbide papers followed by diamond suspension to a mirror finish for microstructural characterization and nano-mechanical tests. Rigaku III Ultima X-ray diffractometer (XRD, Rigaku Corporation, Tokyo, Japan) with 1.54 Å wavelength Cu-Kα radiation was used for crystal structure and phase characterization of the alloys. Microstructure, grain size, and grain orientation of the alloys were characterized by scanning electron microscopy (SEM) using FEI Quanta ESEM (FEI Company, Hillsboro, OR, USA) and electron backscatter diffraction (EBSD) technique.

Nano-indentation creep tests were done using a TI Premier Triboindenter (Bruker, Minneapolis, MN, USA) equipped with XSol600 heating stage for heating the samples up to 600 °C. To avoid oxidation, the tests were performed in a mixture of Ar + 5% H_2 gas environment. A Berkovich sapphire tip was used in all creep tests. The initial tip calibration was done with a standard fused quartz reference sample. For each material, we adopted two different types of tests: (i) static constant load and (ii) dynamic mechanical analyze (DMA). The static creep tests were done by ramping the load to 500 mN and 1000 mN at temperatures of 298 K, 423 K, and 573 K and then held at maximum load for 120 s to determine creep response followed by unloading. The aim was to compare creep performance of the selected alloys as a function of temperature at two different loads. High load was used for the static creep tests to minimize the effect of surface and avoid indentation size effect (ISE). In nano-DMA tests, the selected loads were 50 mN, 100 mN, 500 mN and 1000 mN to study the ISE on dynamic creep behavior at the two temperatures of 298 K and 423 K. The frequency and amplitude were set to 100 Hz and 10% of the peak load, respectively. In all tests, a high loading rate of 20 mN/s was chosen to minimize plastic deformation during the loading segment so creep primarily occurred during the dwell time. Prior to each creep test at elevated temperature, the sample was held at the prescribed set point for at least 20 min for temperature stabilization and the indenter tip was maintained close to the sample to reduce temperature gradient. The thermal drift was automatically corrected by the Triboindenter software and was between 0.05–0.1 nm/s during testing. The effect of drift was also minimized by using large indentation depth and a short hold time [25]. At least 16 indents were done for each condition and the distance between each indent was larger than 100 μm to avoid interaction of their plastic zones.

3. Results

Figure 1 summarizes the microstructural characterization for CoCrNi, CoCrFeMnNi, and pure Ni. The backscattered SEM images of the homogenized samples are shown in Figure 1a,c,e with the insets showing X-ray diffraction patterns for the alloys. All the alloys showed single-phase FCC crystal structure without any secondary phases or precipitates. Figure 1b,d,f show the EBSD images of CoCrNi, CoCrFeMnNi, and Ni, respectively, indicating that all the alloys had equiaxed grains with the inset showing the grain size distribution. The grain sizes for CoCrNi were in the range of 5–50 μm (average

≈ 19 μm) and that for CoCrFeMnNi were in the range of 5–55 μm (average ≈ 22 μm). Ni showed a grain size distribution of 5–65 μm with an average of 30 μm. Significant numbers of annealing twins were observed in the microstructures indicating low stacking fault energy in the MPEAs.

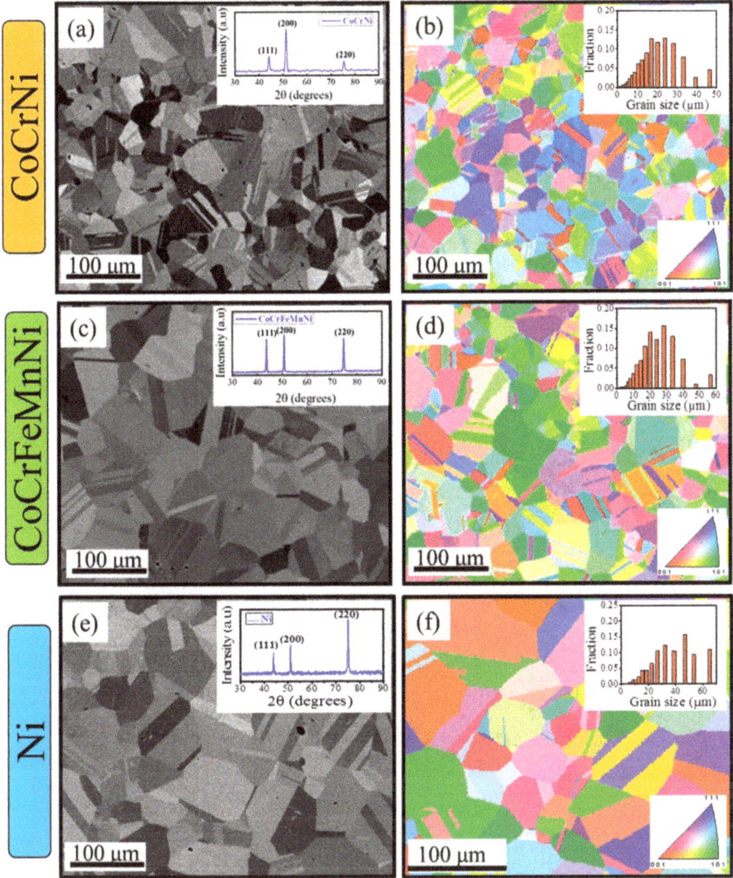

Figure 1. Backscattered SEM images with XRD patterns as insets for (**a**) CoCrNi, (**c**) CoCrFeMnNi, and (**e**) pure Ni. EBSD maps with grain size distribution as insets for (**b**) CoCrNi, (**d**) CoCrFeMnNi, and (**f**) pure Ni.

Creep behavior under static load was evaluated as a function of temperature at two different loads. The load and displacement data recorded during nano-indentation creep of CoCrFeMnNi MPEA is shown in Figure 2a,b as a function of temperature. Figure 2a shows the results at 500 mN load while Figure 2b shows the data at 1000 mN. For each test, an array of 4 × 4 indents was made with 100 μm distance between each indent, therefore covering grains with several different orientations. The plots for Ni and CoCrNi were very similar and not included here. At the same load, indentation depth in CoCrFeMnNi sample increased with increasing temperature from 298 K to 423 K, indicating reduction in hardness. The corresponding change in displacement during hold time was used to analyze the time-dependent deformation behavior of the alloys. This is plotted in Figure 2c,d for CoCrFeMnNi at temperatures of 298 K, 423 K, and 573 K and peak load of 500 mN and 1000 mN, respectively. The creep displacement increased rapidly with time in the beginning followed by slowing down of the rate of increase. Fitting of the experimental data using an empirical relation for creep displacement is shown and elaborated later in the discussion section. Figure 2e,f show the magnitude of creep displacement as a function of temperature for Ni, CoCrNi, and CoCrFeMnNi alloys at the loads of 500 mN and 1000 mN, respectively. The maximum creep displacements were in the range of 50 nm to 400 nm, depending on the alloy, holding load, and temperature. Each data point in Figure 2e,f is the average of at least 16 indents with the error bar including possible influence of crystal orientation. Increase in temperature and peak load increased creep displacement for pure Ni since it is a thermally activated process and diffusion is enhanced at elevated temperature and high load. However, for the MPEAs, distinctly different behavior was seen compared to pure Ni. At 500 mN load, the creep displacement in the case of CoCrNi first increased from 298 K to 423 K and then decreased with further increase of temperature to 573 K. For CoCrFeMnNi, the creep displacement decreased with increasing temperature. This may be attributed to generation of partial dislocations at these intermediate temperatures and/or solute drag effects which restrict dislocation mobility [26–28]. Overall, CoCrNi showed the smallest creep displacement among the three systems for the load and temperature range investigated.

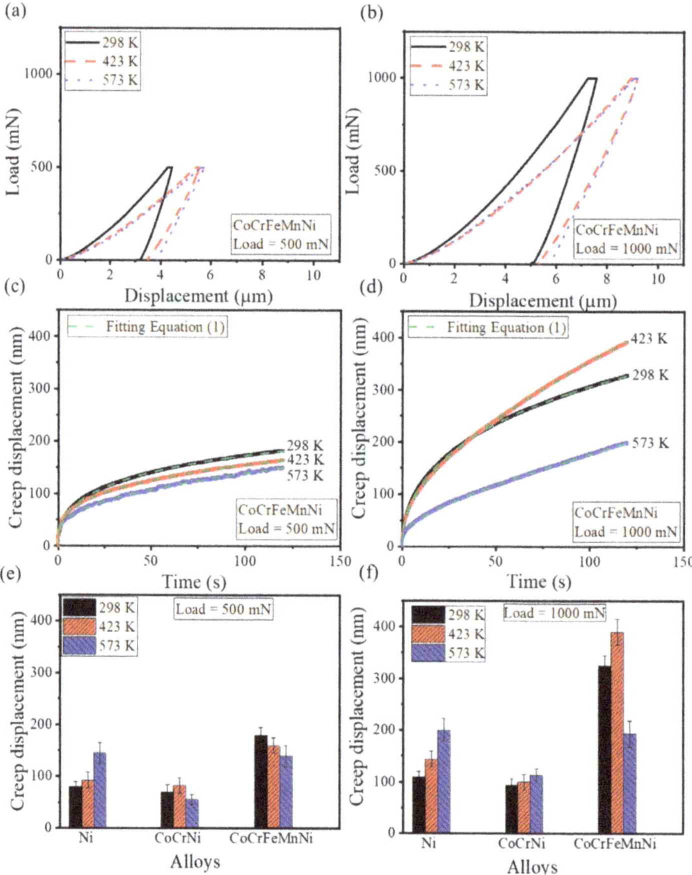

Figure 2. Typical load-displacement plot after nano-indentation creep tests for CoCrFeMnNi alloy as a function of temperature under applied load of (**a**) 500 mN and (**b**) 1000 mN. Creep displacement versus hold time for CoCrFeMnNi alloy at different temperatures under load of (**c**) 500 mN and (**d**) 1000 mN. Total creep displacement as a function of temperature for Ni, CoCrNi, and CoCrFeMnNi alloys at (**e**) 500 mN and (**f**) 1000 mN.

The local time-dependent plastic deformation of CoCrNi and CoCrFeMnNi MPEAs was studied under dynamic load and compared with pure Ni. The samples were loaded to 50 mN, 100 mN, 500 mN, and 1000 mN and held for 120 s under dynamic loads. In each test, the amplitude of load was fixed at 10% of the peak load. Figure 3a,b show the representative creep displacement versus holding time at various loads for CoCrFeMnNi at 298 K and 423 K. The maximum creep depth increased with increasing load from 50 mN to 1000 mN with similar behavior observed for Ni and CoCrNi. Figure 3c,d show the magnitude of maximum creep displacement as a function of load for all studied alloys at 298 K and 423 K. Increasing peak load led to increase in creep displacement due to dislocation activation. Total creep displacement was found to be in the order of CoCrNi < Ni < CoCrFeMnNi, which was in agreement with the data obtained from static loading. The effect of surface oxidation on the measured creep displacement is expected to be minimal due to the use of Ar + 5% H_2 gas environment and higher loads. There was no change in surface appearance of the samples after the high-temperature tests indicating minimal oxidation.

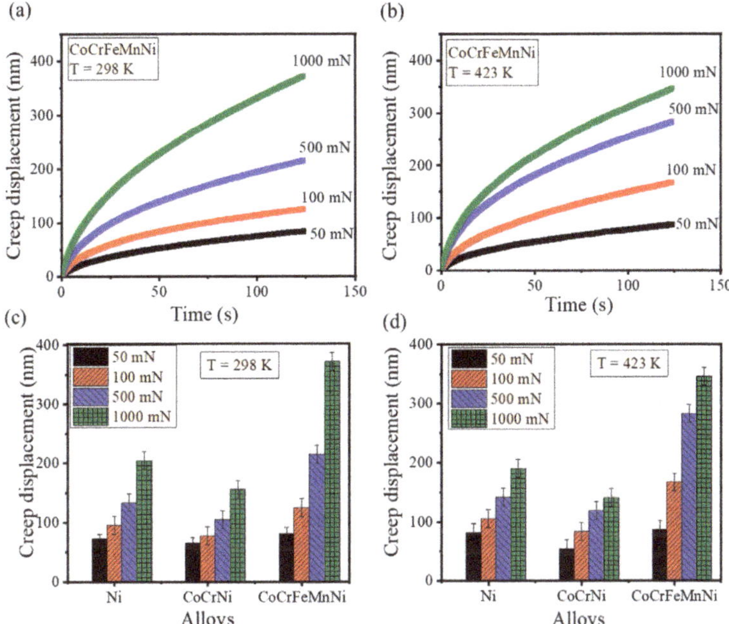

Figure 3. Representative creep displacement versus hold time for CoCrFeMnNi alloy as a function of loads at (**a**) 298 K and (**b**) 423 K. Maximum creep displacement for the three alloys at (**c**) 298 K and (**d**) 423 K showing larger creep depth with increase in load.

4. Discussion

Creep displacement, h, is a function of time, t, and follows the empirical relation [18]:

$$h(t) = h_0 + a(t - t_0)^p + kt \quad (1)$$

where h_0 and t_0 are the initial depth and time during creep period and a, p, and k are fitting constants. The dashed lines in Figure 2c,d represent the fitting curves corresponding to Equation (1) with correlation coefficient $R^2 > 0.95$. For self-similar indentation probe similar to Berkovich used in the current study, indentation strain rate $(\dot{\varepsilon})$ and hardness (H) were obtained as [18]:

$$\dot{\varepsilon} = \frac{1}{h}\frac{dh}{dt} \tag{2}$$

and

$$H = \frac{P}{24.5h_c^2} \tag{3}$$

where $\frac{dh}{dt}$ is the first derivative of the displacement-time curve (i.e., Equation (1)) with respect to t, P is applied load, h_c is contact depth given by $h_c = h_{max} - 0.75P/S$ for Berkovich indenter, and h_{max} and S are maximum penetration depth and material stiffness, respectively. The hardness measured from Equation (3) was divided by 0.9 for sink-in and 1.02 for pile-up [29,30]. The susceptibility for pile-up or sink-in in nano-indentation tests can be found by the ratio of final indentation depth to the maximum indentation depth (h_f/h_{max}) with $h_f/h_{max} \geq 0.7$ indicating pile-up and $h_f/h_{max} < 0.7$ indicating sink-in [31]. Since plastic deformation is a thermally activated process, stress exponent (reciprocal of strain rate sensitivity) provides valuable insight into creep deformation mechanism [18]. The creep stress exponent was calculated for the two MPEAs and pure Ni and its dependence on load and temperature was evaluated.

The creep stress exponent (denoted by n) was calculated from the slope of $\ln\dot{\varepsilon}$ versus $\ln H$ curves [18]. Figure 4 shows the creep stress exponent for the two MPEAs and pure Ni under static load to evaluate the effect of temperature and under dynamic (DMA) mode to study the effect of peak load or indentation depth. Static test analysis was done at two different loads of 500 mN and 1000 mN as shown in Figure 4a,b, respectively. Dynamic test analysis was done at two different temperatures of 298 K and 423 K as shown in Figure 4c,d, respectively. The n value for each condition was obtained from an average of 16 independent indentations. The average n values for the three systems were in the range of 15 to 135 in static tests. Creep stress exponent defines the creep mechanism. Typically, n = 1 is associated with diffusion creep (by lattice or grain boundary diffusion), n = 2 with grain boundary sliding, and $n > 3$ with dislocation creep [23]. Therefore, for all the current alloys, deformation was dominated by dislocation creep (climb or glide). The stress exponent for Ni dropped from 135 at 298 K to 50–60 at 573 K, likely due to thermally activated dislocations at elevated temperature. In a similar study, the stress exponent was found to be in the range of 20–60 for IN-718 at room temperature and decreased to 8–18 at 923 K [32]. The magnitude of stress exponent depends on the balance between generation and annihilation of dislocations [33]. Larger value of stress exponent is typically obtained when more dislocations are generated and involved during deformation [33]. The decrease in n value with increasing temperature may be due to enhancement in dislocation diffusivity and thermal recovery at higher temperature, so annihilation competes with dislocation generation. The negligible change in stress exponent as a function of temperature for CoCrFeMnNi HEA may be attributed to dislocation glide dictated deformation, which is thermally insensitive [34]. Since glide does not require interatomic diffusion (unlike climb), it is the dominant mechanism of deformation even at low temperature and is thermally insensitive [34]. Transmission electron microscopy (TEM) studies of MPEAs showed that in the early stages of plastic deformation, glide of dislocations on {111} planes was the dominant mechanism [35]. Due to low stacking fault energy and partial dislocations in MPEAs, cross slip is hindered, and plasticity occurs by dislocation glide [35]. The temperature independent stress exponent for CoCrFeMnNi may be attributed to sluggish diffusion [1,2], which would favor glide over climb dominated deformation. For pure Ni, deformation was climb dominated over the temperature range studied. For CoCrNi, the sharp drop in creep stress exponent is likely from change in climb dominated creep at low temperatures (similar to pure Ni) to glide dominated creep at high temperatures (similar to CoCrFeMnNi). In summary, the stress exponent for CoCrNi was close to that of pure Ni at room temperature but became similar to CoCrFeMnNi MPEA at the higher temperature. Recent atomistic modeling suggests that the CoCrNi system shows partial chemical ordering at the atomic scale at room temperature while becoming random solid solution at higher temperatures [36]. Partial local ordering may act as barrier for dislocation movement explaining the much higher stress exponent for CoCrNi at 298 K compared to CoCrFeMnNi. Figure 4c,d show the variation in stress exponent for the dynamic tests. As the peak load increased from 50 mN to 1000 mN, the n value increased at both the temperatures of 298 K and 423 K, which is attributed to ISE [37,38]. During the loading process, dislocations are generated in the plastic deformation region beneath the indenter and their density is directly proportional to the load or depth [39]. At low load, dislocation generation rate is slow, and the stress exponent is low. At high load, dislocation generation rate is fast, and they interact and entangle with one another leading to increase in the stress exponent with value up to 100 for pure Ni [19,40]. The observed ISE of n may also be related to the mobility and diffusion of dislocations and how far they are from the free surface [41]. At smaller depth, dislocations are closer to the free surface and therefore have higher mobility and diffusion leading to lower n.

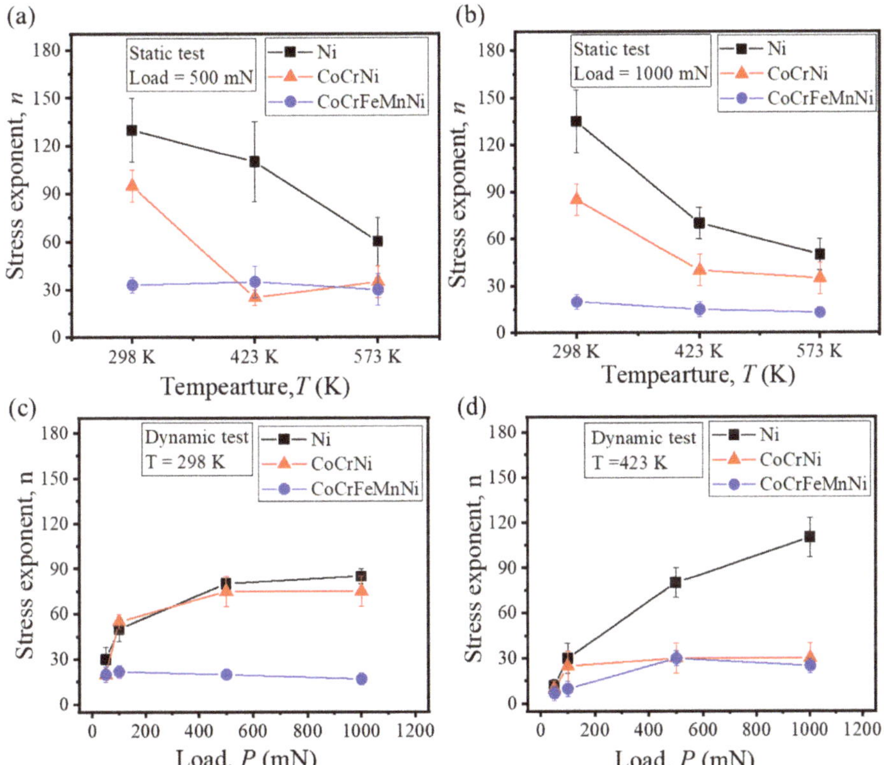

Figure 4. Stress exponent versus temperature from static test for Ni, CoCrNi and CoCrFeMnNi alloys at (**a**) 500 mN and (**b**) 1000 mN showing decrease of stress exponent with increasing temperature for Ni and negligible dependence of stress exponent on temperature for CoCrFeMnNi HEA. Stress exponent versus load from dynamic test for all alloys at (**c**) 298 K and (**d**) 423 K showing indentation size effect for stress exponent.

The high values of stress exponent obtained from nano-indentation of the three alloys may be attributed to the complex stress state underneath the indenter [16,21,38]. The stress exponent is a complex function of microstructure, mobile dislocation density and/or the activation area underneath the indenter [42]. The n value for Ni was significantly larger than those of MPEAs, even though the rate-controlling mechanism was the same for all three systems studied (i.e., dislocation dominated deformation mechanism) with almost identical microstructure. From Figure 4, it was concluded that the stress exponent determined from dynamic tests was slightly lower than that in static test. This may be because of oscillatory load which resulted in better propagation of dislocations and reduction of n.

For further insight into the mechanism of creep, the activation volume (V^*) for the three systems was evaluated as [43]:

$$V^* = \frac{3 \cdot \sqrt{3} \cdot k \cdot T \cdot n}{H} \tag{4}$$

where k is Boltzmann constant, H is hardness, n is stress exponent and T is temperature. The activation volume depends on the stress exponent and hardness. The obtained activation volume for the three systems averaged over the three temperatures (298 K, 423 K, and 573 K) and two loads (500 mN and 1000 mN) were 2.2 ± 0.1 nm^3 (~143b^3), 0.4 ± 0.11 nm^3 (~25b^3) and 0.43 ± 0.1 nm^3 (~26b^3) for Ni, CoCrNi, and CoCrFeMnNi, respectively. Here, the Burgers vector (b) for the FCC alloys was calculated as $b = 1/2a_0$ [110], where a_0 is the lattice parameter with values of 0.352 nm, 0.3567 nm and 0.3597 nm for Ni, CoCrNi, and CoCrFeMnNi, respectively [35,44]. V^* in the range of 100b^3–1000b^3 is associated with dislocation interaction mechanism and in the range of 10b^3–100b^3 with dislocation glide [45]. Therefore, the obtained activation volume for Ni was in the range for dislocation interaction while the lower value of V^* for the two MPEAs indicates dislocation glide mechanism. Activation energy for dislocation nucleation is higher for a MPEA compared to pure metal [10]. Therefore, lower dislocation density in MPEAs favors dislocation glide over dislocation-dislocation interaction. In similar studies, CoCrFeMnNi homogenized sample showed activation volume of 0.2–1.02 nm^3 using stress relaxation tests in the temperature range of 873 K to 1073 K [46] and activation volume of 52b^3 and 209b^3 were reported for coarse grained Ni and Al from nano-indentation tests, respectively [11]. In addition, Ni-based superalloy (IN-718) showed activation volume in the range of 0.05 to 0.1 nm^3 within temperature range of 298 K to 923 K in tensile creep experiment [32].

The temperature dependence of indentation creep rate may be expressed as a power-law relation as [47]:

$$\dot{\varepsilon} = A\sigma^n \exp\left(-\frac{Q}{RT}\right) \approx AH^n \exp\left(-\frac{Q}{RT}\right) \tag{5}$$

where A is a material constant, R is universal gas constant, and Q is the creep activation energy. The slope of $\ln(\dot{\varepsilon}/H^n)$ versus $1/T$ curve gives the value of $-Q/R$ [48,49] as shown in Figure 5 for the three studied systems. The average strain rate and hardness at each temperature over the holding time was selected for analysis. The creep activation energy of Ni, CoCrNi, and CoCrFeMnNi were found to be 200 ± 20 kJ/mol, 400 ± 100 kJ/mol and 25 ± 5 kJ/mol, respectively. The creep activation energy of CoCrNi was estimated to be 400 kJ/mol, while that of creep resistant Ni-based superalloy such as CMSX-2 and IN-X750 calculated from uniaxial tensile test was 230 kJ/mol and 306 kJ/mol, respectively [50,51]. CoCrNi showed the lowest creep displacement and highest activation energy supporting its superior creep resistance. Hardness or strength has a pronounced effect on creep resistance as previously reported [16,18]. CoCrNi showed the highest hardness compared to the other two systems and thus better creep resistance. However, the hardness of Ni and CoCrFeMnNi were similar and we attributed the lower creep resistance of CoCrFeMnNi alloy to its lower stacking fault energy (SFE) compared to Ni [52]. In confined volume deformation as in nano-indentation, lower SFE may result in increased dislocation generation and plastic flow, thus reducing creep resistance during holding time [52].

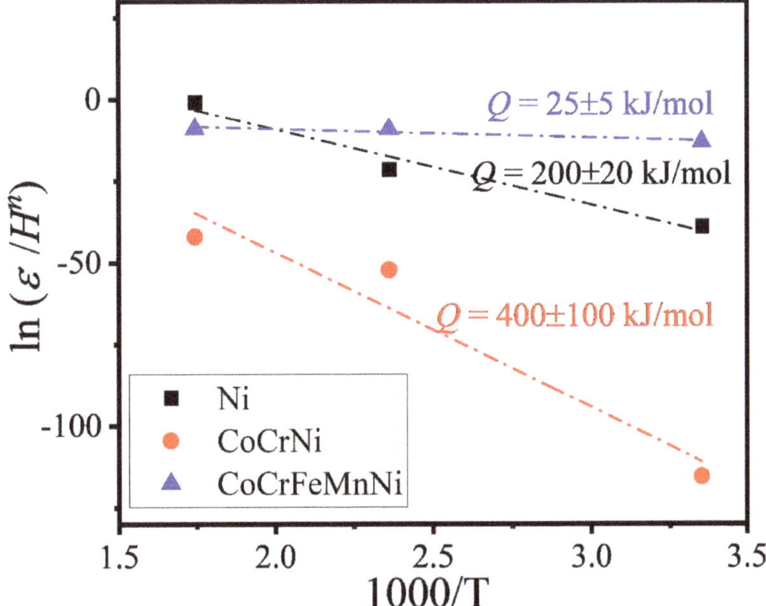

Figure 5. $\ln(\dot{\varepsilon}/H^n)$ versus $1000/T$ with slope giving the activation energy (Q) for Ni, CoCrNi, and CoCrFeMnNi alloys.

5. Conclusions

In summary, nano-indentation creep tests for CoCrNi and CoCrFeMnNi MPEAs were performed under static and dynamic (DMA) loads at 298 K, 423 K, and 573 K. Stress exponent was calculated from steady-state creep. The creep behavior of the alloys was compared in terms of stress exponent, activation volume, and activation energy. The following conclusions may be drawn:

(1) The value of n was in the range of 15 to 135 indicating that time-dependent deformation for all alloys was dislocation dominated.

(2) Activation volume data suggested dislocation glide dominated deformation mechanism for MPEAs and dislocation-dislocation interaction for pure Ni.

(3) The stress exponent decreased with increasing temperature due to thermally activated dislocations. The stress exponent for CoCrFeMnNi was found to be temperature insensitive possibly due to dislocation glide dominated deformation.

(4) The creep stress exponent increased with increasing load (depth) due to higher generation rate of dislocations and their entanglement at greater depth.

(5) CoCrNi showed creep behavior similar to pure Ni at room temperature while it became closer to CoCrFeMnNi at higher temperatures. This may be due to the local chemical ordering in CoCrNi at lower temperatures while becoming more random solid solution like at higher temperatures.

Author Contributions: Conceptualization, M.S. and S.M.; methodology, M.S; validation, M.S. and S.M.; formal analysis, M.S.; investigation, M.S.; data curation, M.S. and S.M.; writing—original draft preparation, M.S.; writing—review and editing, S.M.; visualization, M.S. and S.M.; supervision, S.M.; Project administration, S.M.; All authors have read and agreed to the published version of the manuscript.

Funding: Sundeep Mukherjee acknowledges funding from the National Science Foundation (NSF) under Grant Number 1762545 (CMMI) for part of this work. Any opinions, findings, and conclusions expressed in this paper are those of the authors and do not necessarily reflect the views of the National Science Foundation (NSF).

Acknowledgments: The authors thank the Materials Research Facility (MRF) at University of North Texas for access to the characterization equipment used in this study.

Conflicts of Interest: The authors declare no conflict of interest.

References

1. Miracle, D.B.; Senkov, O.N. A critical review of high entropy alloys and related concepts. *Acta Mater.* **2017**, *122*, 448–511. [CrossRef]
2. Senkov, O.N.; Miracle, D.B.; Chaput, K.J.; Couzinie, J.-P. Development and exploration of refractory high entropy alloys—A review. *J. Mater. Res.* **2018**, *33*, 3092–3128. [CrossRef]
3. Zhang, Y.; Zuo, T.T.; Tang, Z.; Gao, M.C.; Dahmen, K.A.; Liaw, P.K.; Lu, Z.P. Microstructures and properties of high entropy alloys. *Prog. Mater. Sci.* **2014**, *61*, 1–93. [CrossRef]
4. Gludovatz, B.; Hohenwarter, A.; Thurston, K.V.; Bei, H.; Wu, Z.; George, E.P.; Ritchie, R.O. Exceptional damage-tolerance of a medium-entropy alloy CrCoNi at cryogenic temperatures. *Nat. Commun.* **2016**, *7*, 1–8. [CrossRef]
5. Gali, A.; George, E.P. Tensile properties of high-and medium-entropy alloys. *Intermetallics* **2013**, *39*, 74–78. [CrossRef]
6. Nair, R.B.; Arora, H.S.; Mukherjee, S.; Singh, S.; Singh, H.; Grewal, H.S. Exceptionally high cavitation erosion and corrosion resistance of a high entropy alloy. *Ultrason. Sonochem.* **2018**, *41*, 252–260. [CrossRef]
7. Ayyagari, A.; Hasannaeimi, V.; Grewal, H.S.; Arora, H.; Mukherjee, S. Corrosion, erosion and wear behavior of complex concentrated alloys: A review. *Metals* **2018**, *8*, 603. [CrossRef]
8. Xu, X.D.; Liu, P.; Guo, S.; Hirata, A.; Fujita, T.; Nieh, T.G.; Liu, C.T.; Chen, M.W. Nanoscale phase separation in a fcc-based CoCrCuFeNiAl0.5 high-entropy alloy. *Acta Mater.* **2015**, *84*, 145–152. [CrossRef]
9. Sadeghilaridjani, M.; Mukherjee, S. Strain gradient plasticity in multiprincipal element alloys. *JOM* **2019**, *71*, 3466–3472. [CrossRef]
10. Mridha, S.; Sadeghilaridjani, M.; Mukherjee, S. Activation Volume and energy for dislocation nucleation in multi-principal element alloys. *Metals* **2019**, *9*, 263. [CrossRef]
11. Maier, V.; Merle, B.; Göken, M.; Durst, K. An improved long-term nanoindentation creep testing approach for studying the local deformation processes in nanocrystalline metals at room and elevated temperatures. *J. Mater. Res.* **2013**, *28*, 1177–1188. [CrossRef]
12. Sadeghilaridjani, M.; Ayyagari, A.; Muskeri, S.; Hasannaeimi, V.; Salloom, R.; Chen, W.-Y.; Mukherjee, S. Ion irradiation response and mechanical behavior of reduced activity high entropy alloy. *J. Nucl. Mater.* **2020**, *529*, 151955. [CrossRef]
13. Sadeghilaridjani, M.; Muskeri, S.; Hasannaeimi, V.; Pole, M.; Mukherjee, S. Strain rate sensitivity of a novel refractory high entropy alloy: Intrinsic versus extrinsic effects. *Mater. Sci. Eng. A* **2019**, *766*, 138326. [CrossRef]
14. Mridha, S.; Komarasamy, M.; Bhowmick, S.; Mishra, R.S.; Mukherjee, S. Small-scale plastic deformation of nanocrystalline high entropy alloy. *Entropy* **2018**, *20*, 889. [CrossRef]
15. Li, W.H.; Shin, K.; Lee, C.G.; Wei, B.C.; Zhang, T.H.; He, Y.Z. The characterization of creep and time-dependent properties of bulk metallic glasses using nanoindentation. *Mater. Sci. Eng. A* **2008**, *478*, 371–375. [CrossRef]
16. Ma, Y.; Feng, Y.H.; Debela, T.T.; Peng, G.J.; Zhang, T.H. Nanoindentation study on the creep characteristics of high-entropy alloy films: fcc versus bcc structures. *Int. J. Refract. Met. Hard Mater.* **2016**, *54*, 395–400. [CrossRef]
17. Lee, D.-H.; Seok, M.-Y.; Zhao, Y.; Choi, I.-C.; He, J.; Lu, Z.; Suh, J.-Y.; Ramamurty, U.; Kawasaki, M.; Langdon, T.G.; et al. Spherical nanoindentation creep behavior of nanocrystalline and coarse-grained CoCrFeMnNi high-entropy alloys. *Acta Mater.* **2016**, *109*, 314–322. [CrossRef]
18. Ma, Y.; Peng, G.J.; Wen, D.H.; Zhang, T.H. Nanoindentation creep behavior in a CoCrFeCuNi high-entropy alloy film with two different structure states. *Mater. Sci. Eng. A* **2015**, *621*, 111–117. [CrossRef]
19. Wang, Z.; Guo, S.; Wang, Q.; Liu, Z.; Wang, J.; Yang, Y.; Liu, C.T. Nanoindentation characterized initial creep behavior of a high-entropy-based alloy CoFeNi. *Intermetallics* **2014**, *53*, 183–186. [CrossRef]
20. Wang, X.; Gong, P.; Deng, L.; Jin, J.; Wang, S.; Zhou, P. Nanoindentation study on the room temperature creep characteristics of a senary $Ti_{16.7}Zr_{16.7}Hf_{16.7}Cu_{16.7}Ni_{16.7}Be_{16.7}$ high entropy bulk metallic glass. *J. Non-Cryst. Solids* **2017**, *470*, 27–37. [CrossRef]

21. Zhang, L.; Yu, P.; Cheng, H.; Zhang, H.; Diao, H.; Shi, Y.; Chen, B.; Chen, P.; Feng, R.; Bai, J.; et al. Nanoindentation creep behavior of an Al0.3CoCrFeNi high-entropy alloy. *Metall. Mater. Trans. A* **2016**, *47*, 5871–5875. [CrossRef]
22. Jiao, Z.-M.; Ma, S.-G.; Yuan, G.-Z.; Wang, Z.-H.; Yang, H.-J.; Qiao, J.-W. Plastic deformation of Al0.3CoCrFeNi and AlCoCrFeNi high-entropy alloys under nanoindentation. *J. Mater. Eng. Perform.* **2015**, *24*, 3077–3083. [CrossRef]
23. Li, Z.; Zhao, S.; Ritchie, R.O.; Meyers, M.A. Mechanical properties of high-entropy alloys with emphasis on face-centered cubic alloys. *Prog. Mater. Sci.* **2019**, *102*, 296–345. [CrossRef]
24. Mridha, S.; Das, S.; Aouadi, S.; Mukherjee, S.; Mishra, R.S. Nanomechanical behavior of CoCrFeMnNi high-entropy alloy. *JOM* **2015**, *67*, 2296–2302. [CrossRef]
25. Wheeler, J.M.; Armstrong, D.E.J.; Heinz, W.; Schwaige, R. High temperature nanoindentation: The state of the art and future challenges. *Curr. Opin. Solid State Mater. Sci.* **2015**, *19*, 354–366. [CrossRef]
26. Tsao, T.-K.; Yeh, A.-C.; Kuo, C.-M.; Kakehi, K.; Murakami, H.; Yeh, J.-W.; Jian, S.-R. The high temperature tensile and creep behaviors of high entropy superalloy. *Sci. Rep.* **2017**, *7*, 12658. [CrossRef] [PubMed]
27. Kang, Y.B.; Shim, S.H.; Lee, K.H.; Hong, S.I. Dislocation creep behavior of CoCrFeMnNi high entropy alloy at intermediate temperatures. *Mater. Res. Lett.* **2018**, *6*, 689–695. [CrossRef]
28. Sawant, A.; Tin, S.; Zhao, J.-C. High Temperature Nano-Indentation of Ni-Based Superalloys. Proceeding of the 11th International Symposium on Superalloys, TMS (The Minerals, Metals & Materials Society), Champion, PA, USA, 14–17 September 2008; pp. 863–871.
29. Haghshenas, M.; Wang, Y.; Cheng, Y.-T.; Gupta, M. Indentation-based rate-dependent plastic deformation of polycrystalline pure magnesium. *Mater. Sci. Eng. A* **2018**, *716*, 63–71. [CrossRef]
30. McElhaney, K.W.; Vlassak, J.J.; Nix, W.D. Determination of indenter tip geometry and indentation contact area for depth-sensing indentation experiments. *J. Mater. Res.* **1998**, *13*, 1300–1306. [CrossRef]
31. Chen, J.; Shen, Y.; Liu, W.; Beake, B.D.; Shi, X.; Wang, Z.; Zhang, Y.; Guo, X. Effects of loading rate on development of pile-up during indentation creep of polycrystalline copper. *J. Mater. Sci. A* **2016**, *656*, 216–221. [CrossRef]
32. Wang, H.; Dhiman, A.; Ostergaard, H.E.; Zhang, Y.; Siegmund, T.; Kruzic, J.J.; Tomar, V. Nanoindentation based properties of Inconel 718 at elevated temperatures: A comparison of conventional versus additively manufactured samples. *Int. J. Plast.* **2019**, *120*, 380–394. [CrossRef]
33. Zhang, W.-D.; Liu, Y.; Wu, H.; Lan, X.-D.; Qiu, J.; Hu, T.; Tang, H.-P. Room temperature creep behavior of Ti–Nb–Ta–Zr–O alloy. *Mater. Charact.* **2016**, *118*, 29–36. [CrossRef]
34. MorrisJr, J.W. Dislocation-controlled plasticity of crystalline materials: Overview. In *Encyclopedia of Materials: Science and Technology*, 2nd ed.; Elsevier: Amsterdam, The Netherlands, 2001; pp. 2245–2255.
35. Laplanche, G.; Kostka, A.; Reinhart, C.; Hunfeld, J.; Eggeler, G.; George, E.P. Reasons for the superior mechanical properties of medium-entropy CrCoNi compared to high-entropy CrMnFeCoNi. *Acta Mater.* **2017**, *128*, 292–303. [CrossRef]
36. Li, Q.-J.; Sheng, H.; Ma, E. Strengthening in multi-principal element alloys with local-chemical-order roughened dislocation pathways. *Nat. commun.* **2019**, *10*, 3563. [CrossRef]
37. Nix, W.D.; Gao, H. Indentation size effects in crystalline materials: A law for strain gradient plasticity. *J. Mech. Phys. Solids* **1998**, *46*, 411–425. [CrossRef]
38. Li, H.; Ngan, A.H.W. Size effects of nanoindentation creep. *J. Mater. Res.* **2004**, *19*, 513–522. [CrossRef]
39. Almasri, A.H.; Voyiadjis, G.Z. Effect of strain rate on the dynamic hardness in metals. *J. Eng. Mater. Tech.* **2007**, *129*, 505–512. [CrossRef]
40. Liu, Y.; Huang, C.; Bei, H.; He, X.; Hu, W. Room temperature nanoindentation creep of nanocrystalline Cu and Cu alloys. *Mater. Lett.* **2012**, *70*, 26–29. [CrossRef]
41. Zhao, J.; Wang, F.; Huang, P.; Lu, T.J.; Xu, K.W. Depth dependent strain rate sensitivity and inverse indentation size effect of hardness in body-centered cubic nanocrystalline metals. *Mater. Sci. Eng. A* **2014**, *615*, 87–91. [CrossRef]
42. Oikawa, H.; Karashima, S. On the stress exponent and the rate-controlling mechanism of high-temperature creep in some solid solutions. *Metall. Trans.* **1974**, *5*, 1179–1182. [CrossRef]
43. Phani, P.S.; Oliver, W.C. A direct comparison of high temperature nanoindentation creep and uniaxial creep measurements for commercial purity aluminum. *Acta Mater.* **2016**, *111*, 31–38. [CrossRef]

44. Owen, L.R.; Pickering, E.J.; Playford, H.Y.; Stone, H.J.; Tucker, M.G.; Jones, N.G. An assessment of the lattice strain in the CrMnFeCoNi high-entropy alloy. *Acta Mater.* **2017**, *122*, 11–18. [CrossRef]
45. Monclús, M.A.; Molina-Aldareguia, J.M. High temperature nanomechanical testing. In *Handbook of Mechanics of Materials*; Hsueh, C.H., Schmauder, S., Chen, C.-S., Chawla, K.K., Chawla, N., Chen, W., Kagawa, Y., Eds.; Springer: Singapore, 2018.
46. He, J.Y.; Zhu, C.; Zhou, D.Q.; Liu, W.H.; Nieh, T.G.; Lu, Z.P. Steady state flow of the FeCoNiCrMn high entropy alloy at elevated temperatures. *Intermetallics* **2014**, *55*, 9–14. [CrossRef]
47. Li, Y.J.; Mueller, J.; Hoppel, H.W.; Goken, M.; Blum, W. Deformation kinetics of nanocrystalline nickel. *Acta Mater.* **2007**, *55*, 5708–5717. [CrossRef]
48. Shen, L.; Wu, Y.; Wang, S.; Chen, Z. Creep behavior of Sn–Bi solder alloys at elevated temperatures studied by nanoindentation. *J. Mater. Sci.: Mater. Electron.* **2017**, *28*, 4114–4124. [CrossRef]
49. Ginder, R.S.; Pharr, G.M. Characterization of power-law creep in the solid-acid $CsHSO_4$ via nanoindentation. *J. Mater. Res.* **2019**, *34*, 1130–1137. [CrossRef]
50. Rouault-Rogez, H.; Dupeux, M.; Ignat, M. High temperature tensile creep of CMSX-2 Nickel base superalloy single crystals. *Acta Metall. Mater.* **1994**, *42*, 3137–3148. [CrossRef]
51. Picasso, A.C.; Marzocca, A.J. On apparent activation energies of creep in nickel-base superalloys. *Scrp. Mater.* **1999**, *41*, 797–802. [CrossRef]
52. Hu, J.; Sun, G.; Zhang, X.; Wang, G.; Jiang, Z.; Han, S.; Zhang, J.; Lian, J. Effects of loading strain rate and stacking fault energy on nanoindentation creep behaviors of nanocrystalline Cu, Ni-20 wt. %Fe and Ni. *J. Alloys Compd.* **2015**, *647*, 670–680. [CrossRef]

Review

Applications of High-Pressure Technology for High-Entropy Alloys: A Review

Wanqing Dong [1], Zheng Zhou [1], Mengdi Zhang [1], Yimo Ma [1], Pengfei Yu [1,*], Peter K. Liaw [2] and Gong Li [1,*]

1 State Key Laboratory of Metastable Materials Science and Technology, Yanshan University, Qinhuangdao 066004, China

2 Department of Materials Science and Engineering, The University of Tennessee, Knoxville, TN 37996-2200, USA

* Correspondence: ypf@ysu.edu.cn (P.Y.); gongli@ysu.edu.cn (G.L.)

Received: 23 June 2019; Accepted: 26 July 2019; Published: 8 August 2019

Abstract: High-entropy alloys are a new type of material developed in recent years. It breaks the traditional alloy-design conventions and has many excellent properties. High-pressure treatment is an effective means to change the structures and properties of metal materials. The pressure can effectively vary the distance and interaction between molecules or atoms, so as to change the bonding mode, and form high-pressure phases. These new material states often have different structures and characteristics, compared to untreated metal materials. At present, high-pressure technology is an effective method to prepare alloys with unique properties, and there are many techniques that can achieve high pressures. The most commonly used methods include high-pressure torsion, large cavity presses and diamond-anvil-cell presses. The materials show many unique properties under high pressures which do not exist under normal conditions, providing a new approach for the in-depth study of materials. In this paper, high-pressure (HP) technologies applied to high-entropy alloys (HEAs) are reviewed, and some possible ways to develop good properties of HEAs using HP as fabrication are introduced. Moreover, the studies of HEAs under high pressures are summarized, in order to deepen the basic understanding of HEAs under high pressures, which provides the theoretical basis for the application of high-entropy alloys.

Keywords: high-entropy; high pressure; high pressure torsion; diamond anvil cells

1. Introduction

Emerging in recent years, high-entropy alloys (HEAs) are newly developed alloys, with many outstanding properties which have broken the design concept of traditional alloys, and have a variety of principal elements and special crystal structures. Although the composition of the high-entropy alloys varies, the phase composition is very simple. Usually only one or two solid-solution phases are detected by X-ray diffraction, and rare intermetallic compounds are formed. Since there is no principal element in HEAs, the performance of the alloy is affected by the combined influence of the constituent elements. As a result, the performance of HEAs is somewhat unpredictable. The structure and stability of the alloy has crucial influence on its properties. Therefore, it is necessary to study the structures, and the stability of the structure of the different HEAs under various conditions. HEAs are solid solutions composed of many elements which can maintain their stable structures under normal temperatures and atmospheric pressure. However, during long-term or high-temperature annealing, the new phase will be created, which will affect the performance of the alloy [1–3]. The pressure, temperature, and chemical composition are the basic thermodynamic factors that determine the state of a substance. The use and control of temperature and chemical composition are almost synchronous with the development of human civilization. However, due to the limitations of technical conditions, the application of pressure has just begun. High-pressure technology is a relatively-young and emerging discipline, but most of the condensed matter in the universe is under high pressure. The research on high pressures relies heavily on the experimental techniques, and each progression of the method has led to a significant expansion of our basic understanding of material behavior under high pressures. The progress in high-pressure experimental technology has directly promoted the development of the high-pressure science and provided an advanced method for frontier subjects. It has become an important field in modern scientific research. There are few studies on the structures of HEAs under high pressures. Interactions within the materials will change under high pressures and induce the generation of high-pressure phase transitions, thus becoming a new material with special properties. The materials will show many unique properties under high pressures, ones which do not exist under normal conditions, providing a subject for the in-depth study of materials. The behavior of HEAs under high pressures is a potential research direction for the future.

2. High-Entropy Alloys (HEAs)

2.1. Concept

HEAs are kinds of alloys developed in recent years. HEAs are loosely defined as solid-solution alloys that contain more than five principal elements in an equal or near-equal atomic percent (at. %) [4]. These kinds of alloys were defined by Yeh et al. [4] in 2004 as HEAs, and in the same year named by Cantor [5] et al. as the multi-component alloy. The concept of HEAs is based on the development of bulk amorphous alloys in the 1990s, when people were looking for alloys with ultra-high glass-forming ability. According to the well-known confusion principle, the more components of the alloy, the higher the chaos of the liquid alloy. That is, with a high entropy of mixing, it is easy for the alloy to retain the structure of the melt, thus forming a disordered amorphous structure. Figure 1 shows a panoramic view of the materials that humans have used over the past 10,000 years. The picture gives different types of materials, from ceramics to metals, polymers, and most recently composites, the same age at which HEA production is noted. With an increasing understanding of HEAs, the requirements for the content of each element and the number of elements have gradually loosened in the definition of HEA. At present, quaternary equiatomic or part of non-equiatomic quaternary alloys are also defined as HEAs [6–8], and some literature now refers to multi-component alloys containing small amounts of intermetallic compounds as HEAs [9,10].

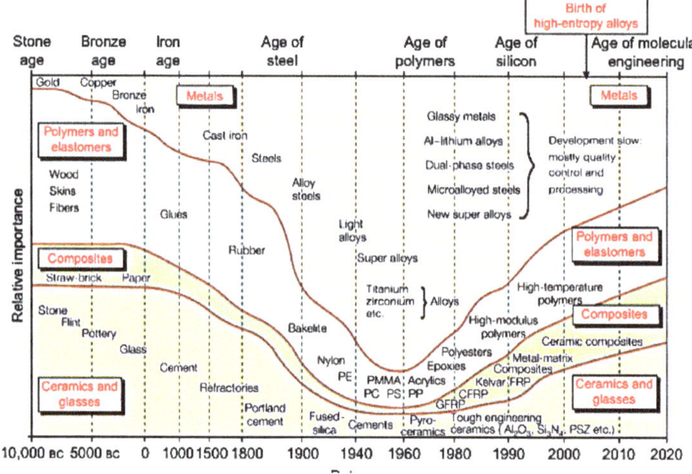

Figure 1. Historical evolution of engineering materials [11].

2.2. Four Core Effects

Yeh [12] summarized the four core effects in HEAs. Those were: (1) High-entropy effects; (2) sluggish diffusion; (3) severe lattice distortion; and (4) cocktail effects.

2.2.1. High-Entropy Effects

This is the most important characteristic of the HEA; i.e., the formation of intermetallic compounds or complex phases is inhibited due to the high entropy of the HEAs, tending to form solid-solution phases [4,5,13,14]. In other words, the high entropy produced by multiple principal elements can inhibit the generation of intermetallic compounds.

2.2.2. Sluggish Diffusion

In traditional alloys, there is a high probability that a less prevalent atom is surrounded by the main element's atoms, which are called the solute atoms. It means that during the atom diffusion, the interaction with the surrounding atoms is essentially constant. However, the atoms that surround an atom in the HEAs are diverse, leading to different activation energies required by various lattice sites during atomic diffusion in HEAs [15]. In comparison with traditional alloys, more energy is needed during the atomic diffusion for HEAs. Although diffusion activation varies greatly with the elements, the trend of the increased diffusion activation energy can be obtained from low entropy to high entropy.

The sluggish diffusion theoretically suppresses the grain growth in HEAs [4] and explains the formation of nano-sized precipitates. On the other hand, the sluggish diffusion increases the phase stability of HEAs, especially at high temperatures; the phase stability and high-temperature strength even exceed some Ni-based superalloys [16].

2.2.3. Severe Lattice Distortion

In HEAs, each atom is surrounded by other kinds of atoms. Due to different compositional atomic sizes in HEAs, that feature can lead to the severe lattice distortion. The severe lattice distortion affects the mechanical, physical, and chemical properties of the material. As shown in Figure 2, the probability that each element in the HEA occupies the position of the lattice is the same, and the atomic radii of different atoms will cause severe lattice distortion.

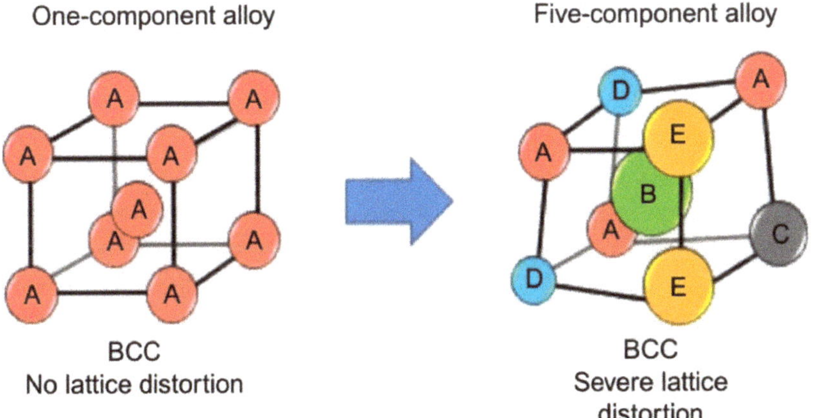

Figure 2. Schematic illustration of lattice distortion.

2.2.4. Cocktail Effect

The cocktail effect for alloys was first mentioned by Ranganathan [17] to describe some basic characteristics of the elements affecting the properties of alloys. However, for the HEAs that break through the traditional design concept, the cocktail effect does not mean that the performance of the alloys is simply a superposition of the properties of each component. There are also interactions between different elements that eventually lead to a composite effect in HEAs [12].

2.3. Research Status

Due to the four core effects of HEAs, many HEAs exhibit mechanical, physical, and chemical properties that are superior to pure metals and other alloys. Many research groups have carried out much research on HEAs. At present, several mature high-entropy systems have been formed, including $Al_xCoCrCuFeNi$ [4], CoCrFeMnNi [5], WNbMoTaV [8], and GdTbDyTmLu [18]. The structures and properties can be regulated through a series of elemental additions or changes in the content, such as the addition of Ti [19], Zr [20], B [20], V [21], and/or Mo [22,23]. Some HEAs have the high glass-forming ability, such as PdPtCuNiP [24], ScCaYbMgZn [25], and ZrHfTiCuFe [26]. The structures of the HEAs include all the three simple metal crystal structures: A face-centered-cubic (fcc) structure represented by CoCrFeNi, a body-centered-cubic (bcc) structure represented by TaZrHfNbTi, and a hexagonal-closed-packed (hcp) structure represented by YGdTbDyLu. The deformation mechanisms of hcp-phase metals have been summarized: Either tensile or compressive strain induced twins or basal and prismatic slips [27–29].

In addition to the excellent mechanical properties of the HEAs at room temperature, the HEAs also have good performance under extreme conditions, such as high and low temperatures. The VNbMoTaW alloy developed by Senkov et al. maintains a yield strength of 800 MPa at 600 to 800 °C, and the yield strength is still more than 400 MPa at 1600 °C [30]. The CoCrFeMnNi single-phase fcc alloy demonstrates the characteristics of "lower temperature and more toughness," and the fracture toughness value is greater than 200 MPa·m$^{1/2}$ at a low temperature (77 K) [31]. For $Co_{1.5}CrFeNi_{1.5}Ti$ and $Al_{0.2}Co_{1.5}CrFeNi_{1.5}Ti$ HEAs, the wear resistance is at least twice that of conventional wear-resistance steels, such as SUJ and SKH51 steel under similar hardness conditions [32]. Moreover, the mechanism of the phase transition in HEAs has received great attention from researchers. At present, the transformation-induced plastic deformation and polymorphism have been studied [7,33]. However, the phase-transition kinetics of the HEAs still need to be further studied. The atomic mechanism and thermodynamic principle of the HEA phase transition are still in dispute [34,35]. Surface energy at the interface may affect the solidification and diffusion of solids, such as precipitation, resulting in phase transitions [36].

3. High Pressure

The rapid development of the high-pressure technology and the interpenetration with other technologies have become a hot topic in science today. High pressure is an important means to control the properties of the material, and is a decisive variable in modern scientific research [37]. Bundy synthesized diamond with technologies employing high pressures and high temperatures, which paved the way for the high-pressure technology for the application of new materials [38].

According to the different high-pressure loading methods or technologies, the high-pressure experimental technology can be divided into the static high-pressure loading technology and dynamic high-pressure loading technology. The duration of the traditional, dynamic high-pressure experiments is very short, typically no more than a few milliseconds, and can reach very high pressures and temperatures. However, the defects are also obvious: The time span is too short to be effectively detected and accurately controlled, and the large amount of extra heat will affect the expected experimental results. In contrast, the static high-pressure technique can perform the non-destructive research on the material under fixed pressure conditions and can obtain good data. Static and dynamic high pressures are complementary to each other, which enhances our basic understanding of the material structure and performance under different high pressures. The dynamic high-pressure loading technology is an experimental technique for subjecting samples to transient high-pressure and high-temperature environments by the shock waves or high-speed physical impact generated from an explosion, or other means. At present, there are many experimental techniques that can achieve dynamic high pressures, which can be simply divided into high-speed dynamic loading and low-speed dynamic loading. The methods implementing the former mainly include the underground nuclear explosion, magnetic flux compression, rapid expansion, or an explosion of air currents. These methods are characterized by fast loading (can realize nanosecond loading), and high-pressure limits. But it is difficult to accurately control the pressure range, relying heavily on experience and repeated experiments. Many high-speed dynamic-loading methods are still in development and inappropriate for research under normal laboratory conditions. In contrast, the experimental techniques of low-speed dynamic loading are relatively mature and have mostly been used in experiments. These loading techniques have a large time span (from hours to milliseconds or even microseconds), high repeatability, high-pressure loading accuracy, and wide adaptation (which can be applied to room temperature, high temperature, low temperature, magnetic field, and other conditions), which can be combined with other test methods and characteristics. It is of great help to study the structures and properties of materials under nonequilibrium thermodynamic conditions, which is an important direction in the high-pressure research.

The static high-pressure loading method is a technique for obtaining high-pressure experimental conditions using a static compression method. The sample is slowly compressed by means of the external mechanical loading device. Since the compression process is slow enough, the heat generated in the process can be fully exchanged with the external environment. Hence, the static high-pressure loading process is an isothermal compression process. At present, the equipment to achieve the static high pressure mainly includes large cavity presses and diamond-anvil-cell presses. Currently, the highest pressure of the former is up to hundreds of thousands GPa, which is the main equipment for the synthesis of high-pressure materials, and it has a unique advantage in the use of high-temperature and high-pressure, synthetic block, superhard materials. Compared to the former, the diamond-anvil cell (DAC) is small in size, requires few samples, and has low experimental costs. Multiple in-situ experiments (including X-ray diffraction, Raman spectrum, etc.) can be performed under a variety of experimental conditions, and higher-pressure limits can be achieved. The hexahedron-anvil-press instrument is the most widely-used static high-pressure device. Figure 3 is the schematic of the instrument. The instrument is composed of six anvil cells, extruding a square-pressure transmitting medium, and the sample is wrapped in the medium. As shown in the figure, the machine applies the pressure from six directions to the center simultaneously to achieve the static equilibrium [39,40].

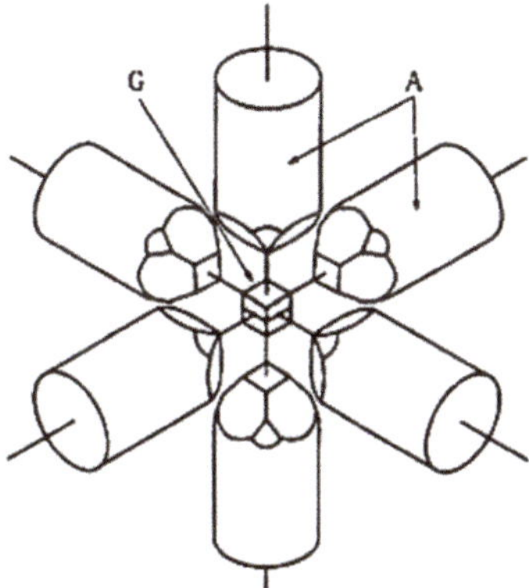

Figure 3. Schematic of the hexahedron anvil press.

The DAC, another device that can achieve the static high-pressure loading, it is smaller in size, and Figure 4 shows the principle of the DAC. DACs are composed of two opposing diamond anvils that squeeze the samples between them to create hydrostatic/non-hydrostatic pressures [41]. In 1977, Bruas combined the DAC technology with synchrotron-radiation X-ray diffraction for the first time [42], which greatly promoted the development of the high-pressure science. In 2015, Dubrovinsky et al. used nano-diamonds in combination with a two-stage pressurized device to obtain a high static pressure of 770 GPa [43]. The core of the DAC is to use the extremely-high hardness of the diamond, with two diamonds pressing on top of each other, to generate high pressures.

The high-pressure torsion (HPT) process was first proposed by Bridgman in 1935 [44]. With the development of microscopy technology, it was not until the 1980s that researchers discovered that ultra-fine crystalline structure could be prepared by high-pressure torsion technology [45]. The principle of HPT is exhibited in Figure 5. The sample is placed between the two anvils and subjected to extremely high pressures of hundreds of GPa. When there is a relative rotation between the two anvils driven by external forces, the friction between the sample and the anvils drives the sample to rotate and causes the shear deformation of the sample. Although the sample is subjected to the large strain plastic deformation, it will not rupture, due to the high pressure. By means of HPT, the size of crystal grains can be significantly reduced, and a dense nano-bulk material can be prepared effectively.

In high-pressure scientific research, the structure-phase transitions, which combine pressure or temperature and pressure as driving forces, can be divided into two basic types, namely reconstructive phase transitions and displacive phase transitions. The classification is based on whether the chemical bonds that form the periodic grid are destroyed after the phase transformation. In the process of the reconstructive phase transition, the main chemical bonds are broken and recombined to form a new structure. There is no clear orientation relationship with the previous phase in crystallography. All the reconstructive phase transitions belong to the first-order phase transition, and the discontinuity of the volume change during the phase transition is obvious. Since there are dynamic barriers between the equilibrium pressures of the adjacent two phases, it often leads to a hysteresis effect in the phase transition. The phase-transition pressure point in the pressurization process is greater than the pressure-relief process. The secondary bonds may break when a displacive phase transition occurs, but the main chemical bonds will not break, usually due to the displacement of the atoms or the tilt of the polyhedral structure. Under a high pressure, many phase transitions are displacive phase transitions, and the degree of the discontinuous volume change is very small. Most of them still belong to the first-order phase transition.

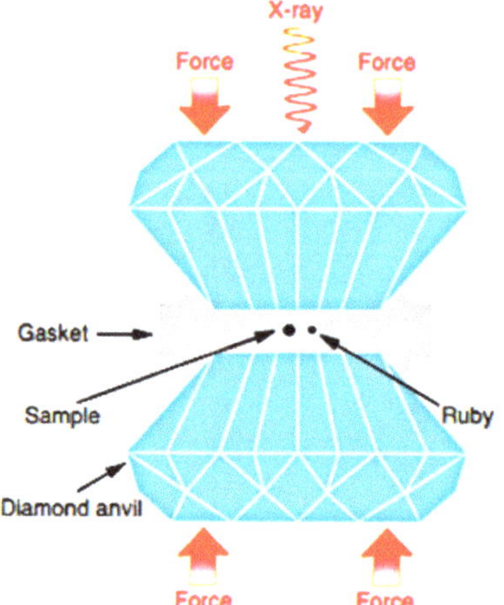

Figure 4. Schematic of the diamond anvil cell (DAC).

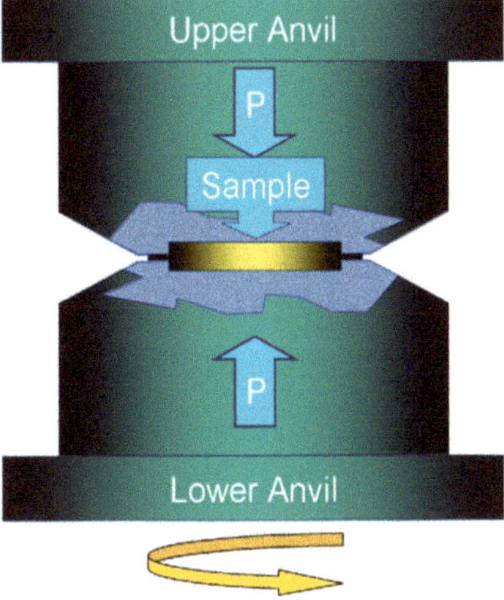

Figure 5. Schematic illustration showing the principles of high-pressure torsion (HPT).

4. HEAs under High Pressure

4.1. Dynamic High Pressure

The HEA is a new kind of alloy between the traditional alloy and the amorphous alloy. The deformation mechanism shows different characteristics from the traditional alloys. Therefore, it is necessary to study the structure and phase-transition process of HEAs under high pressures. The microstructures of HEAs under high pressures is also one of the hot topics in the alloy research. Huang et al. [46] developed a facile, two-step carbothermal shock (CTS) method that employs flash heating and cooling (temperature of 2000 K, shock duration of 55 ms, and ramp rates in the order of 10^5 K/s) of metal precursors on oxygenated-carbon support to produce HEA nanoparticles (HEA-NPs) with up to eight metallic elements. Figure 6 shows the scanning transmission electron microscopy (STEM) elemental maps for PtPdRhRuCe HEA-NPs. The aforementioned is a method of synthesizing HEAs using the high pressure produced by the thermal shock. Nanoparticles are useful in a wide range of applications. Huang et al. [46] developed a method for making HEA nanoparticles, and the "carbothermal shock synthesis" can be tuned to select for the nanoparticle size as well as final structure. These carbothermal shock (CTS) capabilities facilitate a new research area for the materials discovery and optimization.

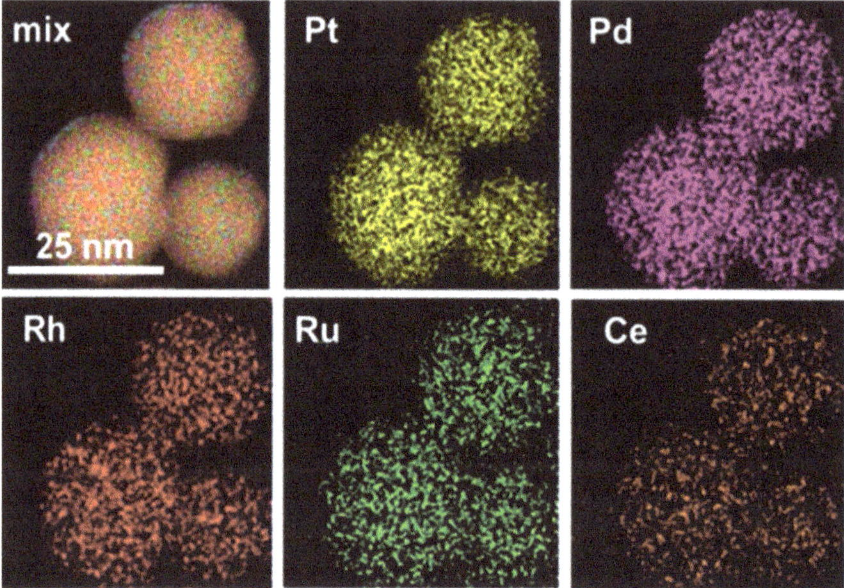

Figure 6. Scanning transmission electron microscope (STEM) elemental maps for PtPdRhRuCe high-entropy-alloy nanoparticles (HEA-NPs), reproduced from [46], with permission from authors.

4.2. Diamond Anvil Cells

The duration of traditional dynamic high-pressure experiments is very short, generally no more than a few milliseconds. Therefore, it is difficult to effectively detect and accurately control, and the large amount of the extra heat will affect the expected experimental results. Thus, the experiment of placing HEAs under high pressures usually adopts a static high-pressure technology. Yu et al. [47] used DAC to pressurize the rare-earth HoDyYGdTb HEA and studied the phase transition under high pressures. Figure 7 shows the pressure-volume relationship of different phases of the HoDyYGdTb HEA measured at room temperature, and the red line indicates the fitting result using the Birch-Murnaghan equation of the state in [48]. The sample was prepared by arc-melting, then scraped, and loaded into the T301-stainless-steel gasket hole with a diameter of 180 μm. The specimen was pressurized up to 60.1 GPa. The HEA is shown to follow the trivalent rare-earth crystal structure sequence of hcp → Sm-type → dhcp → dfcc, which correlated the s → d charge transfer of the HEA. The bulk modulus and atomic volume of the rare-earth HEA agree extremely well with the calculated values with the "additivity law." The pressure-included phase transformation among bcc, fcc, and hcp phases in transition metals were reported previously. What is most noteworthy is that Au changes the structure of HCP under extreme conditions [49]. Zhang et al. [50] studied the phase transition in Ni-based HEAs under high pressures. For high-pressure experiments, the sample was loaded in the chamber, which was indented from the Re gasket with a pair of diamond anvils with a thickness of 40 μm. The single-phase CoCrFeNi HEA alloy was pressurized to observe the phase transition. A pressure-induced fcc-hcp phase transition was found in the CoCrFeNi HEA at the pressure of 13.5 GPa and at an ambient temperature. The hcp structure is recoverable when the pressure is released. The phase transformation is very sluggish and did not finish at 39 GPa

The HEA is essentially an alloy-design concept, and there is no requirement on which elements must be used. Therefore, it also provides a broad scope for the study of HEAs. The transition elements are the most frequently-used components of HEAs. As the major element in our planet, Fe is the most interesting element to be studied, and the phase transformations (including from bcc to hcp structures) under high pressures or high temperature have already been found [51–53]. In ambient conditions, cobalt is in the hcp structure, and a pressure-induced phase transition to the fcc phase was found [54]. Figure 8 shows this pressure-induced phase transition. However, no pressure-included phase transition was found in Ni up to 260 GPa [55]. [50].

Polymorphism is widely observed in many materials, which describes the occurrence of different lattice structures in a crystalline material, and is a potential research direction in the future. The polymorphism in the CoCrFeMnNi HEA is reported in [33]. By employing in situ, high-pressure, synchrotron radiation X-ray diffraction, the polymorphic transition from fcc to hcp structures in the CoCrFeMnNi HEA is observed. The hydrostatic pressure was up to 41 GPa using a DAC. The CoCrFeMnNi HEA has an fcc single–phase structure and remains stable up to 19.5 GPa. When the pressure reaches 22.1 GPa, new peaks appeared, indicating that the phase transition has occurred. All the new peaks revealed an fcc-to-hcp transition under high pressures. During the decompression, the phase transition was irreversible. Huang et al. [56] studied the deviatoric deformation kinetics in the CoCrFeMnNi HEA under hydrostatic compression. The HEA was subjected to a hydrostatic pressure of 20 GPa via a DAC. The main significance of this study was to provide another perspective for studying the deformation mode of the HEA system through high-pressure experiments [56].

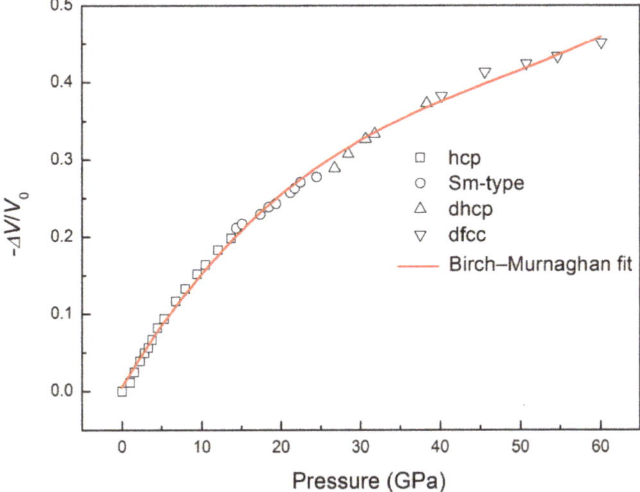

Figure 7. The measured pressure-volume relationship for various phases of the HoDyYGdTb high-entropy alloy (HEA) at up to 60.1 GPa, at room temperature, data from [47].

Gong Li et al. [57] studied the pressure-volume relationship of CoCrFeNiAlCu HEA using in-situ high-pressure energy-dispersive X-ray diffraction with synchrotron radiation at high pressures, and the results show that the CoCrFeNiAlCu HEA keeps a stable fcc + bcc structure in the experimental pressure ranges from 0 to 24 GPa. The equation of the state of the HEA determined by the calculation of the radial distribution function in the non-phase-transitional case is:

$$-\Delta V/V_0 = 2.7P - 0.256P^2 + 0.012P^3 - 2.928 \times 10^{-4}P^4 + 2.907 \times 10^{-6}P^5$$

where V_0 is the volume at zero pressure, $(\Delta V/V_0)$ is the relative volume change.

Cheng et al. [58] studied an ordered, bcc-structured (B2 phase) AlCoCrFeNi HEA using in situ, synchrotron radiation X-ray diffraction up to 42 GPa and non-in transmission electron microscopy. Pressure-induced polymorphic transitions (PIPT) to potentially disordered phase were observed. Yusenko et al. [59] studied the temperature and pressure stability for a hcp $Ir_{0.19}Os_{0.22}Re_{0.21}Rh_{0.20}Ru_{0.19}$

HEA. The sample was loaded in a DAC cell and did not result in phase transition with a maximum pressure of 45 GPa. Ahmad et al. [60] performed in-situ high-pressure and high-temperature XRD measurements on bcc-$Hf_{25}Nb_{25}Zr_{25}Ti_{25}$, fcc-$Ni_{20}Co_{20}Fe_{20}Mn_{20}Cr_{20}$ and hcp-$Re_{25}Ru_{25}Co_{25}Fe_{25}$ HEAs; all of the HEAs remained stable and no phase transition was observed.

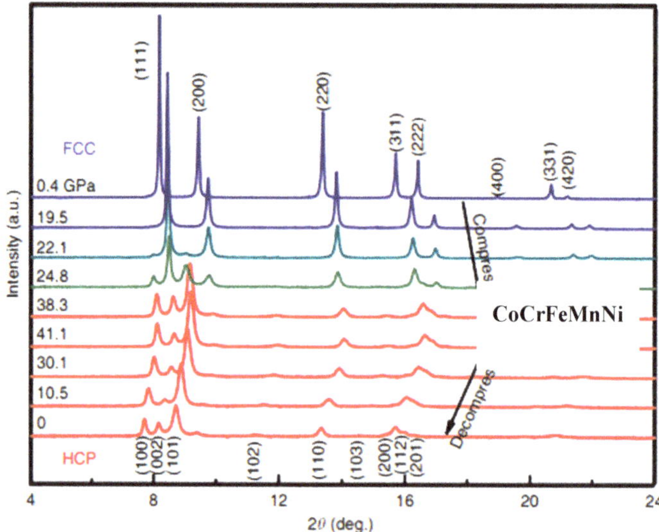

Figure 8. CoCrFeMnNi XRD patterns during compression and decompression, reproduced from [33], with permission from authors.

4.3. *High-Pressure Torsion*

Due to the severe lattice distortion effect and different chemical bonds of their constituent elements, the plastic deformation mechanisms of HEA could be different from that of conventional alloys. The high-pressure torsion (HPT) method is currently used mainly as a severe plastic deformation (SPD) technique for grain refinement, and it has taken a long time to synthesize a metastable phase by HPT processing. High-pressure torsion is generally used to prepare non-porous bulk ultrafine grain samples, which can generate large plastic deformation of the sample through shear stress (γ = 10–100) [61–64]. At present, there have been reports on high-entropy alloys treated by high pressure torsion. These reports are mainly focused on the face centered cubic HEAs. Tang [65] obtained the nano-scale $Al_{0.3}$CoCrFeNi HEA by HPT and studied the strengthening mechanism of the annealing process. Schuh [66] used high pressure torsion to treat the CoCrFeMnNi HEA; the grain can be refined to 50 nm, and the strength and hardness can be increased to 1950 MPa and 520 HV, respectively. After high-pressure torsion of the face centered cubic high-entropy alloy, the grain gradually refines as the strain increases. As shown in Figure 9, the plastic-deformation mechanism mainly includes dislocation slip and twinning.

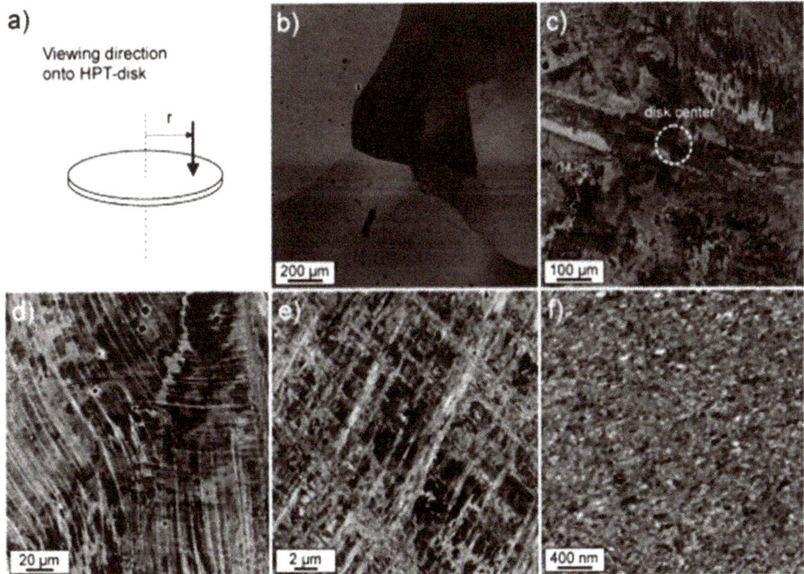

Figure 9. Microstructural evolution in HPT disks investigated with SEM using back-scattered electron contrast, reproduced from [66], with permission from authors.

Yu et al. showed that the plastic-deformation mechanisms of the single-phase fcc $Al_{0.1}CoCrFeNi$ HEA induced by HPT at room temperature [67]. The sample was prepared by arc-melting and then casted as a plate by vacuum induction. The plate was hot-isostatic pressed (HIPed) at 1473 K, and 100 MPa, for 4 h, and then placed in a horizontal tube furnace at 1423 K for 50 h. The samples were compressed at 6 GPa through 1 and 2 turns with the rotation speed of 1 rpm. Processing by HPT produces a very substantial grain refinement in HEAs. The deformation mechanisms include the dislocation slip and twinning at room temperature. As shown in Figure 10, the average Vickers microhardness of the HIPed sample is 135 Hv. After HPT for 1 revolution, the microhardness reaches a saturation of about 482 Hv at the edge of the sample.

A deformation-induced phase transformation in a single-phase fcc $Co_{20}Cr_{26}Fe_{20}Mn_{20}Ni_{14}$ HEA during the cryogenic HPT was reported [68]. The sample was prepared by arc-melting and then homogenizing by heat treatment at 1050 °C for 24 h; the sample was cold-rolled after the heat treatment, followed by annealing at 1050 °C for 1 h with water quenching. The HPT processing was performed at 77 K in the liquid nitrogen (cryo-HPT) and at room temperature (300 K-HPT), the samples were compressed at 5 GPa for 5 revolutions. The thermodynamic calculations indicated that the hcp phase showed a higher stability than the fcc phase. The cryo-HPT providing an extra driving force for the fcc to hcp phase transformation of the $Co_{20}Cr_{26}Fe_{20}Mn_{20}Ni_{14}$ alloy at 77 K. The XRD pattern of the annealed sample shows the FCC phase and the r phase. In contrast, there was no r phase in the XRD pattern of the 300 K-HPT or cryo-HPT treated samples. Only the fcc peak was found. The absence of the r-phase peak after HPT treatment may have been due to grain refinement and residual strain after HPT treatment, which causes the main fcc peak to broaden. Unlike other conditions, the XRD pattern of the low temperature and high pressure treated alloy contains a well-developed hcp peak. This gives us reason to assume that, unlike 300 K-HPT, low temperature HPT induces a phase transition of fcc-to-hcp [68]. Furthermore, the microstructure and thermal stability of the nanocrystalline CoCrFeMnNi HEA after HPT were reported, and the results indicated that the grain size was refined, along with an unusual increase of the strength [66]. The hardness of $Al_{0.3}CoCrFeNi$ was significantly increased processed by HPT [65].

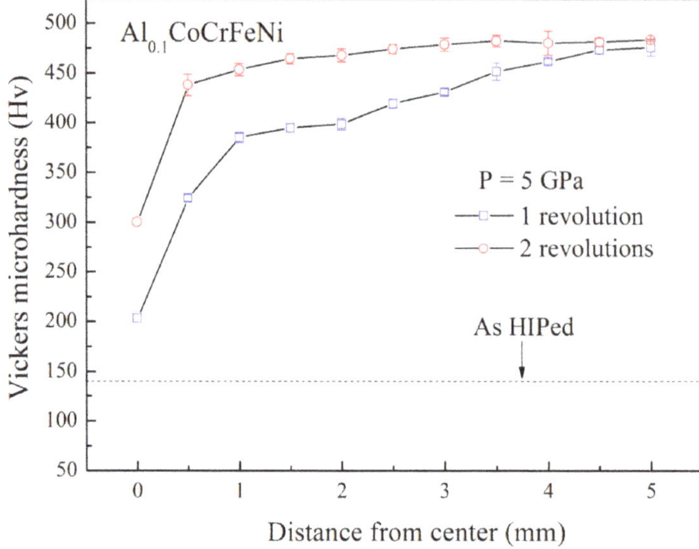

Figure 10. Vickers microhardness plotted against the distance from the center for the sample, data from [67].

4.4. Hexahedron Anvils Press

HEAs with high hardness prepared by high-pressure sintering were reported by Yu et al. [69]. The equiatomic CoCrFeNiCu and CoCrFeNiMn HEAs were prepared by high-pressure sintering (HPS). The elemental powders were developed in a planetary ball mill. Then the powders were sintered in the hexahedron anvil press at 1273 K and 5 GPa for 15 min. A graphite tube was taken as the heating device and the pyrophyllite as the pressure-transmitting medium. The structures and mechanical properties were carefully investigated. It revealed that the structure of the HEA powders have a main fcc phase and a minor bcc phase. After high-pressure sintering, the phase transition from the bcc to fcc structures occurred, and exhibited a simple fcc solid-solution structure. The hardness of the CoCrFeNiCu HEA increased from 133 HV to 494 HV by HPS, and the hardness of the CoCrFeNiMn increased from 300 HV to 587 HV. The increase of hardness is the reason for the decrease of the grain size. The grain size is about 100 nm after HPS. It reveals that the HPS is an effective way to design excellent HEAs.

5. Future Work

HEAs are considered to be one of the three breakthroughs in alloying theory in recent decades (the other two are bulk metallic glass and metal rubber). There are huge developments underway for this unique design concept. The high mixing entropy makes it have an extensive application potential. Since the discovery of HEAs, it has been about two decades. Only in recent years, it has received sufficient attention from researchers and developed rapidly. There is still a great amount of work to be done. The high-pressure technology has been more and more widely used in the fields of fabricating the new material preparation and changing the structures and organizations of materials, such as synthesizing diamonds with the ultra-high-pressure technology, producing ultra-hard and antifriction materials, etc. The constant renewal of the high-pressure technology will enable the development of excellent and more promising materials. Additionally, high pressure science has obvious frontier characteristics, which brings forth many new opportunities and challenges for the development of science and technology. National security is a potentially important driving force of the international attention, in addition to mere intellectual interest and economic interests. HEAs will be an important materials development route for the field of the high-pressure science in the future. The properties of HEAs change when high pressure act on the alloys. The most significant change in the alloys, due to the high pressure, is the phase change during the high-pressure crystallization. The materials show many unique properties under high pressures. The research of high entropy alloy under high pressure has already established a certain foundation, but the industrial applications have not matched the expected results. It is necessary to further broaden the application field, which will play a more important role in future theoretical improvement and new material preparation. With HEAs, as emerging alloys with a great potential for the future development, their structures and properties under high pressures need further studies.

6. Conclusions

HEAs as a new class of alloys have been attracting more and more attention in recent years. With the unique design concept, HEAs show wide application potentials. High pressure is an external condition that is important for structural changes and phase transitions. The unique phenomenon of HEAs under high pressures gives us a new understanding of HEAs, providing a new way forward for the in-depth study of HEAs. Some single-phase HEAs, such as CoCrFeNi, rare-earth HoDyYGdTb, and CoCrFeMnNi phase transition under high pressure, but bcc-$Hf_{25}Nb_{25}Zr_{25}Ti_{25}$, fcc-$Ni_{20}Co_{20}Fe_{20}Mn_{20}Cr_{20}$, and hcp-$Re_{25}Ru_{25}Co_{25}Fe_{25}$ remaine stable and no phase transition is observed under high pressure. These unique structural changes and phase transitions of HEAs under high pressure show wide application potentials. As a new research trend, there is a growing interest in the structure and performance of HEAs under high pressures. The studies of HEAs under high pressures are summarized in this paper, with a hope to deepen the fundamental understanding of HEAs under high pressures.

Author Contributions: W.D. performed the data analyses and wrote the manuscript; Z.Z. helped perform the analysis with constructive discussions; M.Z. and Y.M. helped document retrieval; P.K.L. and P.Y. performed the manuscript review; G.L. contributed to the conception of the study.

Funding: This research received no external funding

Acknowledgments: One of the authors (Gong Li) acknowledges the National Science Foundation of China (Grant No. 11674274). Pengfei Yu acknowledges the National Natural Science Funds of China (Grant No. 51601166). Peter K. Liaw very much appreciates the support of the U.S. Army Research Office Projects (W911NF-13-1-0438 and W911NF-19-2-0049) with the program managers, M.P. Bakas, S. N. Mathaudhu, and D.M. Stepp. Peter K. Liaw thanks the support from the National Science Foundation (DMR-1611180 and 1809640) with the program directors, G. Shiflet and D. Farkas.

Conflicts of Interest: The authors declare no conflict of interest.

References

1. Wang, J.; Niu, S.Z.; Guo, T.; Kou, H.C.; Li, J.S. The FCC to BCC phase transformation kinetics in an Al0.5CoCrFeNi high entropy alloy. *J. Alloy. Compd.* **2017**, *710*, 144–150. [CrossRef]
2. Otto, F.; Dlouhý, A.; Pradeep, K.G.; Kubenov, M.; Raabe, D.; Eggeler, G.; George, E.P. Decomposition of the single-phase high-entropy alloy CrMnFeCoNi after prolonged anneals at intermediate temperatures. *Acta Mater.* **2016**, *112*, 40–52. [CrossRef]
3. Stepanova, N.D.; Yurchenko, N.Y.; Panina, E.S.; Tikhonovsky, M.A.; Zherebtsov, S.V. Precipitation-strengthened refractory Al0.5CrNbTi2V0.5 high entropy alloy. *Mater. Lett.* **2017**, *188*, 162–164. [CrossRef]
4. Yeh, J.W.; Chen, S.K.; Lin, S.J. Nano structured high-entropy alloys with multiple principal elements: Novel alloy design concepts and outcomes. *Adv. Eng. Mater.* **2004**, *6*, 299–303. [CrossRef]
5. Cantor, B.; Chang, I.T.H.; Knight, P. Microstructural development in equiatomic multicomponent alloys. *Mater. Sci. Eng. A Struct. Mater. Prop. Microstruct. Process.* **2004**, *375*, 213–218. [CrossRef]
6. Deng, Y.; Asan, C.C.; Pradeepk, G. Design of a twinning-induced plasticity high entropy alloy. *Acta Mater.* **2015**, *94*, 124–133. [CrossRef]
7. Li, Z.; Pradeepk, G.; Deng, Y. Metastable high-entropy dual-phase alloys overcome the strength-ductility trade-off. *Nature* **2016**, *534*, 227–230. [CrossRef]

8. Senkov, O.N.; Wilks, G.B.; Miracle, D.B.; Chuang, C.P.; Liaw, P.K. Refractory high-entropy alloys. *Intermetallics* **2010**, *18*, 1758–1765. [CrossRef]
9. Yurchenko, N.Y.; Stepanov, N.D.; Shaysultanov, D.G. Effect of Al content on structure and mechanical properties of the $Al_xCrNbTiVZr$ (x = 0; 0.25; 0.5; 1) high-entropy alloys. *Mater. Charact.* **2016**, *121*, 125–134. [CrossRef]
10. Stepanov, N.D.; Yurchenko, N.Y.; Shaysultanov, D.G. Effect of Al on structure and mechanical properties of $Al_xNbTiVZr$ (x = 0, 0.5, 1, 1.5) high entropy alloys. *J. Mater. Sci. Technol.* **2015**, *31*, 1184–1193. [CrossRef]
11. Ashby, M.F. *Materials Selection in Mechanical Design*; Elsevier: Oxford, UK, 2011.
12. Yeh, J.W. Recent progress in high-entropy alloys. *Ann. Chim. Sci. Mater.* **2016**, *31*, 48–633. [CrossRef]
13. Zhang, Y.; Yang, X.; Liaw, P.K. Alloy design and properties optimization of high-entropy alloys. *J. Miner. Met. Mater. Soc.* **2012**, *64*, 830–838. [CrossRef]
14. Zhang, Y.; Zhou, Y.J.; Lin, J.P.; Chen, G.L.; Liaw, P.K. Solid-solution phase formation rules for multi-component alloys. *Adv. Eng. Mater.* **2008**, *10*, 534–538. [CrossRef]
15. Tsai, K.Y.; Tsai, M.H.; Yeh, J.W. Sluggish diffusion in Co-Cr-Fe-Mn-Ni high-entropy alloys. *Acta Mater.* **2013**, *61*, 4887–4897. [CrossRef]
16. Juan, C.C.; Hsu, C.Y.; Tsai, C.W. On microstructure and mechanical performance of $AlCoCrFeMo_{0.5}Ni_x$ high-entropy alloys. *Intermetallics* **2013**, *32*, 401–407. [CrossRef]
17. Ranganathan, S. Alloyed pleasures: Multimetallic cocktails. *Curr. Sci.* **2003**, *85*, 1404–1406.
18. Takeuchi, A.; Amiya, K.; Wada, T. High-entropy alloys with a hexagonal close-packed structure designed by equi-atomic alloy strategy and binary phase diagrams. *J. Miner. Met. Mater. Soc.* **2014**, *66*, 1984–1992. [CrossRef]
19. Chen, J.; Niu, P.; Liu, Y. Effect of Zr content on microstructure and mechanical properties of AlCoCrFeNi high entropy alloy. *J. Mater. Des.* **2016**, *94*, 39–44. [CrossRef]
20. Hsu, C.Y.; Yeh, J.W.; Chen, S.K. Wear resistance and high-temperature compression strength of fcc $CuCoNiCrAl_{0.5}Fe$ alloy with boron addition. *Metall. Mater. Trans. A* **2004**, *35*, 1465–1469. [CrossRef]
21. Chen, M.; Lin, S.; Yeh, J.W. Effect of vanadium addition on the microstructure, hardness, and wear resistance of $Al_{0.5}CoCrCuFeNi$ high-entropy alloy. *Metall. Mater. Trans. A* **2006**, *37*, 1363–1369. [CrossRef]
22. Hsu, C.Y.; Sheu, T.S.; Yeh, J.W. Effect of iron content on wear behavior of $AlCoCrFexMo_{0.5}Ni$ high-entropy alloys. *Wear* **2010**, *268*, 653–659. [CrossRef]
23. Zhu, J.M.; Zhang, H.F.; Fu, H.M. Microstructures and compressive properties of multicomponent $AlCoCrCuFeNiMo_x$ alloys. *J. Alloy. Compd.* **2010**, *497*, 52–56. [CrossRef]
24. Takeuchi, A.; Chen, N.; Wada, T. $Pd_{20}Pt_{20}Cu_{20}Ni_{20}P_{20}$ high-entropy alloy as a bulk metallic glass in the centimeter. *Intermetallics* **2011**, *19*, 1546–1554. [CrossRef]
25. Gao, X.Q.; Zhao, K.; Ke, H.B. High mixing entropy bulk metallic glasses. *J. Non-Cryst. Solids* **2011**, *357*, 3557–3560. [CrossRef]
26. Ma, L.; Wang, L.; Zhang, T. Bulk glass formation of Ti-Zr-Hf-Cu-M (M=Fe, Co, Ni) alloys. *Mater. Trans.* **2002**, *43*, 277–280. [CrossRef]
27. Christian, J.W.; Mahajan, S. Deformation twinning. *Prog. Mater. Sci.* **1995**, *39*, 1–157. [CrossRef]
28. Jain, J.; Poole, W.J.; Sinclair, C.W.; Gharghouri, M.A. Reducing the tension- compression yield asymmetry in a Mg-8Al-0.5Zn alloy via precipitation. *Scr. Mater.* **2010**, *62*, 301–304. [CrossRef]
29. Wang, H.; Lee, S.Y.; Gharghouri, M.A.; Wu, P.D.; Yoon, S.G. Deformation behavior of Mg-8.5wt.%Al alloy under reverse loading investigated by in-situ neutron diffraction and elastic viscoplastic self-consistent modeling. *Acta Mater.* **2016**, *107*, 404–414. [CrossRef]
30. Murty, B.S.; Yeh, J.W.; Ranganathan, A.S. Chapter 2—High-entropy alloys: Basic concepts. *High. Entropy Alloy.* **2014**, 13–35.
31. Gludovatz, B.; Hohenwarter, A.; Catoor, D. A fracture-resistant high-entropy alloy for cryogenic applications. *Science* **2014**, *345*, 1153–1158. [CrossRef]
32. Chuang, M.H.; Tsai, M.H.; Wang, W.R. Microstructure and wear behavior of $Al_xCo_{1.5}CrFeNi_{1.5}Ti_y$ high-entropy alloys. *Acta Mater.* **2011**, *59*, 6308–6317. [CrossRef]
33. Zhang, F.; Wu, Y.; Lou, H.B.; Zeng, Z.D.; Prakapenka, V.B.; Greenberg, E.; Ren, Y.; Yan, J.Y.; Okasinski, J.S.; Liu, X.J.; et al. Polymorphism in a high-entropy alloy. *Nat. Commun.* **2017**, *8*, 15687. [CrossRef] [PubMed]
34. Miracle, D.B.; Senkov, O.N. A critical review of high entropy alloys and related concepts. *Acta Mater.* **2017**, *122*, 448–511. [CrossRef]

35. Zhang, Y.; Zuo, T.T.; Tang, Z.; Gao, M.C.; Dahmen, K.A.; Liaw, P.K.; Lu, Z.P. Microstructures and properties of high-entropy alloys. *Prog. Mater. Sci.* **2014**, *61*, 1–93. [CrossRef]
36. Porter, D.A.; Easterling, K.E.; Sherif, M.Y. *Phase Transformations in Metals Andalloys*; CRC Press: Boca Raton, FL, USA, 2009.
37. Hemley, R.J.; Ashcroft, N.W. The revealing role of pressure in the condensed matter sciences. *Phys. Today* **1998**, *8*, 26–27. [CrossRef]
38. Bundy, F.P.; Hall, H.T.; Strong, H. Man-made diamonds. *Nature* **1955**, *176*, 51–55. [CrossRef]
39. Chen, K.W.; Jian, S.R.; Wei, P.J. A study of the relationship between semi-circular shear bands and pop-ins induced by indentation in bulk metallic glasses. *Intermetallics* **2010**, *18*, 1572–1578. [CrossRef]
40. Faupel, F.; Rätzke, K.; Ehmler, H. Diffusion in metallic glasses and supercooled melts. *Rev. Mod. Phys.* **2000**, *644*, 237–280. [CrossRef]
41. Zhang, F.; Lou, H.B.; Cheng, B.Y.; Zeng, Z.D.; Zeng, Q.S. High-pressure induced phase transitions in high-entropy alloys: A review. *Entropy* **2019**, *21*, 239. [CrossRef]
42. Buras, B.; Olsen, J.S.; Gerward, L. X-ray energy-dispersive diffractometry using synchrotron radiation. *J. Appl. Crystallogr.* **1977**, *10*, 431–438. [CrossRef]
43. Dubrovinsky, L.; Dubrovinskaia, N.; Bykova, E. The most incompressible metal osmium at static pressures above 750 gigapascals. *Nature* **2015**, *525*, 226. [CrossRef] [PubMed]
44. Bridgman, P.W. Effects of high shearing stress combined with high hydrostatic pressure. *Phys. Rev.* **1935**, *48*, 825–847. [CrossRef]
45. Zhilyaev, A.P.; Langdon, T.G. Using high-pressure torsion for metal processing: Fundamentals and applications. *Prog. Mater. Sci.* **2008**, *53*, 893–979. [CrossRef]
46. Yao, Y.G.; Huang, Z.N.; Xie, P.F.; Lacey, S.D.; Jacob, R.J.; Xie, H.; Chen, F.; Nie, A.; Pu, T.; Rehwoldt, M. Carbothermal shock synthesis of high-entropy-alloy nanoparticles. *Science* **2018**, *359*, 1489–1494. [CrossRef] [PubMed]
47. Yu, P.F.; Zhang, L.J.; Ning, J.L. Pressure-induced phase transitions in HoDyYGdTb high-entropy alloy. *Mater. Lett.* **2014**, *196*, 137–140. [CrossRef]
48. Francis, B. Finite strain isotherm and velocities for single-crystal and polycrystalline NaCl at high pressures and 300 K. *J. Geophys. Res. Solid Earth* **1978**, *83*, 1258–1263.
49. Dubrovinsky, L.; Dubrovinskaia, N.; Crichton, W.A.; Mikhaylushkin, A.S.; Simak, S.I.; Abrikosov, I.A.; Almeida, J.S.; Ahuja, R.; Luo, W.; Johansson, B. Noblest of all metals is structurally unstable at high pressure. *Phys. Rev. Lett.* **2007**, *98*, 045503. [CrossRef] [PubMed]
50. Zhang, F.X.; Zhao, S.J.; Jin, K.; Bei, H.; Popov, D.; Park, C.; Neuefeind, J.C.; Weber, W.J.; Zhang, Y.W. Pressure-induced fcc to hcp phase transition in Ni-based high entropy solid solution alloys. *Appl. Phys. Lett.* **2017**, *110*, 011902. [CrossRef]
51. Takahashi, T.; Bassett, W.A. High-pressure polymorph of iron. *Science* **1964**, *145*, 483–486. [CrossRef] [PubMed]
52. Saxena, S.K.; Shen, G.; Lazor, P. Experimental evidence for a new iron phase and implications for earth's core. *Science* **1993**, *260*, 1312–1314. [CrossRef] [PubMed]
53. Andrault, D.; Fiquet, G.; Kunz, M.; Visocekas, F.; Hausermann, D. The orthorhombic structure of iron: An in situ study at high-temperature and high-pressure. *Science* **1997**, *278*, 831–834. [CrossRef]
54. Yoo, C.S.; Cynn, H.; Soderlind, P.; Iota, V. New beta (fcc)—Cobalt to 210 GPa. *Phys. Rev. Lett.* **2000**, *84*, 4132. [CrossRef] [PubMed]
55. Sergueev, I.; Dubrovinsky, L.; Ekholm, M.; Yekilova, O.Y.; Chumakov, A.I.; Zajac, M.; Potapkin, V.; Kantor, I.; Bornemann, S.; Ebert, H.; et al. Hyperfine splitting and room-temperature ferromagnetism of Ni at multimegabar pressure. *Phys. Rev. Lett.* **2013**, *111*, 157–601. [CrossRef] [PubMed]
56. Huang, E.W.; Lin, C.M.; Juang, J.Y.; Chang, Y.J.; Chang, Y.W.; Wu, C.S.; Tsai, C.W.; Yeh, A.C.; Shieh, S.R. Deviatoric deformation kinetics in high entropy alloy under hydrostatic compression. *J. Alloy. Compd.* **2019**, *349*, 50113. [CrossRef]
57. Li, G.; Xiao, D.H.; Yu, P.F.; Zhang, L.J.; Liaw, P.K.; Li, Y.C.; Liu, R.P. Equation of state of an AlCoCrCuFeNi high-entropy alloy. *JOM* **2015**, *67*, 2310–2313. [CrossRef]
58. Cheng, B.Y.; Zhang, F.; Lou, H.b.; Chen, X.H.; Liaw, P.K.; Yan, J.Y.; Zeng, Z.D.; Ding, Y.; Zeng, Q.S. Pressure-induced phase transition in the AlCoCrFeNi high-entropy alloy. *Scr. Mater.* **2019**, *161*, 82–92. [CrossRef]

59. Yusenko, K.V.; Riva, S.; Carvalho, P.A.; Yusenko, M.V.; Arnaboldi, S.; Sukhikh, A.S.; Hanfland, M.; Gromilov, S.A. First hexagonal close packed high-entropy alloy with outstanding stability under extreme conditions and electrocatalytic activity for methanol oxidation. *Scr. Mater.* **2017**, *138*, 22–27. [CrossRef]
60. Ahmad, A.S.; Su, Y.; Liu, S.Y.; Ståhl, K.; Wu, Y.D.; Hui, X.D.; Ruett, U.; Gutowski, O.; Glazyrin, K.; Liermann, H.P.; et al. Structural stability of high entropy alloys under pressure and temperature. *J. Appl. Phys.* **2017**, *121*, 235901. [CrossRef]
61. Valiev, R.Z.; Estrin, Y.; Horita, Z. Producing bulk ultrafine-grained materials by severe plastic deformation. *J. Miner. Met. Mater. Soc.* **2010**, *58*, 33–39. [CrossRef]
62. Zhilyaev, A.P.; Lee, S.; Nurislamova, G.V. Microhardness and microstructural evolution in pure nickel during high-pressure torsion. *Scr. Mater.* **2001**, *44*, 2753–2758. [CrossRef]
63. Révész, A.; Hóbor, S.; Lábár, J.L. Partial amorphization of a Cu–Zr–Ti alloy by high pressure torsion. *J. Appl. Phys.* **2006**, *100*, 103522. [CrossRef]
64. Sabirov, I.; Pippan, R. Formation of a W-25%Cu nanocomposite during high pressure torsion. *Scr. Mater.* **2005**, *52*, 1293–1298. [CrossRef]
65. Tang, Q.H.; Huang, Y.; Huang, Y.Y. Hardening of an $Al_{0.3}CoCrFeNi$ high entropy alloy via high-pressure torsion and thermal annealing. *Mater. Lett.* **2015**, *151*, 126–129. [CrossRef]
66. Schuh, B.; Mendez-Martin, F.; Völker, B.; George, E.P.; Clemens, H.; Pippan, R.; Hohenwarter, A. Mechanical properties, microstructure and thermal stability of a nanocrystalline CoCrFeMnNi high-entropy alloy after severe plastic deformation. *Acta Mater.* **2015**, *96*, 258–268. [CrossRef]
67. Yu, P.F.; Cheng, H.; Zhang, L.J.; Zhang, H. Effects of high pressure torsion on microstructures and properties of an $Al_{0.1}CoCrFeNi$ high-entropy alloy. *Mater. Sci. Eng.* **2016**, *655*, 283–291. [CrossRef]
68. Moon, J.; Qi, Y.; Tabachnikova, E.; Estrin, Y.; Choi, W.M. Deformation-induced phase transformation of $Co_{20}Cr_{26}Fe_{20}Mn_{20}Ni_{14}$ high-entropy alloy during high-pressure torsion at 77 K. *Mater. Lett.* **2017**, *202*, 86–88. [CrossRef]
69. Yu, P.F.; Zhang, L.J.; Cheng, H.; Zhang, H.; Ma, M.Z.; Li, Y.C.; Li, G.; Liaw, P.K.; Liu, R.P. The high-entropy alloys with high hardness and soft magnetic property prepared by mechanical alloying and high-pressure sintering. *Intermetallics* **2016**, *70*, 82–87. [CrossRef]

Article

Tribological Behavior of As-Cast and Aged $AlCoCrFeNi_{2.1}$ CCA

Fevzi Kafexhiu *, Bojan Podgornik and Darja Feizpour

Institute of Metals and Technology, Lepi pot 11, 1000 Ljubljana, Slovenia; bojan.podgornik@imt.si (B.P.); darja.feizpour@imt.si (D.F.)
* Correspondence: fevzi.kafexhiu@imt.si; Tel.: +386-1-4701-931

Received: 31 December 2019; Accepted: 28 January 2020; Published: 1 February 2020

Abstract: In the present study, wear behavior as a function of aging time was evaluated for the $AlCoCrFeNi_{2.1}$ eutectic complex, concentrated alloy (CCA) consisting of B2 (BCC), and $L1_2$ (FCC) lamellae in the as-cast state. By aging the material at 800 °C up to 500 h, precipitation of a fine, evenly dispersed micro-phase inside the $L1_2$ takes place. From 500 h to 1000 h of aging, precipitates coarsen by the Ostwald ripening mechanism. Reciprocating wear tests were characterized by a prevailing abrasive wear mechanism, while adhesive and delamination wear components change with aging conditions. The $L1_2$ phase with lower hardness in the as-cast material preferentially deformed during the wear test, which was not the case after aging the material, i.e., with the presence of precipitates. Aging-induced changes show a similar trend for the coefficient of friction and $L1_2$ + precipitates phase fraction, whereas changes in specific wear rate are in a good agreement with changes in B2 phase fraction. In general, aging the $AlCoCrFeNi_{2.1}$ CCA at 800 °C up to 500 h decreases its coefficient of friction due to reduced adhesive wear component and enhances its wear performance through precipitation strengthening.

Keywords: $AlCoCrFeNi_{2.1}$; CCA; HEA; aging; precipitates; wear; tribology

1. Introduction

Multi-principal element alloys (MPEAs) have become an attractive topic in the materials science community because of their great potential in discovering and developing new materials of scientific significance and practical benefit. In contrast to conventional alloys, these materials usually contain at least five elements in equimolar or near-equimolar proportions, which in the case of fulfilling the Hume-Rothery rules can have microstructures of single-phase solid solution with high configurational entropy of mixing—high-entropy alloys (HEAs), or they can have complex microstructures that contain multiple phases of solid solution and/or intermetallic compounds—complex, concentrated alloys (CCAs). The research interest on CCAs is ever increasing as they offer an excellent combination of advantages of single-phase solid solution in HEAs, and secondary-phase strengthening effects of well-established alloys. This combination may as well diminish the strength-ductility competition, which is a well-known issue in conventional alloys.

However, the application of these materials in structural engineering requires, among others, a good understanding of their surface degradation mechanisms including corrosion, erosion, and wear behavior [1]. In this respect, some research has been carried out by different authors on different HEAs and CCAs as reviewed by Ayyagari et al. [2]. Some highlights worth mentioning include the work of Wu et al. [3] who found out that by increasing aluminum content of the $Al_xCoCrCuFeNi$, both the volume fraction of the BCC phase and the hardness value increase and thus the wear coefficient decreases. On the other hand, Tong et al. [4] reported that the wear resistance of the $Al_xCoCrCuFeNi$ was similar to that of ferrous alloys at the same hardness level. Both these authors correlated the high wear resistance to the higher hardness coming from the solid solution strengthening of single-phase HEAs. Hsu et al. [5] in their work discovered that the major wear mechanism of the $AlCoCrFe_xMo_{0.5}Ni$ HEA is abrasion. Also, by performing the oxidation test at the pin/disk interface flash temperature, 500 °C, they came to the conclusion that the oxidation rate of $Fe_{2.0}$ markedly exceeds that of $Fe_{1.5}$, indicating more oxides abrade the surface, resulting in lower wear resistance. Oxidative wear was also encountered by Du et al. [6] who studied the tribological behavior of the $Al_{0.25}CoCrFeNi$ HEA with a simple FCC phase and hardness of 260 HV from room temperature to 600 °C. They found out that below 300 °C, with increasing temperature, the wear rate increased due to high temperature softening. The wear rate remained stabilized above 300 °C due to the anti-wear effect of the oxidation film on the contact interface. The dominant wear mechanism of HEA changed from abrasive wear at room temperature to delamination wear at 200 °C, then delamination wear and oxidative wear at 300 °C and became oxidative above 300 °C. Excellent anti-oxidation property and resistance to thermal softening were reported for the $Co_{1.5}CrFeNi_{1.5}Ti$ and $Al_{0.2}Co_{1.5}CrFeNi_{1.5}Ti$ alloys, which are the main reasons for the outstanding wear resistance, which is at least two times better than that of conventional wear-resistant steels with similar hardness, such as SUJ2 and SKH51 [7].

Löbel et al. [8] investigated the wear behavior of $AlCoCrFeNiTi_{0.5}$ HEA produced by powder metallurgy under reciprocating wear conditions from room temperature to 900 °C. They found out that with increasing temperature up to 650 °C, initially, a slight decrease in wear resistance occurred, whereas a further increase in test temperature resulted in a distinct increase in wear resistance and a decrease in coefficient of friction. Their investigations prove the suitability of the $AlCoCrFeNiTi_{0.5}$ HEA for high-temperature applications, as the formation of protective oxides improves the wear performance.

A novel $AlCoCrFeNi_{2.1}$ eutectic CCA firstly studied by Lu et al. [9], is an alloy with promising properties in as-cast condition [10,11] due to the contribution of both ductile FCC ($L1_2$) and harder BCC (B2) phases resulting in a combination of good strength and ductility. The alloy has also shown to have excellent thermomechanical processing capability by severe cold rolling [12–15] or rotary friction welding [16]. Excellent work on surface wear and corrosion behavior of the $AlCoCrFeNi_{2.1}$ has been performed by Hasannaeimi et al. [17], where a transition from adhesive to oxidative wear was observed as the duration of reciprocating wear test increased.

The purpose of the present research is to study the tribological behavior of $AlCoCrFeNi_{2.1}$ eutectic CCA as a function of microstructure evolution by aging the material at 800 °C for 100, 500, and 1000 h, both for the sake of fundamental scientific understanding and the application worthiness of this particular alloy. Although the precipitation is the main process taking place during aging at 800 °C, the focus of the present research will be only on the effect these precipitates have on the wear behavior of the $AlCoCrFeNi_{2.1}$. Detailed analysis of precipitate kinetics, thermodynamics, and structure/morphology will be published elsewhere.

2. Materials and Methods

The eutectic $AlCoCrFeNi_{2.1}$ CCA was synthesized by vacuum induction melting and ingot casting using commercially available elements with a purity of 99.9%. Before melting, the furnace was evacuated and a subsequent argon gas atmosphere of 300 mbar was created. The molten material was kept for at least 20 min at 1500 °C under induction current, which provided sufficient agitation to ensure the homogeneous distribution of elements in the melt. The melt was cast inside the furnace (Ar atmosphere) in a cast-iron crucible and further cooled down in the air. In order to determine the melting point of the alloy and to find out what reactions might take place during the heating/cooling process, differential scanning calorimetry (DSC) analysis using STA 449 C Jupiter Thermo-microbalance (Netzsch-Gerätebau GmbH, Selb, Germany) was subsequently performed in the temperature range of 30–1425 °C under the dynamic atmosphere of Ar (15 mL min^{-1}), heating and cooling rate of 20 K min^{-1}, using an Al_2O_3 crucible with approximately 219.6 mg of sample material. As indicated by the DSC curves in Figure 1, besides melting and solidification peaks with onset temperatures 1343.9 °C and 1346.3 °C, respectively, a transformation reaction revealed in form of weak peaks can be seen on the heating and cooling curves with onset temperatures around 800 °C and 822 °C, respectively. Therefore, in order to be able to characterize the phase transformation developing around 800 °C, dilatometry technique using Bähr DL 805A/D dilatometer (TA Instruments, Inc, New Castle, DE, USA) was used to heat a cylindrical sample with a diameter of 4 mm and a length of 10 mm to a temperature of 800 °C at which it was held for 20 h, then quenched by streaming N_2 gas to ensure a controlled cooling rate of 10 K s^{-1}. Afterward, the sample was longitudinally cut in half and a metallographic specimen was prepared by hot mounting in Bakelite, then mechanically ground with silicon carbide emery paper from a grade 180 down to 1200, and finally polished with 3 µm and 1 µm abrasive diamond suspensions. Scanning electron microscopy (SEM; JSM-6500F, Jeol, Tokyo, Japan) with back-scattered (BSE) detector and energy dispersive X-ray spectroscopy (EDS) were used for imaging and chemical composition determination. In addition, a lamella for transmission electron microscopy (TEM; JEM-2100 HR, Jeol, Tokyo, Japan, operated at 200 kV) was prepared for detailed microstructural characterization. A TEM lamella was first prepared by cutting a small sample of around 3 mm length, 1 mm height, and 1 mm thickness from a bulk specimen, then coarse and fine ground and polished to around 100 µm thickness using SiC papers from 800 to 4000 grit sizes, and additionally thinned by argon ion-slicing (IonSlicer, EM-09100IS, Jeol, Tokyo, Japan) with ion milling at 6 kV for around 6 h and around 15 min of fine ion milling at 2 kV to reach electron transparency. Scanning transmission electron microscopy (STEM) unit with a bright-field (BF) detector and EDS (JED-2300T, Jeol, Tokyo, Japan) were used for chemical composition determination and elemental mapping. Based on the characterization results, it was decided for the present work to isothermally age the material at 800 °C in a simple furnace (air atmosphere) at three different durations, 100, 500, and 1000 h and subsequently quench it in water.

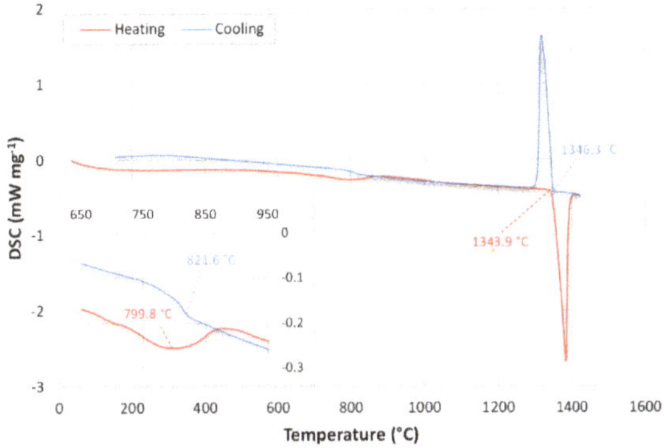

Figure 1. Heating and cooling curves from DSC analysis on the $AlCoCrFeNi_{2.1}$ CCA.

On as-cast and aged material, metallographic samples of cuboid shape with dimensions 20 × 20 × 10 mm^3 were machined, ground and polished applying the same method as described above. SEM-BSE imaging and EDS analyses on at least 5 randomly chosen regions at both 1k and 3k magnifications were acquired. For more representative analysis, images acquired at 1k magnification were used for quantitative evaluation of phase fractions, whereas those acquired at 3k were used for quantitative evaluation of precipitates. Images were digitally analyzed using FIJI (ImageJ, ver. 1.52p, Bethesda, MD, USA) [18,19] with an appropriate size filter and color threshold, which enabled a separate analysis of the three distinct phases. As a result, the surface area of the darker phase, as well as surface area and distribution (*x* and *y* coordinates) of each precipitate could be obtained. Having these data and a known surface area of BSE images, the area fraction of all phases could be easily determined, where average values with standard deviations were derived out of five analyzed images. Furthermore, on the metallographic samples (cuboid 20 × 20 × 10 mm^3), HV10 hardness measurements were performed at room temperature with at least three indentations and a holding time of 14 s using Instron Tukon 2100B instrument (Buehler-Illinois Tool Works (ITW), Lake Bluff, IL, USA). In addition, three repetitions of reciprocating ball-on-plate wear tests using hardened DIN 100Cr6 bearing steel ball with a diameter of 20 mm as a counter-body were performed. The hardened bearing steel ball was used due to its high hardness (~700 HV) thus concentrating the major amount of wear on the investigated alloy's surface, as well as due to the similarity to many machine component applications with the prevailing metal-metal contact. Furthermore, the use of the steel counter-body enables the study of not only abrasive but also adhesive wear mechanisms.

Wear tests were performed using the in-house designed ball-on-plate reciprocating sliding device (Figure 2), as typically found in many tribological investigations, with a stationary ball (counter-body) being loaded against a moving flat-surface specimen (cuboid 20 × 20 × 10 mm^3). Tests were performed at ambient temperature conditions (21 ± 2 °C) with a sliding frequency of 15 Hz at a stroke of 4 mm, resulting in a maximal sliding velocity of 0.12 m s^{-1} and a total duration of 833 s, which corresponds to a 100 m of sliding. The normal load of 20 N was applied, corresponding to 1 GPa of mean Hertzian contact pressure. All tests were performed in dry sliding conditions.

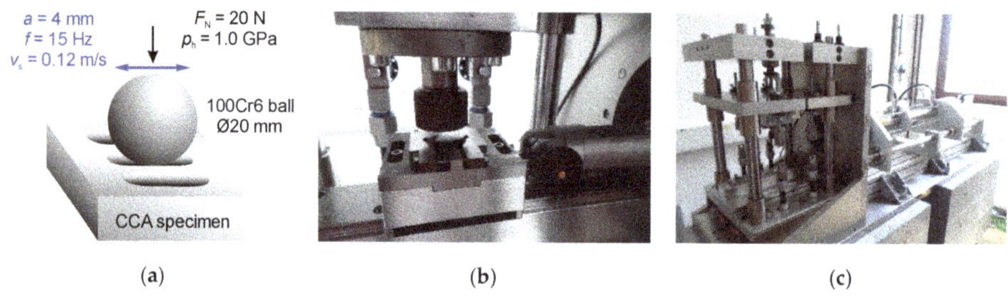

(a) (b) (c)

Figure 2. Ball-on-plate configuration used for wear tests: (**a**) Schematics with test parameters; (**b**) Configuration in the actual testing rig; (**c**) The device used for wear tests.

Worn specimens and counter-bodies were characterized using the 3D surface measurement system (Alicona InfiniteFocus, Alicona Imaging GmbH, Raaba, Austria), being able this way to accurately measure the worn-out material volume V (mm^3) and calculate the specific wear rate coefficient (k) according to the equation

$$k = V \cdot F^{-1} \cdot s^{-1} \quad (1)$$

where F (N) is the maximum load applied, and s (m) is the total sliding distance.

Finally, SEM imaging and EDS analyses were performed on wear tracks in order to shed some light on wear mechanisms.

3. Results and Discussion

3.1. Microstructure and Hardness

The microstructure of the as-cast alloy with typical lamellar/dendritic morphology is shown in the backscattered SEM image in Figure 3a. Precipitation of a fine darker phase inside $L1_2$ (FCC) (light gray area) after 100, 500, and 1000 h of aging at 800 °C is shown in backscattered SEM images in Figure 3b–d, respectively. Besides precipitation and a slight coarsening/decomposition of the B2 lamellae, there are no major morphological changes of the lamellar/dendritic structure with aging. However, elemental concentrations within both phases slightly change with aging time, as shown in Figure 4.

Figure 4 shows the change in at% concentration of elements in all three phases, obtained by EDS analysis. A decrease of Al and Ni concentrations and an increase of the concentrations of Co, Cr, and Fe in the $L1_2$ (FCC) phase (Figure 4a) is more pronounced in the first 100 h of aging at 800 °C, as the precipitation kinetics is the fastest in this time span. The opposite can be seen in the B2 (BCC) phase (Figure 4b). This is in good agreement with changes in phase fraction and precipitation process (Figure 5), where precipitates size and area fraction increase at the expense of the area fraction of $L1_2$

phase, while the B2 area fraction also shows a slight increase. Furthermore, there is a noticeable decrease in at% concentration of Co, Cr, and Fe in the precipitates after 1000 h of aging at 800 °C (Figure 4c) and an increase of Al and Ni concentrations. Note also the similarity of chemical composition between the B2 (BCC) phase after aging (Figure 4b) and the precipitates phase after 1000 h of aging (Figure 4c). This is a clear indication of similarity of B2 and precipitates phase also from the viewpoint of the crystal lattice structure, which needs to be confirmed by additional analysis using XRD, selected area diffraction in TEM, or even high-resolution TEM for a detailed analysis of the interface between precipitates and $L1_2$ phase. This is beyond the scope of the present study and will be investigated separately.

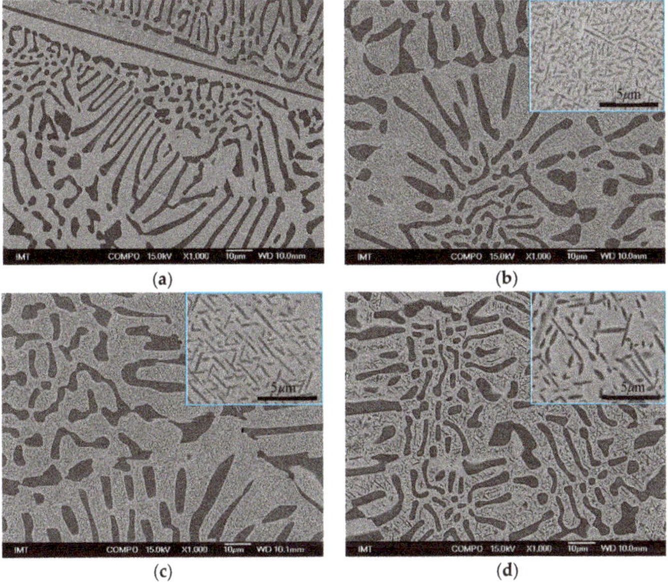

Figure 3. SEM-BSE images of the microstructure of: (**a**) As-cast $AlCoCrFeNi_{2.1}$ CCA with B2 (BCC) lamellae—dark grey and $L1_2$ (FCC) phase—light grey; Precipitation and coarsening of fine B2-like phase inside $L1_2$ after aging at 800 °C for: (**b**) 100 h; (**c**) 500 h; (**d**) 1000 h.

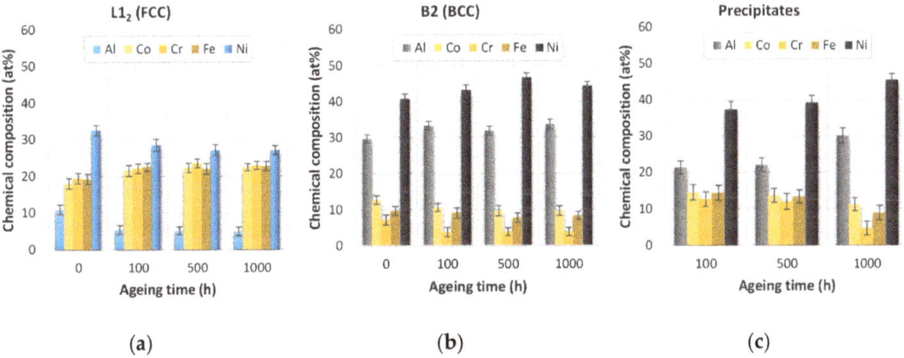

Figure 4. Changes of elemental concentration in at% with aging at 800 °C in: (**a**) $L1_2$ (FCC) phase; (**b**) B2 (BCC) phase; (c) Precipitates.

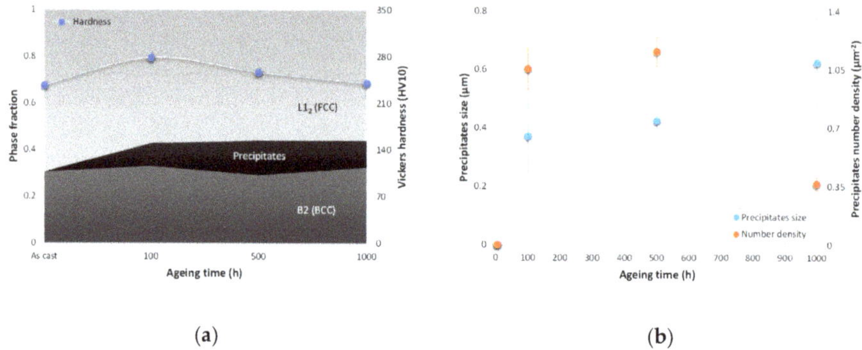

(a) (b)

Figure 5. Microstructure evolution and hardness change as a function of aging at 800 °C: (**a**) Phase fractions and Vickers hardness; (**b**) Precipitates size and their number density.

Changes in phase fraction and Vickers hardness with aging are shown in Figure 5a, whereas precipitate size and number density are given in Figure 5b. The initial phase fractions of B2 and $L1_2$ phases is around 0.3 and 0.7, respectively. After the first 100 h of aging, the precipitates phase fraction (0.1) increases at the expense of the $L1_2$ phase that drops to 0.57, while B2 phase fraction also shows a slight increase (0.33). Precipitates average size expressed as equivalent circle diameter (ECD) at this stage is 0.37 ± 0.1 μm. From this point up to the 1000 h of aging, the $L1_2$ phase fraction remains virtually unchanged, whereas the phase fraction of precipitates slightly fluctuates between 0.15 and 0.12 in accordance with the B2 phase after 500 and 1000 h of aging, respectively. However, this does not mean that the $L1_2$ phase has reached its solid solution equilibrium state because the precipitation from the $L1_2$ phase continues even after 100 h of aging. The precipitate size continues to increase to 0.42 ± 0.1 μm and 0.62 ± 0.2 μm after 500 and 1000 h of aging, respectively. The precipitation rate is the highest in the first 100 h of aging, as represented by the number of precipitates per unit area or number density of precipitates, which at this stage is around 1.05 ± 0.13 μm^{-2}. The precipitation continues up to 500 h of aging but a with much lower rate, as the existing precipitates at this stage continue growing. From this point up to 1000 h of aging, the number of precipitates drops down to 0.37 ± 0.04 μm^{-2}. Since at the same time precipitates size increases while their number density decreases, it means that by aging the $AlCoCrFeNi_{2.1}$ at 800 °C from 500 h onward, the Ostwald ripening process develops, where larger particles coarsen at the expense of dissolving smaller ones.

The slight increase in hardness of $AlCoCrFeNi_{2.1}$ after the first 100 h of aging at 800 °C can be attributed to precipitation of the fine B2-like phase at the expense of the softer $L1_2$ phase, the fraction of which decreases at this stage (Figure 5a). From this point up to 1000 h of aging, the hardness decreases linearly back to the initial value (as-cast condition), which can be attributed to the precipitate coarsening within the softer $L1_2$ phase and continuous impoverishing of the solid solution from solute atoms such as Al and Ni, which diffuse towards the precipitates and the B2 phase.

STEM-EDS elemental analysis and mapping shown in Figure 6b and summarized in Table 1, indicate the different composition of the B2 phase in spectra 1 and 2, and the $L1_2$ phase in spectrum 3 [20]. Spectrum 2 in Table 1 and the elemental map for Cr, at the lower right corner shows, increased Cr signal, which is coming from the Cr-rich nano-precipitates, as reported by Gao et al. [11]. Similar to the B2 phase (spectra 1 and 2), Al and Ni concentration in precipitates (spectra 4, 5, and 6) is higher as compared to the $L1_2$ phase (spectrum 3). A detailed analysis of these precipitates from the viewpoint of their crystal lattice structure, kinetics, thermodynamics of precipitation, etc., is beyond the scope of the present work, therefore it will be studied separately elsewhere.

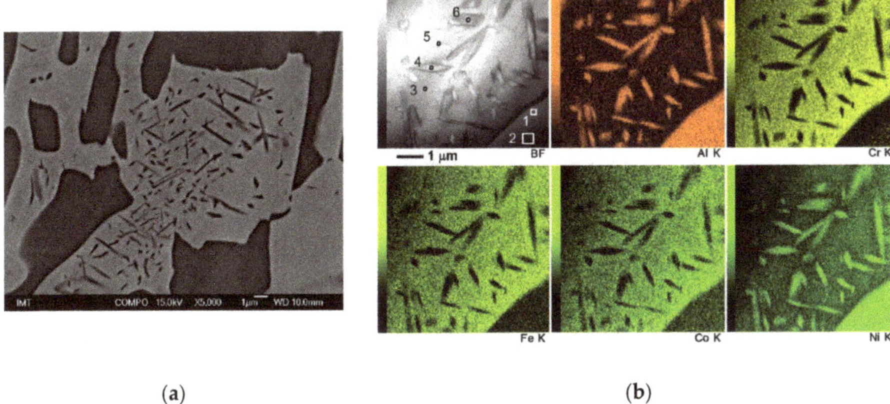

(a) (b)

Figure 6. Electron microscopy characterization of the dilatometry sample (20 h at 800 °C): (**a**) SEM-BSE image showing partial precipitation; (**b**) STEM bright-field image of precipitates with EDS point analyses and mapping.

Table 1. Elements concentration in at% from EDS point analyses in Figure 6.

Spectrum	Al	Co	Cr	Fe	Ni	Total
1	13.4	13.8	4.9	9.4	58.4	100
2	13.2	12.7	8.7	10.8	54.6	100
3	3.5	18.3	22.6	20.4	35.3	100
4	19.8	10.1	3.7	8.7	57.8	100
5	18.6	11.1	5.2	9.7	55.3	100
6	18.6	11.1	4.2	9.5	56.6	100

3.2. Wear Tests

A representative example of 1 out of 3 wear tracks of tested samples is shown in Figure 7 in form of 3D pseudo-color depth images, 2D optical images, and cross-sectional depth profiles of all three wear tracks taken at the same locations shown with the red line across the representative wear tracks in Figure 7a–d. Visual differences in shape or length/width/depth of wear tracks depending on materials condition (aging time) are minor, however, detailed volume measurements could reveal their difference.

The coefficient of friction (COF) shown in Figure 8 is characterized by both short- and long-range fluctuations. The short-range (high frequency) fluctuations can be attributed to the difference in coefficient of friction between the harder B2 phase and the softer $L1_2$ phase. This is supported by the findings of Hasannaeimi et al. [17] who evaluated a small-scale phase-specific scratch behavior and found the variation in COF across the scratch line, where the softer FCC phase shows lower COF as compared to the harder BCC one. Long-range fluctuations, however, could be coming from the different morphology of the new contacting surface, which is continuously uncovered by wearing out the top-most layer of the material in dry sliding contact.

The trend at which the average coefficient of friction of the as-cast $AlCoCrFeNi_{2.1}$ CCA changes with aging time as shown in Figure 9, is in a good correlation with the changing trend of $L1_2$ + precipitates phase fraction. This suggests that the plastic deformation of the softer $L1_2$ matrix with precipitates results in higher COF than the harder B2 phase. In general, by aging the $AlCoCrFeNi_{2.1}$ CCA at 800 °C, COF of decreases with time.

A trend opposite to the one in Figure 9 is shown in Figure 10 for the specific wear rate of the specimen and counter-body, as well as the B2 phase fraction. In as-cast material, the wear rate of counter-body is lower compared to the wear rate of the material. After 100 h of aging at 800 °C, the wear rates of both specimen and counter-body increase and become almost equal. At this point, the B2 phase fraction and hardness also increase slightly, whereas COF decreases. After 500 h of aging, the specific wear rate of the specimen decreases almost three times more than the wear rate of counter-body (Figure 10), whereas the COF remains virtually unchanged (Figure 9). At this point, there is a slight decrease in B2 fraction and a slight increase in precipitate size, fraction, and number density (Figure 5). With further aging up to 1000 h, the wear rate of specimen rises almost to the value of the as-cast state but still remains lower than the wear rate of the counter-body. From this analysis, it can be concluded that adequate aging of $AlCoCrFeNi_{2.1}$ CCA at 800 °C not only decreases the coefficient of friction but also improves the material's wear performance.

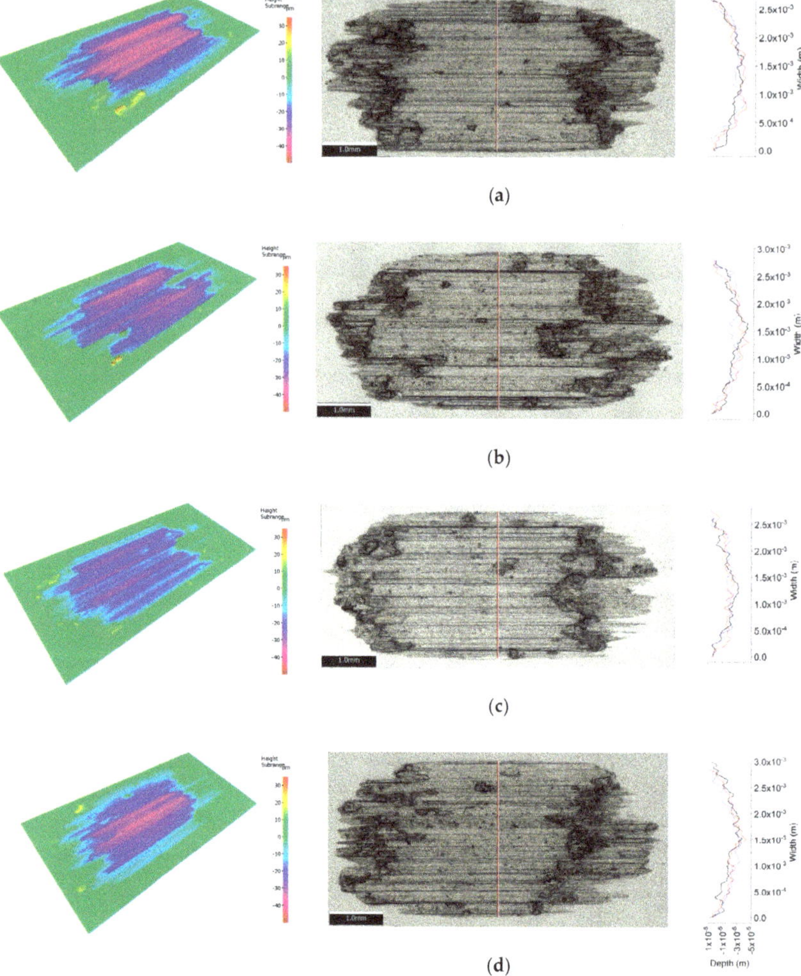

Figure 7. Pseudo-color 3D depth images, optical images, and depth profiles of three wear tracks in (**a**) As-cast material; Material aged at 800 °C for (**b**) 100 h; (**c**) 500 h; (**d**) 1000 h.

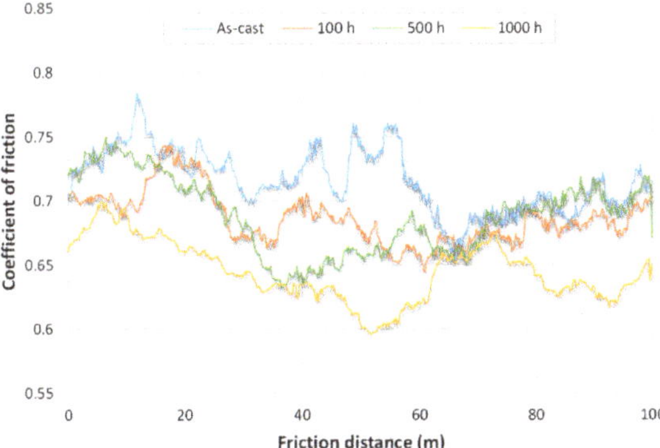

Figure 8. Coefficient of friction throughout the wear test on as-cast $AlCoCrFeNi_{2.1}$ and after aging at 800 °C at three different durations.

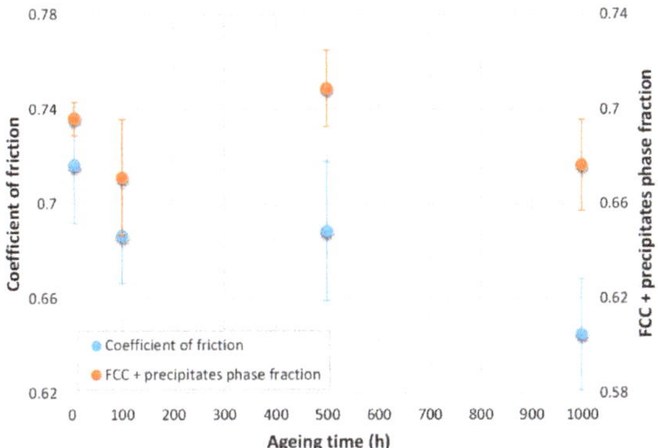

Figure 9. Coefficient of friction and FCC ($L1_2$) + precipitates phase fraction of as-cast $AlCoCrFeNi_{2.1}$ and after aging at 800 °C at three different durations.

SEM-BSE images of the wear track of as-cast $AlCoCrFeNi_{2.1}$ CCA shown in Figure 11a reveals the presence of patches of re-deposited material caused by adhesive wear, erosion grooves from abrasive wear, and severe disintegration and some delamination of the lamellar morphology inside the wear track. An SEM-BSE image at higher magnification in Figure 11b indicates extensive plastic deformation through the planar slip mechanism of the $L1_2$ phase. When the planar slip lines in the $L1_2$ phase reach the semi-coherent interface between the latter and B2 phases, a step-like shape is formed, as shown in Figure 11b.

Inside the wear track, both phases deform uniformly and no inter-phase detaching or void formation is observed. Gao et al. [11] reported that the $L1_2$ phase can accommodate several arrays of parallel mobile dislocations and deform by planar slip. They also reported that the B2 phase in the eutectic fails in a brittle manner, while the $L1_2$ phase shows ductility and necking, leading to dual-mode fracture in this alloy. Hasannaeimi et al. [17] reported that during wear test of the $AlCoCrFeNi_{2.1}$, $L1_2$ and B2 phases deformed simultaneously, while the B2 phase accommodates medium density of dislocations. This was attributed to the 3D back-stress acting on the $L1_2$ phase which can maintain synchronous deformation in heterogeneous systems. This back stress is further enhanced by semi-coherent boundaries between B2 and $L1_2$ phases, and the lower fraction of B2 lamellae. The accumulative effect of these conditions resulted in the activation of dislocation in the brittle B2 phase facilitated by a high density of dislocation pile-up at the phase boundaries, and this modified the brittle behavior of B2 phase to accommodate deformation.

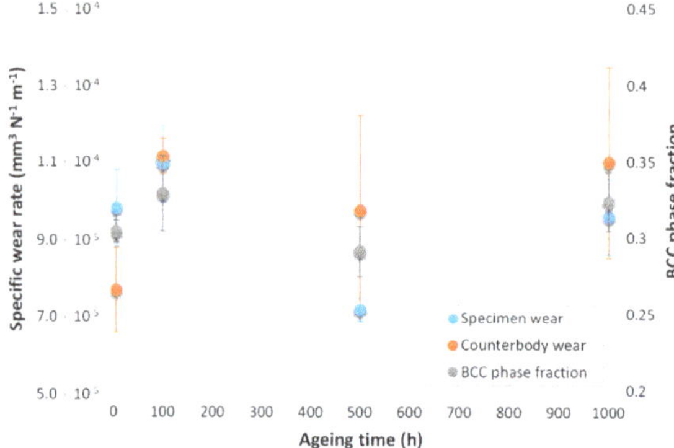

Figure 10. The specific wear rate of specimens and counter-body, and BCC (B2) phase fraction of as-cast $AlCoCrFeNi_{2.1}$ and after aging at 800 °C at three different durations.

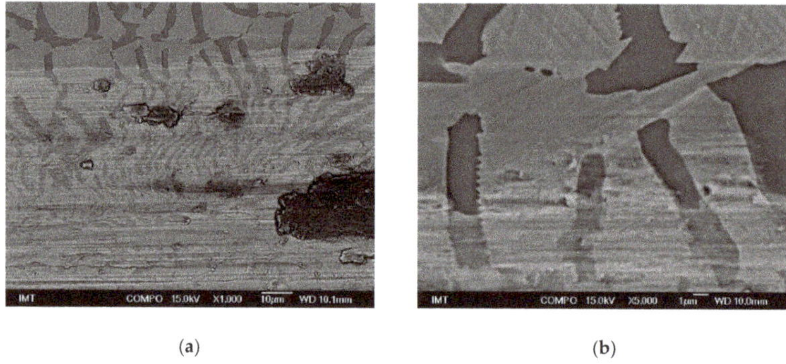

(**a**) (**b**)

Figure 11. SEM-BSE images of wear tracks of $AlCoCrFeNi_{2.1}$ CCA in as-the cast state: (**a**) Wear track edge with patches of re-deposited material; (**b**) Slip lines in the vicinity of wear track.

SEM-EDS analyses in Figure 12a summarized in Table 2, indicate that inside the wear track, the main wear mechanism is abrasion with little patches of oxidized redeposited material (self-adhesion) and some surface delamination. Figure 12b shows EDS analyses of the piled-up oxides of counter-body and self-adhered material, as summarized in Table 3, where Si and Mn, which are present in the counter-body material (DIN 100Cr6) are also detected.

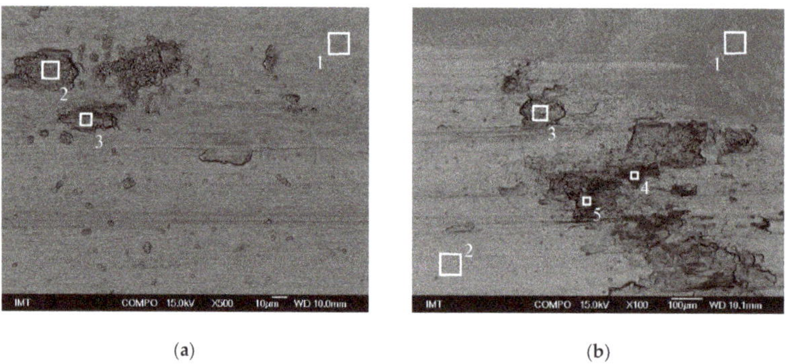

(**a**) (**b**)

Figure 12. SEM-BSE images with EDS analyses on the wear track of $AlCoCrFeNi_{2.1}$ in the as-cast state: (**a**) Inside the wear track; (**b**) At the edge of the wear track.

Similar oxide pile-up phenomenon but in reduced amount is present in all durations of the aged material, as shown in Figure 13a for 100 h of aging. The difference between the aged material and the one in as-cast condition is the fraction of $L1_2$ and B2 phases, resulting in different hardness level, as well as the presence, size, and number density of precipitates that strengthened the $L1_2$ phase, resulting in no plastic deformation by planar slip, as seen in the as-cast material. This is shown in Figure 13b for 100 h of aging and is similar to the material aged for 500 and 1000 h. Differences observed in tribological behavior obtained by aging can be ascribed to these changes, causing alterations in wear mechanism. In the case of as-cast alloy without precipitates, due to lower B2 phase fraction as compared to the $L1_2$, abrasive wear is predominantly combined with adhesive wear, where delamination is largely prevented by dislocation pile-up and planar slip at the phase boundaries [11], thus modifying the brittle behavior of B2 phase [17]. Intensified adhesive wear component results in higher friction but low counter-body wear. After 100 h of aging, the adhesive wear component is reduced due to lower $L1_2$ phase fraction and increased hardness, also indicated by a small drop in friction. However, increased B2 phase fraction, absence of planar slip, and precipitation strengthening result in increased brittleness and intensified delamination. This amplifies the formation of hard wear particles, which remain in the reciprocating sliding contact and lead to increased wear of the material and counter-body (Figure 10). Prolonged aging time (500 h) results in a slightly reduced B2 phase and increased $L1_2$ phase fractions but a high level of strengthening with precipitates (Figure 5b), thus providing the best combination of high hardness and reduced brittleness with minimal wear. In this case, the main wear mechanism is abrasive wear with minimized components of adhesion and delamination. Furthermore, over-aging for 1000 h leads to a decrease in the number density of precipitates and an increase in their size, which results in decreased hardness, as shown in Figure 5. Lower hardness means increased abrasive wear, with a smaller number of large precipitates representing reduced resistance to sliding and a thus lower coefficient of friction (Figure 9).

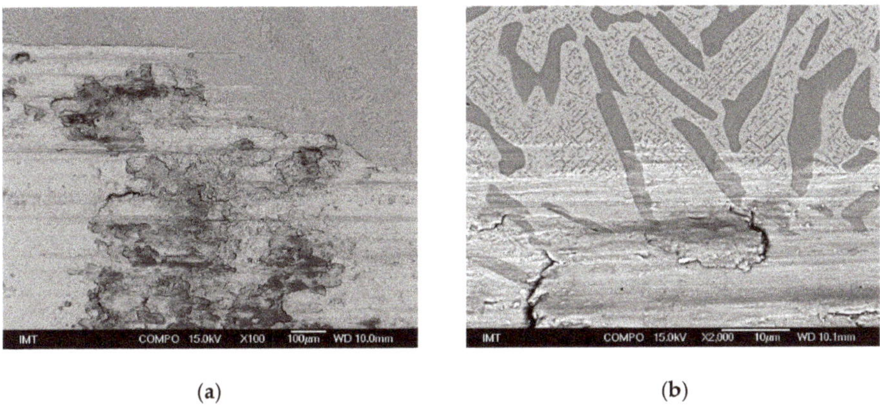

(**a**) (**b**)

Figure 13. SEM-BSE images of the wear track edge after 100 h of aging at 800 °C: (**a**) Oxide pile-up at the edge of the wear track; (**b**) Edge of the wear track showing no deformation by planar slip inside the $L1_2$ phase.

Table 2. Elements concentration in at% from EDS point analyses in Figure 12a.

Spectrum	O	Al	Cr	Fe	Co	Ni	Total
1	0.0	24.1	12.3	13.0	12.7	37.8	100
2	29.0	9.6	11.6	20.2	9.4	20.1	100
3	23.9	8.7	13.4	24.9	9.3	19.9	100

Table 3. Elements concentration in at% from EDS point analyses in Figure 12b.

Spectrum	O	Al	Si	Cr	Mn	Fe	Co	Ni	Total
1	1.9	16.2	0.0	15.1	0.0	15.9	16.6	34.3	100
2	0.0	15.6	0.0	15.9	0.0	16.6	17.1	34.8	100
3	15.0	3.3	0.6	16.3	0.9	48.9	3.5	11.5	100
4	47.6	3.0	0.4	9.9	0.5	26.4	3.8	8.4	100
5	44.1	4.2	0.5	10.7	0.4	25.7	4.4	10.0	100

4. Conclusions

In the present study, wear behavior as a function of aging time was evaluated for the $AlCoCrFeNi_{2.1}$ eutectic complex, concentrated alloy consisting of B2 (BCC) lamellae, and $L1_2$ (FCC) phase in as-cast state, as well as fine evenly distributed precipitates inside the $L1_2$ phase after aging at 800 °C for 100, 500, and 1000 h. Between 0 and 500 h of aging, both precipitates number and size increase, while from 500 h to 1000 h of aging, precipitates coarsen by Ostwald ripening mechanism. Abrasive wear prevailed in reciprocating wear analysis, with adhesive wear and delamination component changing depending on the aging conditions. The $L1_2$ phase without precipitates and lower hardness in the as-cast material preferentially deformed during the wear test, resulting in intensified adhesive wear component but minimum delamination, hindered by planar slip between B2 and $L1_2$ phases. This was not the case after aging the material, i.e., with the presence of precipitates. In general, aging the $AlCoCrFeNi_{2.1}$ alloy at 800 °C decreases its coefficient of friction due to reduced adhesive wear component and enhances its wear performance through precipitation strengthening. However, under-aging (100 h) results in increased material brittleness and thus increased delamination, while over-aging (1000 h) results in precipitates coarsening, and decreased hardness and abrasive wear resistance. The best performance, combining high hardness and reduced brittleness with minimal wear, is achieved with an intermediate aging of 500 h.

Author Contributions: Conceptualization, F.K. and B.P.; Methodology, F.K.; Formal analysis, F.K.; Investigation, F.K. and D.F.; Writing—original draft preparation, F.K.; Writing—review and editing, B.P. and D.F.; Visualization, F.K.; Project administration, F.K.; Funding acquisition, F.K. All authors have read and agreed to the published version of the manuscript.

Funding: This research was funded by Javna Agencija za Raziskovalno Dejavnost Republike Slovenije, grant number Z2-9220.

Acknowledgments: Acknowledgements go to the Institute of Metals and Technology (IMT) in Ljubljana where research was performed. Also, authors are thankful to all who were involved in the experimental part of this research.

Conflicts of Interest: The authors declare no conflict of interest. The funders had no role in the design of the study; in the collection, analyses, or interpretation of data; in the writing of the manuscript, or in the decision to publish the results.

References

1. Qiu, Y.; Thomas, S.; Gibson, M.A.; Fraser, H.L.; Birbilis, N. Corrosion of high entropy alloys. *npj Mater Degrad* **2017**, *1*, 15. [CrossRef]
2. Ayyagari, A.; Hasannaeimi, V.; Grewal, H.; Arora, H.; Mukherjee, S. Corrosion, Erosion and Wear Behavior of Complex Concentrated Alloys: A Review. *Metals* **2018**, *8*, 603. [CrossRef]
3. Wu, J.-M.; Lin, S.-J.; Yeh, J.-W.; Chen, S.-K.; Huang, Y.-S.; Chen, H.-C. Adhesive wear behavior of AlxCoCrCuFeNi high-entropy alloys as a function of aluminum content. *Wear* **2006**, *261*, 513–519. [CrossRef]
4. Tong, C.-J.; Chen, M.-R.; Yeh, J.-W.; Lin, S.-J.; Chen, S.-K.; Shun, T.-T.; Chang, S.-Y. Mechanical performance of the AlxCoCrCuFeNi high-entropy alloy system with multiprincipal elements. *Metall Mat Trans A* **2005**, *36*, 1263–1271. [CrossRef]

5. Hsu, C.-Y.; Sheu, T.-S.; Yeh, J.-W.; Chen, S.-K. Effect of iron content on wear behavior of AlCoCrFexMo0.5Ni high-entropy alloys. *Wear* **2010**, *268*, 653–659. [CrossRef]
6. Du, L.M.; Lan, L.W.; Zhu, S.; Yang, H.J.; Shi, X.H.; Liaw, P.K.; Qiao, J.W. Effects of temperature on the tribological behavior of Al0.25CoCrFeNi high-entropy alloy. *J. Mater. Sci. Technol.* **2019**, *35*, 917–925. [CrossRef]
7. Chuang, M.-H.; Tsai, M.-H.; Wang, W.-R.; Lin, S.-J.; Yeh, J.-W. Microstructure and wear behavior of AlxCo1.5CrFeNi1.5Tiy high-entropy alloys. *Acta Mater.* **2011**, *59*, 6308–6317. [CrossRef]
8. Löbel, M.; Lindner, T.; Pippig, R.; Lampke, T. High-Temperature Wear Behaviour of Spark Plasma Sintered AlCoCrFeNiTi0.5 High-Entropy Alloy. *Entropy* **2019**, *21*, 582. [CrossRef]
9. Lu, Y.; Dong, Y.; Guo, S.; Jiang, L.; Kang, H.; Wang, T.; Wen, B.; Wang, Z.; Jie, J.; Cao, Z.; et al. A promising new class of high-temperature alloys: Eutectic high-entropy alloys. *Sci. Rep.* **2014**, *4*. [CrossRef] [PubMed]
10. Lu, Y.; Gao, X.; Jiang, L.; Chen, Z.; Wang, T.; Jie, J.; Kang, H.; Zhang, Y.; Guo, S.; Ruan, H.; et al. Directly cast bulk eutectic and near-eutectic high entropy alloys with balanced strength and ductility in a wide temperature range. *Acta Mater.* **2017**, *124*, 143–150. [CrossRef]
11. Gao, X.; Lu, Y.; Zhang, B.; Liang, N.; Wu, G.; Sha, G.; Liu, J.; Zhao, Y. Microstructural origins of high strength and high ductility in an AlCoCrFeNi 2.1 eutectic high-entropy alloy. *Acta Mater.* **2017**, *141*, 59–66. [CrossRef]
12. Bhattacharjee, T.; Wani, I.S.; Sheikh, S.; Clark, I.T.; Okawa, T.; Guo, S.; Bhattacharjee, P.P.; Tsuji, N. Simultaneous Strength-Ductility Enhancement of a Nano-Lamellar $AlCoCrFeNi_{2.1}$ Eutectic High Entropy Alloy by Cryo-Rolling and Annealing. *Sci. Rep.* **2018**, *8*, 1–8. [CrossRef] [PubMed]
13. Bhattacharjee, T.; Zheng, R.; Chong, Y.; Sheikh, S.; Guo, S.; Clark, I.T.; Okawa, T.; Wani, I.S.; Bhattacharjee, P.P.; Shibata, A.; et al. Effect of low temperature on tensile properties of $AlCoCrFeNi_{2.1}$ eutectic high entropy alloy. *Mater. Chem. Phys.* **2018**, *210*, 207–212. [CrossRef]
14. Wani, I.S.; Bhattacharjee, T.; Sheikh, S.; Lud, Y.P.; Chatterjee, S.; Bhattacharjeea, P.P.; Guo, S.; Tsujib, N. Ultrafine-grained AlCoCrFeNi2.1 eutectic high-entropy alloy. *Mater. Res. Lett.* **2016**, *4*, 174–179. [CrossRef]
15. Wani, I.S.; Bhattacharjee, T.; Sheikh, S.; Bhattacharjee, P.P.; Guo, S.; Tsuji, N. Tailoring nanostructures and mechanical properties of $AlCoCrFeNi_{2.1}$ eutectic high entropy alloy using thermo-mechanical processing. *Mater. Sci. Eng. A* **2016**, *675*, 99–109. [CrossRef]
16. Li, P.; Sun, H.; Wang, S.; Hao, X.; Dong, H. Rotary friction welding of $AlCoCrFeNi_{2.1}$ eutectic high entropy alloy. *J. Alloys Compd.* **2020**, *814*, 152322. [CrossRef]
17. Hasannaeimi, V.; Ayyagari, A.V.; Muskeri, S.; Salloom, R.; Mukherjee, S. Surface degradation mechanisms in a eutectic high entropy alloy at microstructural length-scales and correlation with phase-specific work function. *npj Materials Degradation* **2019**, *3*, 16. [CrossRef]
18. Schindelin, J.; Arganda-Carreras, I.; Frise, E.; Kaynig, V.; Longair, M.; Pietzsch, T.; Preibisch, S.; Rueden, C.; Saalfeld, S.; Schmid, B.; et al. Fiji: An open-source platform for biological-image analysis. *Nat. Meth.* **2012**, *9*, 676–682. [CrossRef]
19. Rueden, C.T.; Schindelin, J.; Hiner, M.C.; DeZonia, B.E.; Walter, A.E.; Arena, E.T.; Eliceiri, K.W. ImageJ2: ImageJ for the next generation of scientific image data. *BMC Bioinf.* **2017**, *18*, 529. [CrossRef] [PubMed]
20. Wani, I.S.; Bhattacharjee, T.; Sheikh, S.; Clark, I.T.; Park, M.H.; Okawa, T.; Guo, S.; Bhattacharjee, P.P.; Tsuji, N. Cold-rolling and recrystallization textures of a nano-lamellar AlCoCrFeNi2.1 eutectic high entropy alloy. *Intermetallics* **2017**, *84*, 42–51. [CrossRef]

Article

Gradient Distribution of Microstructures and Mechanical Properties in a FeCoCrNiMo High-Entropy Alloy during Spark Plasma Sintering

Mingyang Zhang [1], Yingbo Peng [2], Wei Zhang [1,*], Yong Liu [1], Li Wang [3], Songhao Hu [4] and Yang Hu [5]

1 Powder Metallurgy Research Institute, Central South University, Changsha 410083, China; hugezmy123@gmail.com (M.Z.); yonliu@csu.edu.cn (Y.L.)
2 College of Engineering, Nanjing Agricultural University, Nanjing 210031, China; ybpengnj@njau.edu.cn
3 Department Metal Physics, Helmholtz-Zentrum Geesthacht, 21502 Geesthacht, Germany; li.wang1@hzg.de
4 Henan Huanghe Whirlwind Co., Ltd., Xuchang 461500, China; husonghao2008@outlook.com
5 Yuanmeng Precision Technology (Shenzhen) Institute, Shenzhen 518055, China; yanghu_hust@126.com
* Correspondence: waycsu@csu.edu.cn; Tel.: +86-731-8887-7669

Received: 31 January 2019; Accepted: 15 March 2019; Published: 19 March 2019

Abstract: A novel graded material of a high-entropy alloy (HEA) FeCoCrNiMo was fabricated by spark plasma sintering (SPS) processing. After SPS, the HEA specimens consisted of a single face-centred cubic (FCC) phase in the center, but dual FCC and a tetragonal structure σ phase near the surface. Surprisingly, the sintering pressure was sufficient to influence the proportion of phases, and thus the properties of HEA samples. The hardness of the specimens sintered under the pressures of 30, 35, and 40 MPa increased gradually from 210 $HV_{0.2}$, which is the single FCC phase in the center, to the maximum value near the surface as a result of the gradual increase in the fraction of the transformed σ phase. The σ phase, being a complex hard and brittle intermetallic particle to manipulate the properties of FCC-type HEA systems, which could be influenced by pressure, indicated a major possibility for designing gradient HEA materials.

Keywords: high-entropy alloy; spark plasma sintering; pressure; microstructure; mechanical properties

1. Introduction

Because of the rapid development of modern engineering and manufacturing industries, high-performance alloys urgently need to be developed. High-entropy alloys (HEAs) constitute a unique class of alloys exhibiting high strength and hardness, decent wear and corrosion resistance, and other attractive mechanical properties for both scientific research and practical applications [1–4]. For further studies on HEAs, phase transformations could be crucial for controlling their microstructures to obtain superior properties [5–7]. A significant amount of research on HEAs has been performed to study phase transformation using vacuum arc melting [8]; this procedure was usually restricted to laboratory settings. Using spark plasma sintering (SPS) to consolidate mechanically alloyed HEA powders is a promising method to obtain high-performance bulk HEAs [9–12]. The sintering pressure is one of the most important parameters in the SPS method [13]. Moreover, pressure will influence the equilibrium between the gaseous and liquid phases. The influence of pressure is typically neglected for the equilibrium of two or more solid phases. Standard phase diagrams involve only composition and temperature as the relevant variables [14]. This approach is based on phase transformations involving only solid phases not being associated with significant volume changes, unlike changes from liquid to gas. According to this theory, if the pressure could be controlled and further influence the phase transformation, it would represent a breakthrough in the field of controlling the structure and performance of HEAs.

In the present study, a FeCoCrNiMo HEA was successfully prepared using the SPS method under different pressures. The gradient distribution of microstructures and mechanical properties in the SPS samples were investigated. The face-centered cubic (FCC) to a tetragonal structure σ phase transformation was also studied. There is the possibility of the σ intermetallic compounds [15]. Consequently, the microstructure of the alloy can be varied by careful control of thermal treatment. Moreover, this phase transformation of FCC to σ phase, which can be influenced by SPS pressure, indicates a major possibility to design a novel gradient material of HEAs used in tools and dies.

2. Experimental

The investigated alloy with a nominal composition $Fe_{24.1}Co_{24.1}Cr_{24.1}Ni_{24.1}Mo_{3.6}$ (in at. %) was prepared using powder metallurgy (99.9%, Vilory new materials Co. Ltd, Xuzhou, China). Powders consisting of particles under 200 mesh in size were prepared using gas atomization and were mechanically milled using conventional planetary-milling equipment. Next, the powder with a particle size under 200 mesh was mechanically milled using conventional planetary milling equipment. The weight ratio between the powder and the stainless-steel balls was 1:10 and ethanol was added as the milling medium. The milling time was 20 h and the milling speed was 300 rev min^{-1}. The milled powders were then added into a graphite die 40 mm in diameter and consolidated using an HPD 25/3 SPS equipment under reduced pressure (10^{-3} Pa). The sintering temperature was 1150 °C and the pressures were 30, 35, and 40 MPa. After a holding time of 480 s, the sintered billets were cooled down to room temperature in the furnace. Samples were prepared by mechanical grinding using 1200 to 4000 grit SiC papers followed by a final polishing step (size: φ 40 × 2 mm^3). The transverse fracture strength of the samples (size: 12 × 2 × 30 mm^3) was determined by an Instron 3369 mechanical testing facility (Instron, Norwood, MA, USA) using the three-point method (span length: 25 mm, test speed: 2 mm/min), and tested twice for each treatment. A FEI Quanta FEG 250 scanning electron microscope (SEM, FEI, Hillsboro, OR, USA) equipped with an energy-dispersive X-ray (EDX) analyzer was used to investigate the microstructure and chemical compositions of the sintered specimens (20 kV, using spot analysis and backscattering mode). A Philips CM 200 transmission electron microscopy (TEM, Royal Philips, Amsterdam, The Netherlands) operating at 200 kV was used to identify the structure of the precipitates by selected area electron diffraction (SAED) analysis. The TEM specimen was prepared by a crossbeam workstation AURIGA 40 (Zeiss, Oberkochen, Germany) equipped with a focused ion beam (FIB) column and scanning electron microscopy (SEM). The phase constitution of the specimens used a Rigaku Rapid IIR (Rigaku, Tokyo, Japan) micro-area X-ray diffractometer (XRD, 40 kV, from 20° to 100°, a circular region with a diameter of 30 μm, PDF database 2009, using 20 minutes for each position) equipped with a 2D detector (phi: 360°, omega: −15°~150°) utilizing Cu Kα radiation. The phase transition temperature was analyzed by differential scanning calorimeter (DSC) using a NETZSCH STA 449C thermal analyzer (RT~1300 °C, 40 K/min, Ar atmosphere; Netzsch, Selb, Germany). The hardness of the alloy was determined using Buehler 5104 hardness tester (Buehler, Lake Bluff, IL, USA) under a 200 g load for 15 s and was averaged from three measurements. The indentation profile was obtained by NanoMap 500 DLS 3D surface profiler (Aep Technology, Santa Clara, CA, USA).

3. Results and Discussion

3.1. Microstructure

The morphology and phase composition of the powders before and after ball milling are shown in Figure 1a,b. The previously spherical particles of HEA powders were crushed to form irregular shape particles; the specific surface area of powders increased. As the specific surface area of the particles increases, a higher sintering driving force can be obtained, and the degree of sintering densification is improved. According the XRD pattern in Figure 1c, both powders exhibit an FCC structure. This means that the ball milling process does not lead to phase transition.

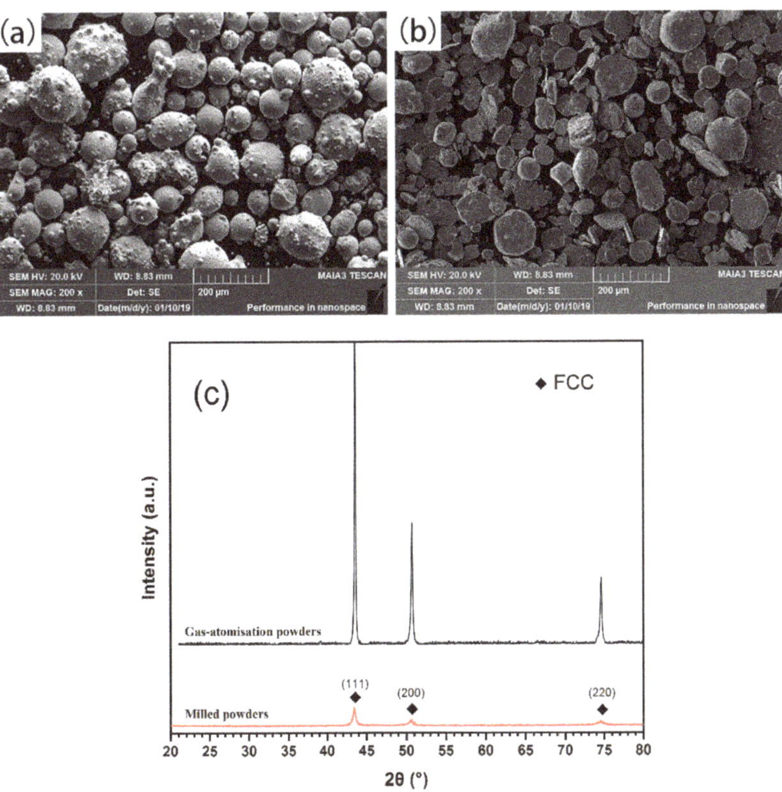

Figure 1. Microstructure of FeCoCrNiMo high-entropy alloy (HEA) powders before (**a**) and after (**b**) ball milling, (**c**) X-ray diffractometer (XRD) pattern of two powders. FCC—face-centered cubic.

Figure 2a–c shows the longitudinal cross-sections of the HEAs samples sintered at 1150 °C under different pressures. Judging from the contrast images, the microstructure did not consist of a single phase, as did those obtained after lower temperature sintering in our previous study [16].

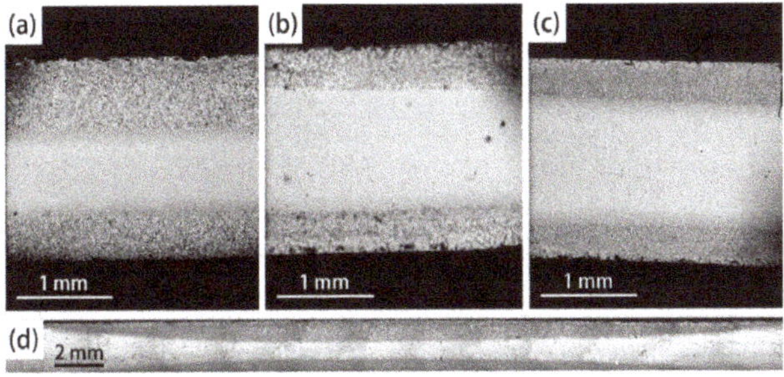

Figure 2. Longitudinal cross-sections of the same area in the FeCoCrNiMo HEA samples sintered at 1150 °C under different pressures: (**a**) 30, (**b**) 35, and (**c**) 40 MPa, as well as (**d**) the macrostructure of the 30 MPa sintered sample.

The microstructure gradually changed along the radial direction, towards the center of the samples. Obviously, the phase transformation occurred on the upper and lower surfaces of the three samples during SPS. The thickness of the phase transformed region decreased gradually along the radial direction of the cylinder towards both ends, as shown in Figure 2d. The reason for this phenomenon is the inhomogeneous distribution of the temperature field along the radial direction in the cylindrical graphite die during SPS. By simulating the temperature field of graphite die in the SPS process, the sintering temperature at the center is about 1700 °C and the border of the sample and the die can be as high as 450 °C [17]. Thus, because of the inhomogeneous distribution of temperature field, the volume fraction of phase transition is less than that of the middle part.

Under the sintering pressure of 30 MPa, the microstructure exhibited a distinct gradient distribution, as illustrated in Figure 3. It can be seen that multiple phase structures were successfully synthesized during SPS. The volume fraction of the transformed phase increased gradually from the center to edge of the specimen. As shown in Figure 3b, the volume fraction of the transformed phase decreased as the depth increased, which shows that sintering pressure directly affects the degree of phase transformation and presents a gradient distribution. For example, when the sintering pressure is 30 MPa, the volume fraction of the transformed phase is reduced from about 27% at the edge to 14% at a depth of 550 microns, and the volume fraction is reduced by half. When the sintering pressure is 40 MPa, the volume fraction of the transformed phase is sharply reduced from about 19% at the edge to 1% at a depth of 550 μm, which is about 20 times lower. It can be seen that the sintering pressure has a significant influence on the transformed phase volume fraction and the distribution of the transformed phase along the thickness direction. By controlling the sintering pressure, the gradient distribution of the multiple phase along the pressure direction can be realized and regulated.

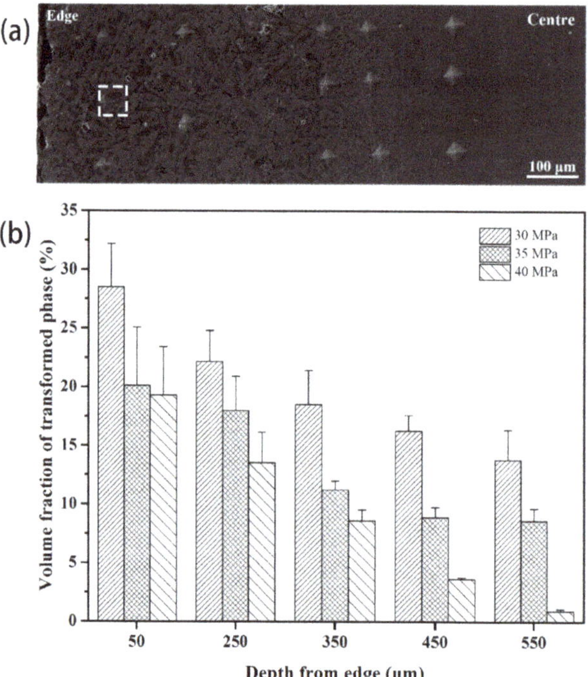

Figure 3. Scanning electron microscopy (SEM) image (**a**) and volume fraction of the transformed phase (**b**) of the FeCoCrNiMo HEA after sintering at 1150 °C and different sintering pressures. The mark shows the approximate position of the energy-dispersive X-ray (EDX) map.

3.2. Phase Identification

To understand the phase transformation, the phase constitution of the specimens was investigated using a combination of SEM, EDX, and XRD techniques.

As shown in Figure 4, the distribution of the elements Fe, Co, Cr, Ni, and Mo could be clearly identified in the samples. The precipitated phase was a Cr-rich phase, containing relatively low amounts of Co and Ni. No intermetallic compound was formed; therefore, the precipitate was a hard, Cr-rich σ phase. Powder metallurgy with a fast cooling rate was employed to reduce the preferred orientation effects observed in the cast alloys [18], the result of which could be used for accurately measuring the local lattice distortion. The distribution of the FCC and σ phase structures in the specimen can be identified using XRD analysis, as shown in Figure 5. The center of the specimen that underwent SPS under 30 MPa of pressure was a single FCC structure. While gradually transitioning towards the surface of the specimen, the {111} FCC diffraction peak became less intense than that of the center. However, the peaks corresponding to the {110} and {200} planes could be identified, indicating the formation of the σ phase structure. In addition, EDX analyses (Table 1) showed that the FCC matrix was rich in Mo and Ni, whereas the precipitated σ phase was rich in Cr, indicating that the matrix was represented by the FCC phase and the precipitate was the ordered σ phase. The XRD peak intensities are in a good accordance with the polycrystalline powders and the chemical compositions. Previous studies on the phase transformation of HEAs caused by SPS or the vacuum arc melting method reported that the body-centered cubic (BCC) phases or a tetragonal structure σ phase were composed of a spinodally modulated matrix, and precipitates exhibiting a near-equiaxed shape were distributed uniformly throughout the HEA. The FCC phases exhibiting net-like structure were located at the boundaries of the BCC phases [19,20]. However, the distribution of the σ phase and FCC structures observed was significantly different than those described in these previous studies. Interestingly, the microstructure presented a significant gradient distribution as the volume fraction of the σ structure increased. The mixed structures presented neither a net-shaped nor a dendritic form. This suggested that the FCC phase primarily formed during sintering, while the σ phase precipitated. The σ phase is an intermetallic precipitate, which dispersed and distributed in the FCC matrix.

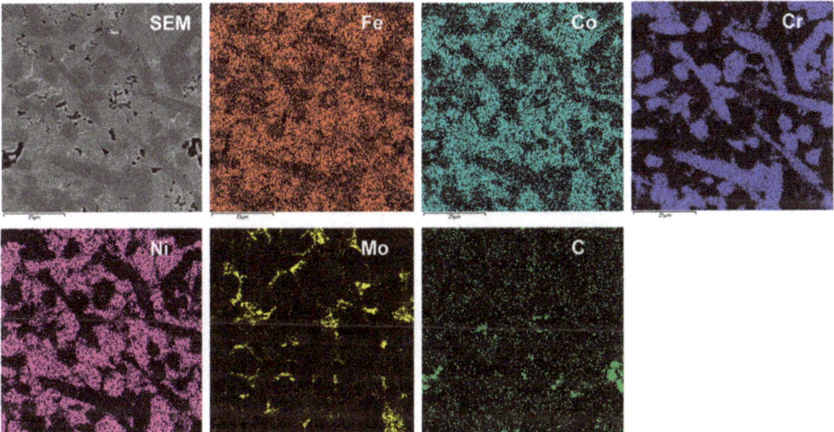

Figure 4. SEM image and EDX maps of Fe, Co, Cr, Ni, Mo, and C of marked area in Figure 3a of HEA after spark plasma sintering (SPS) at 1150 °C and 30 MPa.

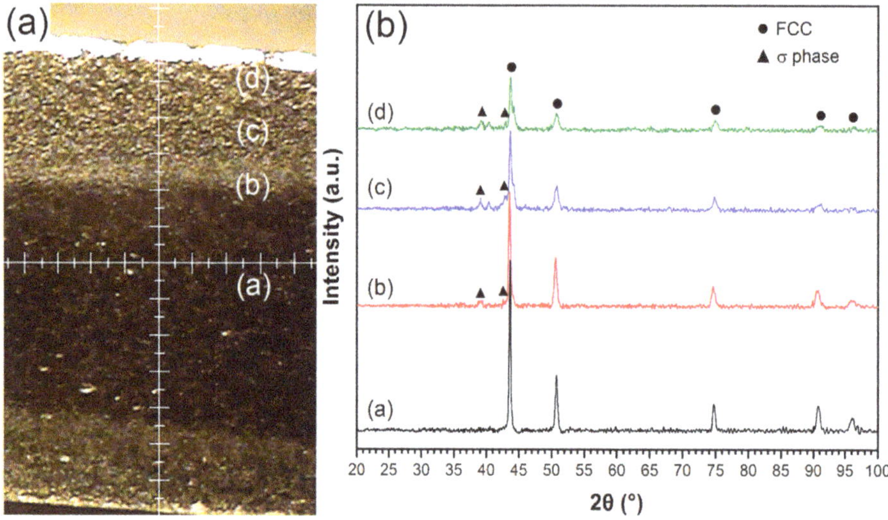

Figure 5. XRD pattern of different areas in the sample of HEA after SPS at 1150 °C and 30 MPa, (**a**) the macrostrucutre of the 30 MPa sintered sample, and (**b**) XRD pattern of the corresponding position in (**a**).

Table 1. Chemical composition of FeCoCrNiMo after spark plasma sintering (SPS) at 1150 °C and 30 MPa. FCC—face-centered cubic.

Chemical Composition (at. %)	Cr	Fe	Co	Ni	Mo
σ phase	50.8	19.1	17.3	10.1	2.6
FCC	20.0	23.9	24.8	22.5	6.7

In addition, there is no segregation of C in both the FCC and σ phases, which excludes the possibility of carbide formation (C diffusion of graphite die during SPS). There are several places where C element is segregated in the micropore or porosities; it is certain that these micropore or porosities are inevitable in sintering process, but have nothing to do with and are not the result of the phase transition from FCC to σ phase.

In Table 1, the Mo content of the σ phase is only 2.6%, whereas that of the FCC phase is approximately 6.7%. As the Mo content of nominal composition is 4.34 at.%, the FCC phase was enriched in Mo. This was the result of the good inter-solubility of Cr, Fe, Co, and Ni, while the solubility of Mo in the other elements was poor. Therefore, during the dissolution process of the solid solution, Mo, as a solute element, was repelled towards the FCC phase and redistributed along with Cr, while Cr entered the solid solution. The precipitated σ phase became rich in Cr and poor in Mo.

According to the (CoFeNi)–Cr–Mo pseudo ternary phase diagram at 900 °C and the (CoFeCrNi)–Mo pseudo-binary phase diagram in the work of [21], the microstructures of the dual phase (FCC and σ phase) structure (marked area in Figure 3) are examined and presented in Figure 6. The P1 area exhibits a single-phase polycrystalline structure, which can be identified in Figure 3. With the proceeding to the center of the specimen, the microstructures evolve to a dendritic structure, and Mo starts to segregate in the interdendritic areas and gradually shows a mixed two-phase structure. The TEM images and the selected-area diffraction (SAED) pattern in Figure 6a,c, showing a large precipitate embodied in the FCC matrix, clearly confirm that the precipitate consists of a mixture of the FCC and the σ phase, which are enriched with Cr and Mo elements (Table 1). It is noted that the FCC phase particles were precipitated out from the σ phase particle during cooling from a high temperature by solid-phase transformation. Thus, this complex structure can be simply referred to as a "double precipitation" during cooling, which can be explained by the pseudo binary diagram.

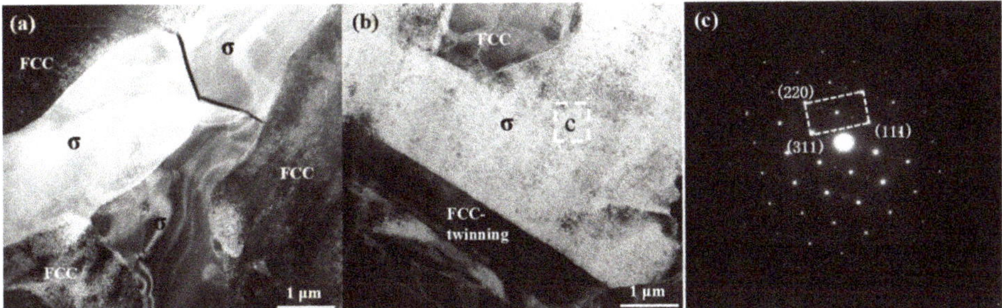

Figure 6. Transmission electron microscopy (TEM) image (**a**), (**b**) of the marked area in Figure 3a of FeCoCrNiMo HEA after sintering at 1150 °C and 30 MPa, and the selected-area diffraction (SAED) pattern (**c**) of the marked area in (**b**).

The phases formed in the SPS-processed HEA are essentially metastable—they are indeed formed solid-state phases during SPS, which are then kept at the ambient temperature because of the sluggish diffusion kinetics of HEAs. The TEM images are shown in Figure 6b. It was identified that an FCC phase re-precipitated in the σ phase. It can be concluded that the pseudo phase diagram quite successfully predicts all the structural features in the current alloy system. However, the solubility of Mo in the FCC matrix at SPS temperatures should correspond to the "FCC + σ" zone in the pseudo phase diagram in the work of [21], so there was no Mo-rich μ phase observed. Also, FCC twinning has also been observed in Figure 6b, which may be the result of phase transformation induction. This twinning structure will have a beneficial effect on the mechanical properties of HEA [22]. It is noted that the FCC matrix has intrinsically low stacking fault energy [23] and alloying of Mo could further reduce such energy, all of which promotes both twinning and widely dissociated and reactive dislocations.

According to the above analysis, the precipitates were identified as the σ phases that were transformed from the initial FCC structure during SPS. As can be seen in Figure 2, the thickness of the σ phase precipitates decreased gradually as the SPS pressure increased from 30 to 40 MPa. In addition, the FCC to σ phase transition temperature of FeCoCrNiMo HEA was about 1260 °C, which was measured by DSC (Figure 7). The first exothermic peak is at about 960 °C, which represents the beginning of the sintering reaction and the formation of the FCC phase. However, phase transition occurs at 1150 °C under the SPS condition, which is attributed to the effect of sintering pressure in the SPS process. When pressure existed, the phase transformation became easier, which reduced the transformation temperature and shortened the time of transformation. Thus, the sintering pressure significantly reduced the phase transformation temperature. Moreover, the σ phase is an ordered tetragonal structure as an intermetallic compound. From the point of view of atomic stacking density, the change of phase-volume is sensitive to pressure, and the larger the pressure, the smaller the trend of phase volume grown up. This theory has been proven in TiAl-based alloys, which are also intermetallic compounds. Under HIP (Hot Isostatic Pressing) conditions, the phase volume is sensitive to sintering pressure [14]. This explained why when sintering pressure rose to 35 and 40 MPa, the volume fraction of the transformed σ structure decreased. Accordingly, the gradient distribution of microstructure is caused by the sintering pressure and by adjusting the sintering pressure, the volume fraction of phase transformation can be changed to obtain the required gradient materials.

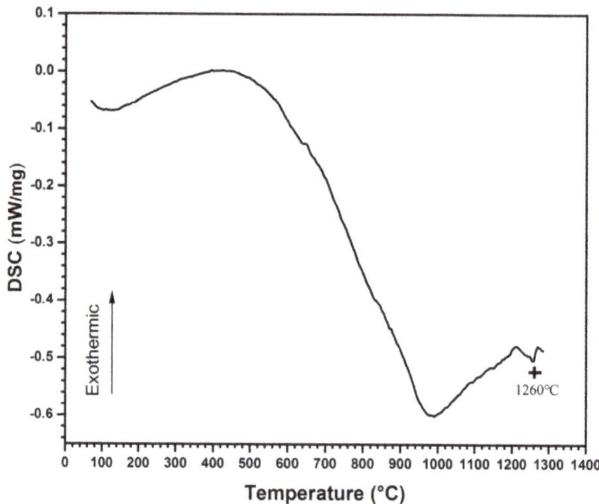

Figure 7. Differential scanning calorimeter (DSC) of the FeCoCrNiMo HEA sample processed by SPS.

3.3. Mechanical Properties

Bending and microhardness tests were used to analyze the effect of the distribution of the FCC and σ phases on the mechanical properties of HEA.

Figure 8 shows the bending curve of the HEA sample SPS processed under different pressures. According to Figure 8, when the sintering pressure is 40 MPa, the sample has the highest transverse strength of 1004 MPa and the highest fracture strain of 2.3%. With the decrease of sintering pressure, the strength and strain also decrease. When the sintering pressure is 30 MPa, the bending strength is 779 MPa. Obviously, the transverse strength decreases with the increase of the volume fraction of the σ phase. This is because the σ phase is an intermetallic compound with intrinsic brittleness. In contrast, the FCC phase has higher toughness. Comprehensively, when the volume fraction of the σ phase is small, the transverse strength of the FCC phase is more reflected. When sintering pressure is 30 MPa, the σ phase increases significantly, which leads to the decrease of transverse strength.

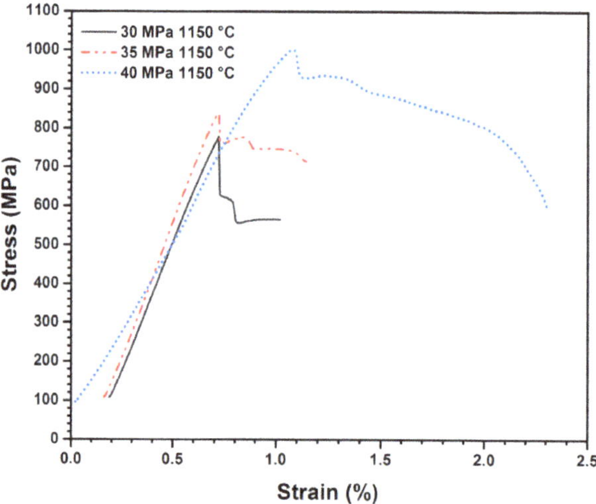

Figure 8. Bending curve of the HEA sample SPS processed under different pressure.

In addition, the results of the bending tests demonstrated that the SPS sample, which contained a larger volume fraction of the σ phase, could resist greater bending stress. This observation agreed with the work of Tsai et al., who found that the hardness of the $Al_{0.3}CrFe_{1.5}MnNi_{0.5}$ HEA nearly tripled after the σ phase formed [24], as the σ phase is a very hard phase [25]. Although, because of the different composition, it is different compared with the σ phase in FeCoCrNi–Mo HEA, other chemical elements may result in different properties. However, the σ phase is a typical ordered, tetragonal structure. Whatever the elements formed, its essence is a hard phase compared with the FCC phase.

The sample exhibited a gradual microhardness change similar to the gradually distribution of the σ phase. The center of all the three specimens consisted of a single FCC phases structure with a constant microhardness value of approximately 210 $HV_{0.2}$. As an increasing amount of the new structure precipitated, the hardness gradually increased as the distance from the surface decreased, and the hardness increased to the maximum value at 50 μm from the surface, as shown in Figure 9. The highest hardness could be attributed to the effect of the large volume fraction of hard σ phases. It also illustrates that the sintering pressure directly affects the gradient distribution of the FCC to σ phase transformation in the thickness direction. When the sintering pressure is 30 MPa, the hardness distribution shows a steep trend owing to the transformation of the FCC phase to the σ phase. However, at 40 MPa, the overall hardness distribution tends to be flat owing to the thinner phase transition area.

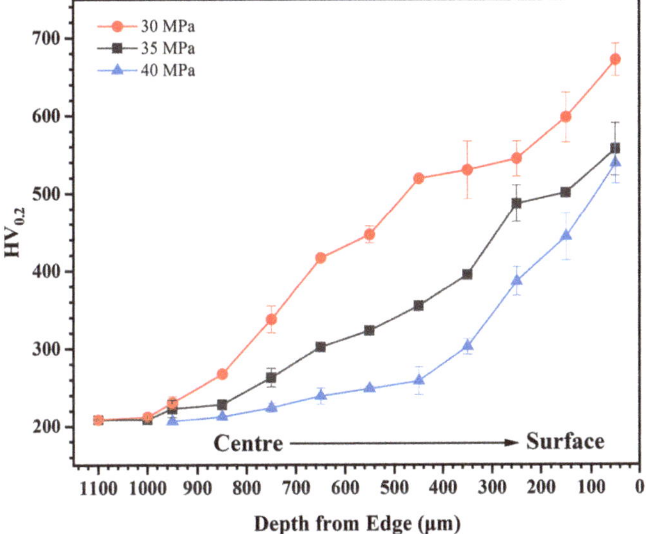

Figure 9. Microhardness distribution with depth from edge of the FeCoCrNiMo HEA after sintering at 1150 °C and 30, 35, and 40 MPa.

Moreover, from the indentation morphology of different depths, taking the sintering of 30 MPa as an example, the volume fraction of σ phase at the edge is larger, and the indentation depth is shallower, as shown in Figure 10. As the depth increases to the middle, the volume fraction of the σ phase decreases, the FCC phase increases, and the indentation area and depth increase. This further proves that the σ phase is a hard and brittle intermetallic phase, and the FCC phase is softer than the σ phase.

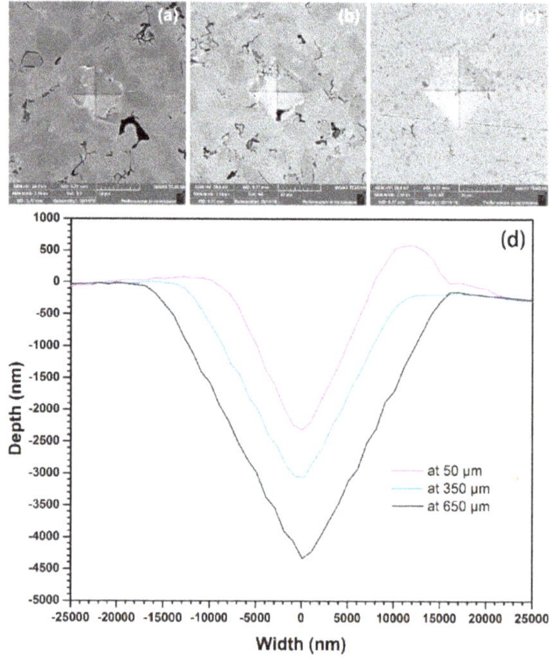

Figure 10. Indentation morphology of different depths from edge (**a**) 50 μm, (**b**) 350 μm, (**c**) 650 μm, and (**d**) indent profile of the FeCoCrNiMo HEA after sintering at 1150 °C and 30 MPa.

4. Conclusions

The mechanical properties of the FeCoCrNiMo HEA exhibited a significantly gradual change because of gradient distribution of the mixed FCC and σ phase structure. This gradient distribution was mainly assumed to be caused by the sintering pressure, which was under the specific sintering temperature of 1150 °C. It was found that the alloying of Mo into the CoCrFeNi HEA system generated the precipitation of the hard and brittle σ phase in the FCC matrix. The hard, Cr-rich σ phase was homogeneously distributed throughout the FCC matrix, and the volume fraction of the σ structure increased from the center to the surface of the HEA sample, like it would for gradient materials. The volume fraction of transformed σ structure can easily be adjusted by sintering pressure. The sintering pressure directly affects the gradient distribution of the FCC to the σ phase transformation in the thickness direction of the HEA samples. The implication for control of properties via changing the phase balance in HEAs will provide a strong technical base for the tool and dies of a novel gradient material.

Author Contributions: Conceptualization, W.Z. and Y.L.; Methodology, W.Z.; Validation, Y.P.; Formal Analysis, W.Z. and Y.P.; Investigation, M.Z. and L.W.; Resources, Y.L. and S.H.; Data Curation, M.Z.; Writing-Original Draft Preparation, M.Z.; Writing-Review & Editing, W.Z. and Y.P.; Project Administration, W.Z. and Y.L.; Funding Acquisition, Y.L., S.H. and Y.H.

Funding: This research was funded by the project of Innovation and Entrepreneur Team Introduced by Guangdong Province, grant number 201301G0105337290, and the Special Funds for Future Industrial Development of Shenzhen, grant number HKHTZD20140702020004.

Acknowledgments: The authors wish to acknowledge the financial support of State Key Laboratory of Powder Metallurgy (CSU 621011808), XUCHANG Fellowship Program [grant number: XW2017-40], the project of Innovation and Entrepreneur Team Introduced by Guangdong Province (201301G0105337290), and the Special Funds for Future Industrial Development of Shenzhen (No. HKHTZD20140702020004).

Conflicts of Interest: The authors declare no conflict of interest.

References

1. Yeh, J.W.; Chen, S.K.; Lin, S.J.; Gao, M.C.; Dahmen, K.A.; Liaw, P.K.; Lu, Z.P. Nanostructured high-entropy alloys with multiple principal elements: Novel alloy design concepts and outcomes. *Adv. Eng. Mater.* **2004**, *6*, 299–303. [CrossRef]
2. Zhang, Y.; Zuo, T.T.; Tang, Z.; Gao, M.C.; Dahmen, K.A.; Liaw, P.K.; Lu, Z.P. Microstructures and properties of high-entropy alloys. *Prog. Mater. Sci.* **2014**, *61*, 1–93. [CrossRef]
3. Tsai, M.H.; Yeh, J.W. High-entropy alloys: a critical review. *Mater. Res. Lett.* **2014**, *2*, 107–123. [CrossRef]
4. Chuang, M.H.; Tsai, M.H.; Wang, W.R.; Lin, S.J.; Yeh, J.W. Microstructure and wear behavior of $AlxCo_{1.5}CrFeNi_{1.5}Tiy$ high-entropy alloys. *Acta Mater.* **2011**, *59*, 6308–6317. [CrossRef]
5. He, J.Y.; Wang, H.; Huang, H.L.; Xu, X.D.; Chen, M.W.; Wu, Y.; Liu, X.J.; Nieh, T.G.; An, K.; Lu, Z.P. A precipitation-hardened high-entropy alloy with outstanding tensile properties. *Acta Mater.* **2016**, *102*, 187–196. [CrossRef]
6. Shun, T.T.; Du, Y.C. Microstructure and tensile behaviors of FCC $Al_{0.3}CoCrFeNi$ high entropy alloy. *Alloy. Compd.* **2009**, *479*, 157–160. [CrossRef]
7. Li, Z.; Pradeep, K.G.; Deng, Y.; Raabe, D.; Tasan, C.C. Metastable high-entropy dual-phase alloys overcome the strength–ductility trade-off. *Nature* **2016**, *534*, 227–230. [CrossRef] [PubMed]
8. Singh, S.; Wanderka, N.; Murty, B.S.; Glatzel, U.; Banhart, J. Decomposition in multi-component AlCoCrCuFeNi high-entropy alloy. *Acta Mater.* **2011**, *59*, 182–190. [CrossRef]
9. Praveen, B.S.; Murty, R.S. Alloying behavior in multi-component AlCoCrCuFe and NiCoCrCuFe high entropy alloys. *Mater. Sci. Eng. A* **2012**, *534*, 83–89. [CrossRef]
10. Fu, Z.Q.; Chen, W.P.; Wen, H.M.; Zhang, D.L.; Chen, Z.; Zheng, B.L.; Zhou, Y.Z.; Lavernia, E.J. Microstructure and strengthening mechanisms in an FCC structured single-phase nanocrystalline $Co_{25}Ni_{25}Fe_{25}Al_{7.5}Cu_{17.5}$ high-entropy alloy. *Acta Mater.* **2016**, *107*, 59–71. [CrossRef]
11. Gwalani, B.; Pohan, R.M.; Lee, J.; Lee, B.; Banerjee, R.; Ryu, H.J.; Hong, S.H. High-entropy alloy strengthened by in situ formation of entropy-stabilized nano-dispersoids. *Sci. Rep.* **2018**, *8*, 14085. [CrossRef]
12. Pohan, R.M.; Gwalani, B.; Lee, J.; Alam, T.; Hwang, J.Y.; Ryu, H.J.; Banerjee, R.; Hong, S.H. Microstructures and mechanical properties of mechanically alloyed and spark plasma sintered $Al_{0.3}CoCrFeMnNi$ high entropy alloy. *Mater. Chem. Phys.* **2018**, *210*, 62–70. [CrossRef]
13. Mamedov, V. Spark plasma sintering as advanced PM sintering method. *Powder Metall.* **2002**, *45*, 322–328. [CrossRef]
14. Huang, A.; Hu, D.; Loretto, M.H.; Mei, J.; Wu, X.H. The influence of pressure on solid-state transformations in Ti–46Al–8Nb. *Scr. Mater.* **2007**, *56*, 253–256. [CrossRef]
15. Liu, W.H.; Lu, Z.P.; He, J.Y.; Luan, J.H.; Wang, Z.J.; Liu, B.; Liu, Y.; Chen, M.W.; Liu, C.T. Ductile CoCrFeNiMox high entropy alloys strengthened by hard intermetallic phases. *Acta Mater.* **2016**, *116*, 332–342. [CrossRef]
16. Zhang, M.; Zhang, W.; Liu, Y.; Liu, B.; Wang, J. FeCoCrNiMo high-entropy alloys prepared by powder metallurgy processing for diamond tool applications. *Powder Metall.* **2018**, *61*, 123–130. [CrossRef]
17. Wang, Y.; Fu, Z. Study of temperature field in spark plasma sintering. *Mater. Sci. Eng.* **2002**, *B90*, 34–37.
18. Yeh, J.W.; Chang, S.Y.; Hong, Y.D.; Chen, S.K.; Lin, S.J. Anomalous decrease in X-ray diffraction intensities of Cu–Ni–Al–Co–Cr–Fe–Si alloy systems with multi-principal elements. *Mater. Chem. Phys.* **2007**, *103*, 41–46. [CrossRef]
19. Wang, J.; Niu, S.Z.; Guo, T.; Kou, H.C.; Li, J.S. The FCC to BCC phase transformation kinetics in an $Al_{0.5}CoCrFeNi$ high entropy alloy. *Alloy. Compd.* **2017**, *710*, 144–150. [CrossRef]
20. Zhang, A.; Han, J.; Meng, J.; Su, B.; Li, P. Rapid preparation of AlCoCrFeNi high entropy alloy by spark plasma sintering from elemental powder mixture. *Mater. Lett.* **2016**, *181*, 82–85. [CrossRef]
21. Wu, Q.; Wang, Z.; He, F.; Li, J.; Wang, J. Revealing the Selection of σ and μ Phases in CoCrFeNiMox High Entropy Alloys by CALPHAD. *J. Phase Equilibria Diffus.* **2018**, *39*, 446–453. [CrossRef]
22. Liu, T.K.; Wu, Z.; Stoica, A.D.; Xie, Q.; Wu, W.; Gao, Y.F.; Bei, H.; An, K. Twinning-mediated work hardening and texture evolution in CrCoFeMnNi high entropy alloys at cryogenic temperature. *Mater. Des.* **2017**, *131*, 419–427. [CrossRef]
23. Zaddach, A.J.; Niu, C.; Koch, C.C.; Irving, D.L. Mechanical properties and stacking fault energies of NiFeCrCoMn high-entropy alloy. *JOM* **2013**, *65*, 1780–1789. [CrossRef]

24. Tsai, K.C.; Chang, J.; Li, H.; Tsai, R.C.; Cheng, A.H. A second criterion for sigma phase formation in high-entropy alloys. *Mater. Res. Lett.* **2016**, *4*, 1–6. [CrossRef]
25. Tsai, M.H.; Yuan, H.; Cheng, G.M. Significant hardening due to the formation of a sigma phase matrix in a high entropy alloy. *Intermetallics* **2013**, *33*, 81–86. [CrossRef]

Article

Microstructure and Properties of High-Entropy $Al_xCoCrFe_{2.7}MoNi$ Alloy Coatings Prepared by Laser Cladding

Minghong Sha †, Chuntang Jia †, Jun Qiao, Wenqiang Feng, Xingang Ai, Yu-An Jing, Minggang Shen and Shengli Li *

School of Materials and Metallurgy, University of Science and Technology Liaoning, Anshan 114051, China; s13591200297@163.com (M.S.); jia19950608@163.com (C.J.); juncourses@163.com (J.Q.); wqfeng14b@imr.ac.cn (W.F.); aixingang@126.com (X.A.); jyachina@163.com (Y.-A.J.); lnassmg@163.com (M.S.)

* Correspondence: hongsmh_116@163.com; Tel.: +86-156-4125-9199

† These authors contributed equally to this work.

Received: 25 October 2019; Accepted: 17 November 2019; Published: 20 November 2019

Abstract: High-entropy $Al_xCoCrFe_{2.7}MoNi$ (x = 0, 0.5, 1.0, 1.5, 2.0) alloy coatings were prepared on pure iron by laser cladding. The effects of Al content on the microstructure, hardness, wear resistance and corrosion resistance of the coatings were studied. The results showed that the crystal phases of the $Al_xCoCrFe_{2.7}MoNi$ coatings changed from Mo-rich BCC1 + FCC to (Al, Ni)-rich BCC2 + Mo-rich BCC1 when x increased from 0 to 0.5, and the phase changed to an (Al, Ni)-rich BCC2 + (Mo, Cr)-rich σ phase as x increased further. The hardness of the coatings increased as the Al content increased. The $Al_{2.0}CoCrFe_{2.7}MoNi$ coating exhibit best wear resistance. Addition of Al increased the corrosion potential in a 3.5 wt.% NaCl solution, and the coating with x = 1.0 exhibited the highest corrosion resistance.

Keywords: high-entropy alloys; laser cladding; corrosion resistance; wear resistance; microstructure

1. Introduction

Yeh et al. [1,2] put forward the concept of a high-entropy alloy (HEA) in 2004, and changed the traditional concept of alloy design. An HEA is defined as an alloy with a configurational entropy larger than 1.5R in the random solution state [3,4], where R is the gas constant. Owing to their special composition and structure, an HEA exhibits high phase stability, wear resistance, and corrosion resistance [5–12]. Zhang et al. [13] prepared an HEA coating of FeCoCrAlNi on 304 stainless steel by laser cladding. The results showed that the coating exhibits better corrosion resistance and pitting resistance than uncoated 304 stainless steel in a 3.5 wt.% NaCl solution. Niu et al. [14] studied the effect of Al content in an $Al_xFeCoCrNiCu$ (x = 0.25, 0.5, 1.0) HEA on its corrosion resistance in a 1 mol/L H_2SO_4 solution and a 1 mol/L HCl solution, respectively. The corrosion resistance and pitting resistance in the 1 mol/L H_2SO_4 solution increased when the Al content was less than 0.5, while they decreased when the Al content reached 1.0. Kao et al. [15] studied the corrosion resistance of an $Al_xCoCrFeNi$ HEA and found that the corrosion potential (E_{corr}) and corrosion current (I_{corr}) are independent of Al content. Therefore, the effects of Al content on corrosion resistance of HEAs are still not fully understood. An AlCoCrFeNi HEA has been extensively studied for its uncomplicated FCC and BCC phases [16–19]. Some researchers added Ti, Nb, and other elements to the alloy to obtain the desired microstructures, hardness and wear resistance [20–22]. Mo has small thermal expansion coefficient, high strength at high temperatures, high hardness, strong corrosion resistance and high thermal conductivity [23]. It is shown that the addition of Mo increases the strength of $AlCrFeNiMo_x$ and $CoCrFeNiMo_x$ HEAs due to the formation of the sigma (σ) phase [24,25]. The σ phase is a hard, brittle phase commonly found in superalloys and can significantly change the mechanical properties of the alloy [26–28]. Its effect on corrosion resistance has not been reported. The effects of Mo content on the structure and properties of AlCoCrFeNiMo$_x$ HEAs are being investigated in another of our studies. Due to a higher Fe content in the coating, it is helpful to improve the coating's bond with a steel substrate; Al$_x$CoCrFe$_{2.7}$NiMo HEAs were determined as the coating materials to be studied in our current work.

As a new technology, laser cladding has many advantages over traditional cladding technologies, providing coatings with minimum dilution, minimum deformation, and high surface quality. The effects of Al content on microstructure, hardness, wear resistance and corrosion resistance of Al$_x$CoCrFe$_{2.7}$MoNi coatings prepared by laser cladding were evaluated in this study.

2. Materials and Methods

Pure iron was selected as the base material in order to eliminate the effects of other elements. Its high purity allows accurate analysis and characterization of the structure and properties of HEAs. Table 1 shows the chemical composition of the base material measured by chemical analysis.

The pure metal powders used in the experiments are the atomized powders produced by BGRIMM (Beijing, China). Pure powders of Al, Co, Cr, Fe, Ni, and Mo (>99.9%, wt.%) with an average particle size of 75 µm were used as raw cladding materials. The powders were weighed and mixed according to the proportions listed in Table 2 (Fe$_{2.7}$ was achieved by appropriate laser parameters determined through multiple test attempts to control the dilution ratio of the coating and the base material). Then, the mixed alloy powder was put into a stainless-steel tank and thoroughly dry blended for 5 h in a planetary ball mill with a rotating speed of 300 r/min. After sieving, placed in a vacuum dries oven to prevent oxidation. A mixed powder layer with 1 mm thickness was placed on the base material and radiated by the laser in an argon atmosphere. Single-pass laser cladding was used to deposit coatings of Al$_x$CoCrFe$_{2.7}$MoNi at 1350 W, 980 nm wavelength, 20 mm/s scanning rate, and 3 mm spot diameter (Laserline LDF 4000, Laserline GmbH, Mülheim-Kärlich, Germany). The radiated samples were then annealed at 900 °C for 5 h to relieve thermal stress and prevent microcrack formation. The structures of the samples were analyzed using X-ray diffraction (XRD, PANalytical X-pert Power, Malvern Panalytical Ltd., Worcestershire, UK) with a line detector (X'Celerator) at 2θ ranging from 15° to 90° in 0.065° increments with Cu K$_\alpha$ radiation. High Score Plus software and PDF-2004 database (JCPDS, Newtown Square, PA, USA) were used to analyze the diffraction pattern. The specimens were eroded in aqua regia for 5–10 s, and the morphologies and compositions of the coatings were analyzed using a scanning electron microscope (SEM, Hitachi S-3400N, Hitachi, Ltd., Tokyo, Japan) with an energy dispersive spectrometer (EDS, TEAM PEGASUS2040), and with a transmission electron microscope (TEM, FEI Talos F200, FEI Co. Ltd, Hillsboro, OR, USA) with an EDS (FEI Super X). The size of the investigated area used for the measurements of the overall compositions of coatings in Table 3 is 1300 µm × 265 µm. Microhardness was measured from the bond zone to the coating surface using a microhardness tester (Qness Q10A, Qness GmbH, Golling, Austria) with a 9.81 N loading force and 15 s loading time. The wear resistance was tested with friction and wear test equipment (UMT TriboLab, Bruker Corporation, Billerica, MA, USA) with a pair of ceramic balls. A 13 N normal load, 100 mm/s reciprocating speed, a 10 mm reciprocating straight line distance and 1800 mm total wear distance were used in the wear tests. The weight of the samples before and after wear tests was weighed with a balance (0.01 mg precision). The E_{corr} and I_{corr} were measured with an electrochemical workstation (Autolab PGSTAT302N) from −1.2 to 1.2 V and 1.0 mV/s scanning speed in a 3.5 wt.% NaCl aqueous solution. The E_{corr} and I_{corr} of the coatings were obtained by Tafel linear extrapolation. A platinum electrode, saturated AgCl electrode and the specimen were used as the auxiliary, reference, and working, respectively. The chemical valence states of metal elements in passive films formed on the surfaces of the Al$_{1.0}$ and Al$_{2.0}$ coatings were measured using X-ray photoelectron spectroscopy (XPS, Escalab 250Xi, Thermo Fisher Scientific, Waltham, MA, USA) with monochromatic Al Kα excitation.

Table 1. Pure iron content for matrix (wt.%).

Element	Fe	Al	S	P	Mn	Si	C
Content	99.457	0.22	0.014	0.011	0.120	0.150	0.028

Table 2. Composition of the mixed powder (wt.%).

x	Al	Co	Cr	Fe	Mo	Ni
0	0	18.34	16.18	17.38	29.85	18.26
0.5	4.03	17.60	15.52	16.68	28.65	17.53
1.0	7.74	16.92	14.92	16.03	27.54	16.85
1.5	11.18	16.28	14.37	15.43	26.51	16.22
2.0	14.38	15.70	13.85	14.88	25.56	15.64

3. Experimental Results

3.1. Crystal Structure

X-ray diffraction patterns from the $Al_xCoCrFe_{2.7}NiMo$ coatings are shown in Figure 1. The coatings are mainly composed of simple solid solutions and intermetallic compounds. The phase structure is composed of both BCC and FCC solid solutions when $x = 0$, while BCC1 and BCC2 solid solutions appear and the peak value is more intense when $x = 0.5$. Tiny Bragg peaks corresponding to the σ and BCC phases are visible in the XRD pattern when $x = 1.0$, and no new phase appears in the XRD pattern as the Mo content increases ($x = 1.5$ and 2.0).

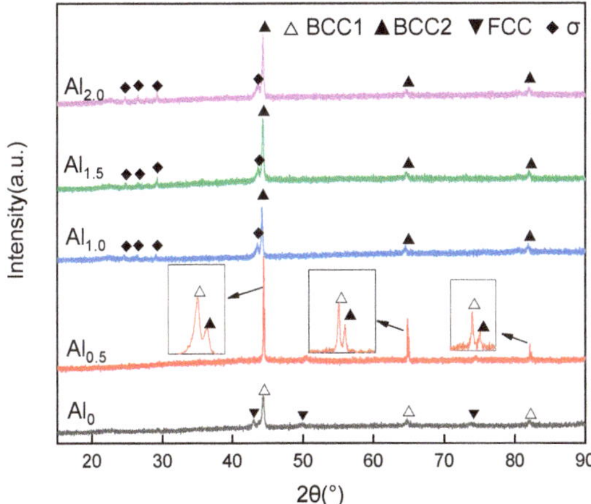

Figure 1. XRD pattern from $Al_xCoCrFe_{2.7}MoNi$ coatings.

3.2. Microstructure

SEM images of the microstructure of the $Al_xCoCrFe_{2.7}MoNi$ ($x = 0, 0.5, 1.0, 1.5, 2.0$) coatings are shown in Figure 2a–c, e, and g. In view of the fineness of microstructure and the limitations of the SEM, TEM images of the microstructures of $Al_{1.0}$, $Al_{1.5}$, and $Al_{2.0}$ HEAs are presented in Figure 2d, f and h. The target composition and actual composition of the coatings measured by EDS are listed in Table 3. The chemical compositions in different micro-regions of $Al_xCoCrFe_{2.7}MoNi$ are shown in Table 4. A few precipitates containing Fe and Cr appear in region A of the coating without Al, as shown in Figure 2a. Figure 2b shows that the $Al_{0.5}$ coating consists of dendrites. Figure 2c,e,g shows that $Al_{1.0}$, $Al_{2.0}$, and $Al_{3.0}$ alloys have fine microstructures, respectively. The light and dark phases appear in $Al_{1.0}$, indicated by D and C, respectively, as shown in Figure 2d. The EDS results and analysis of the diffraction spots show that the C phase is a Mo-rich σ phase, and the D phase is an (Al, Ni)-rich BCC2 phase. Figure 2f shows that two kinds of dark areas (granules and sheets) and one bright area can be seen in the $Al_{1.5}$ coating, indicated by C, C_1, and D, respectively. An analysis of the diffraction spots show that the C and C_1 phases belong to the (Mo, Cr)-rich σ phase, while the D phase is an (Al, Ni)-rich BCC2 phase. The dark strip disappears as Mo content increases, as shown in Figure 2h, and the microstructure is composed of a granular (Mo, Cr)-rich σ phase and (Al, Ni)-rich BCC2 phase.

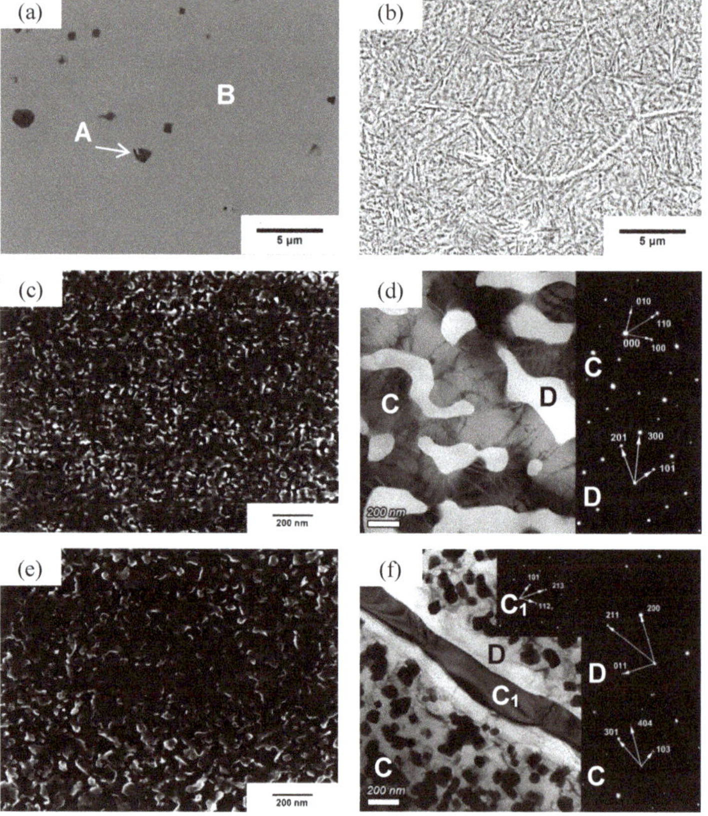

Figure 2. *Cont.*

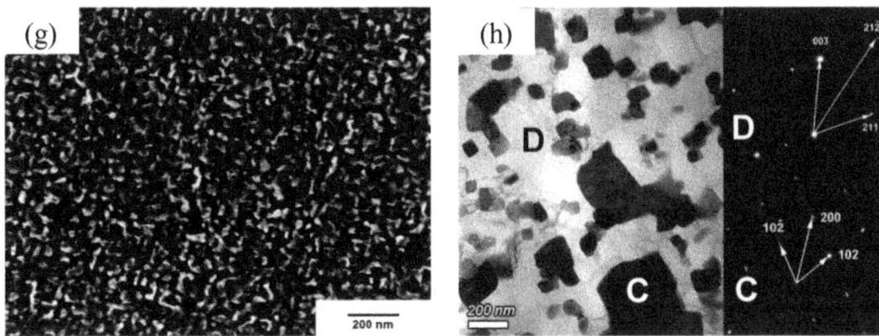

Figure 2. Microstructures of the $Al_xCoCrFe_{2.7}MoNi$ coatings. (**a–c**,**e**,**g**) show SEM images; (**d**,**f**,**h**) show TEM images.

Table 3. Composition of coatings measured by EDS (at.%).

x	Type	Al	Cr	Fe	Co	Ni	Mo
0	Actual	0	14.52	40.39	15.04	16.56	13.49
	Target	0	14.93	40.30	14.93	14.93	14.93
0.5	Actual	5.92	13.73	37.87	14.88	12.82	14.79
	Target	6.94	13.89	37.50	13.89	13.89	13.89
1.0	Actual	12.71	13.25	36.33	12.96	12.76	11.98
	Target	12.99	12.99	35.06	12.99	12.99	12.99
1.5	Actual	16.36	13.93	32.77	12.19	12.81	11.94
	Target	18.29	12.20	32.93	12.20	12.20	12.20
2.0	Actual	19.25	11.34	33.95	11.45	11.69	12.32
	Target	22.99	11.49	31.03	11.49	11.49	11.49

Table 4. Composition of coatings micro-regions (at.%).

Alloy	Region	Al	Co	Cr	Fe	Ni	Mo
$Al_{0.0}$	A region	0	0	79.98	20.02	0	0
	B region	0	17.71	13.85	53.28	13.61	7.55
$Al_{0.5}$	D region	5.25	11.91	8.59	53.41	9.24	11.60
$Al_{1.0}$	C region	1.76	10	18.2	49.44	2.46	17.75
	D region	28.91	17.04	1.96	22.17	29.57	0.32
$Al_{1.5}$	C region	1.01	8.98	13.72	58.11	2.01	34.89
	D region	25.60	11.63	3.84	38.55	19.20	1.18
	E region	6.34	7.93	15.87	43.97	3.17	22.68
$Al_{2.0}$	C region	3.53	7.87	15.21	53.58	2.16	17.34
	D region	26.68	12.58	3.84	36.52	19.76	0.58

3.3. Microhardness

The microhardness measurements from different positions in the $Al_xCoCrFe_{2.7}MoNi$ coatings are shown in Figure 3. The $CoCrFe_{2.7}NiMo$ alloy has the lowest average hardness (272 HV), which can be attributed to generation of the FCC phase. The microhardness increases as Al content increases, and $Al_{2.0}CoCrFe_{2.7}NiMo$ has the highest average hardness (1142 HV). Hardness test results show that the formation of the BCC2 phase increases the hardness of the coating.

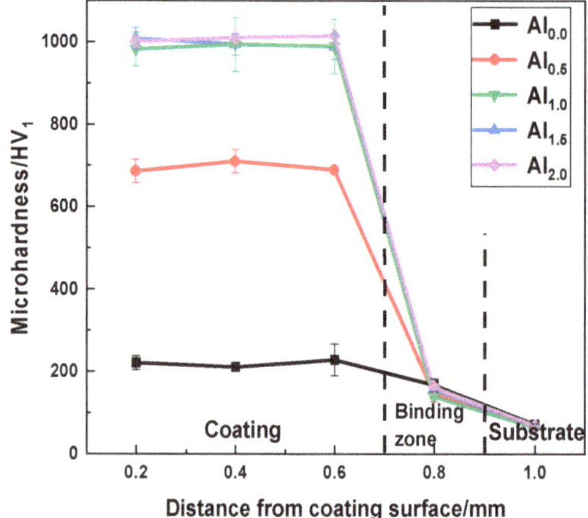

Figure 3. Microhardness of $Al_xCoCrFe_{2.7}MoNi$ coatings.

3.4. Wear Resistance

Wear in a material is related to its structure and external environment. The wear resistance of samples $Al_{1.0}$, $Al_{1.5}$, and $Al_{2.0}$ was analyzed in this paper. The morphology of worn $Al_xCoCrFe_{2.7}MoNi$ (x = 1.0, 1.5, 2.0) coatings is shown in Figure 4a$_1$–c$_1$ are enlarged partial details of Figure 4a–c, respectively. There is a convex scaly plastic deformation layer on the friction surface of the $Al_{1.0}$ sample, as shown in Figure 4a,a$_1$, which resulted from repeated grinding during the wear test. An oxide developed at the junction of the flaky furrow and the scaly deformed layer because of severe friction and high temperature during the wear test. The wear mechanism is mainly adhesive wear and oxidative wear. The wear of the $Al_{1.5}$ sample is with a few flake furrows, in which oxides were found, as shown in Figure 4b,b$_1$, indicating the wear occurs via oxidation, slight adhesion wear and slight abrasive wear. Sample $Al_{2.0}$ also exhibits a scaly plastic deformation layer with oxides and a flaky furrow in Figure 4c,c$_1$. The wear mechanism in sample $Al_{2.0}$ is adhesive wear and oxidative wear. The measured weight losses from the coatings due to wear are listed in Table 5; sample $Al_{2.0}$ exhibited the least wear of 0.1 mg.

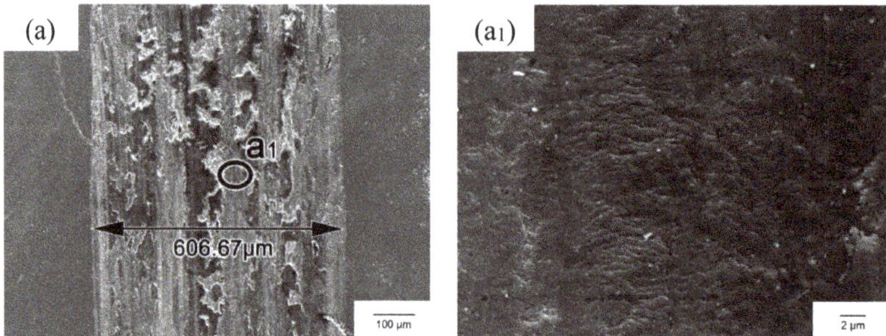

Figure 4. *Cont.*

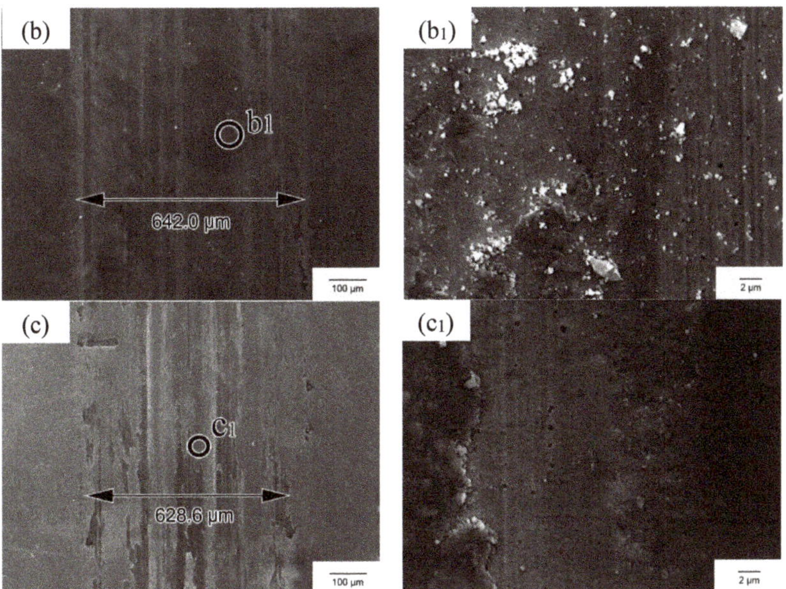

Figure 4. Wear morphology of the $Al_xCoCrFe_{2.7}MoNi$ (x = 1.0, 1.5, 2.0) coatings; (**a**,**a1**) x = 1.0; (**b**,**b1**) x = 1.5; (**c**,**c1**) x = 2.0.

Table 5. Weight loss from the $Al_xCoCrFe_{2.7}MoNi$ (x = 1.0, 1.5, 2.0) coatings.

Alloy	Before Abrasion/g	After Abrasion/g	Abrasion Weight Loss/mg
$Al_{1.0}$	10.0048	10.0044	0.4
$Al_{1.5}$	8.4143	8.4141	0.2
$Al_{2.0}$	10.1304	10.1303	0.1

3.5. Corrosion Resistance

Potentiodynamic polarization curves of the $Al_xCoCrFe_{2.7}MoNi$ coatings in the 3.5 wt.% NaCl solution are shown in Figure 5. The E_{corr} and I_{corr} of the coatings were obtained by Tafel linear extrapolation, as shown in Table 6. These results show that, except for sample $Al_{1.0}$, the self-corrosion potential of the other coatings increases as the Al content increases. The self-corrosion current density with sample $Al_{1.0}$ is the least, while the self-corrosion current density with samples $Al_{1.5}$ and $Al_{2.0}$ are slightly larger. Sample $Al_{0.0}$ exhibits the lowest self-corrosion potential and higher self-corrosion current density, indicating that it has the greatest corrosion tendency, highest corrosion rate and worst corrosion resistance. Figure 6 shows XPS results from passive films on the $Al_xCoCrFe_{2.7}NiMo$ (x = 1.0, 2.0) coatings after the corrosion experiments in the 3.5 wt.% NaCl solution. The composition of the passive film is Al_2O_3, CoO, Co_2O_3, Cr_2O_3, Fe_2O_3, MoO_3, and NiO when x = 1.0, while the passive film is primarily composed of Al_2O_3, CoO, Cr_3O_4, Fe_2O_3, MoO_3, and NiO when x = 2.0. Al_2O_3, Cr_2O_3, CoO, MoO_3, and NiO were detected on the surfaces of all coatings, which can provide certain protection in a corrosive environment. The compositions (in relative at.%) of the passive films determined from XPS measurements are summarized in Figure 7. This shows that the content of Al_2O_3 is higher than that of other metal oxides, and the relative contents of Ni, Co and Fe oxides decrease as the Al content increases. This is primarily due to the fact that Al is active and oxidizes easier than the other elements listed here.

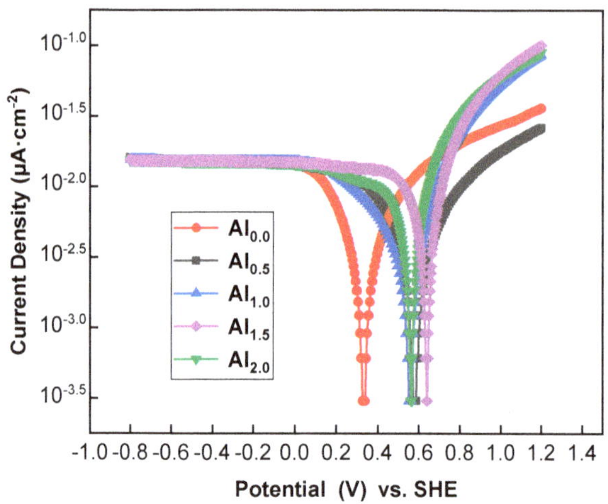

Figure 5. Polarization curves of $Al_xCoCrFe_{2.7}MoNi$ coatings in a 3.5 wt.% NaCl solution.

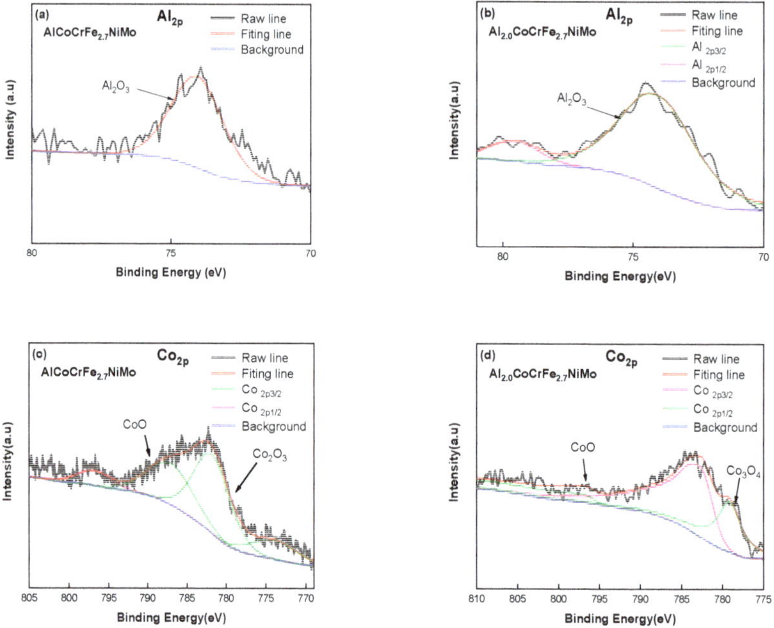

Figure 6. *Cont.*

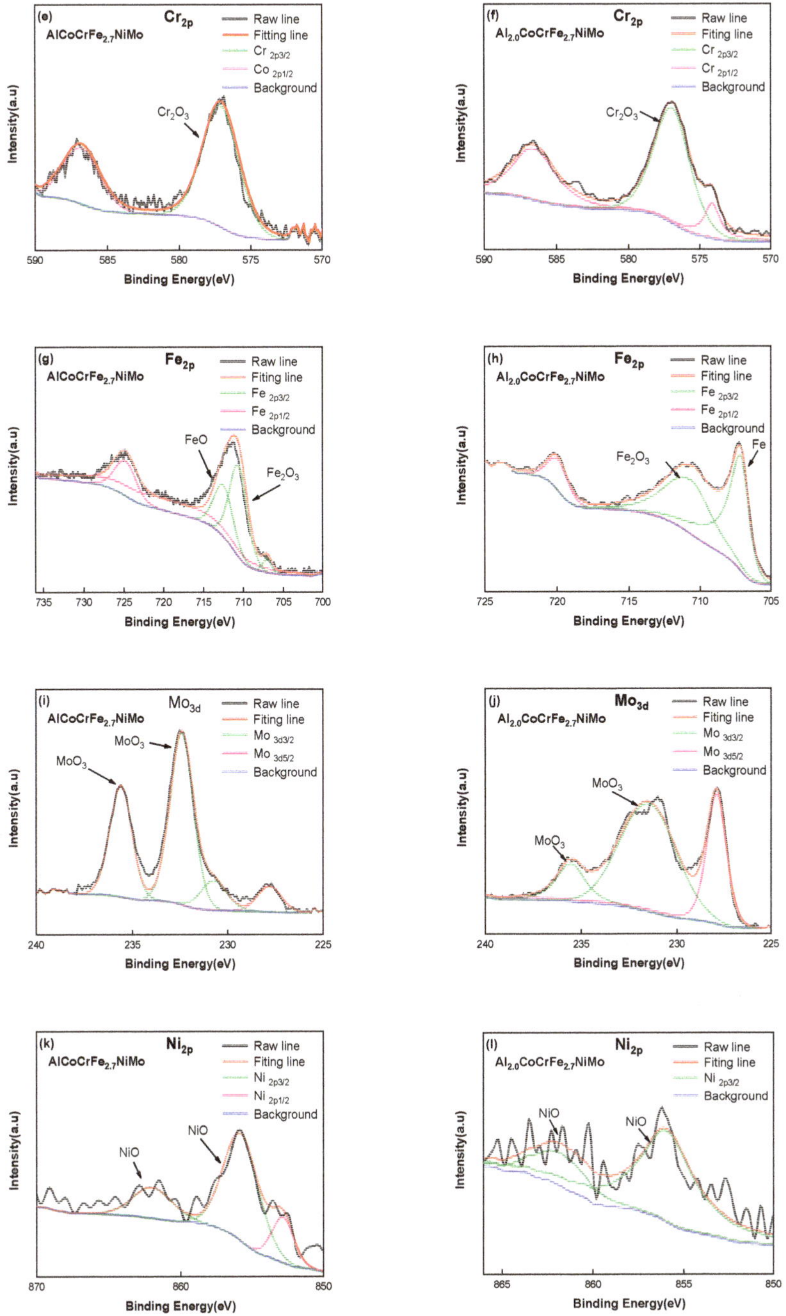

Figure 6. XPS spectra from the passive film formed on the surfaces of the coatings. (**a**,**c**,**e**,**g**,**i**,**k**) $x = 1.0$; (**b**,**d**,**f**,**h**,**j**,**l**) $x = 2.0$.

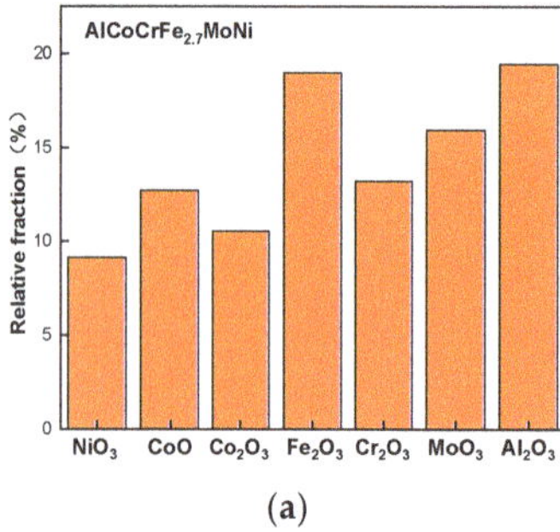

(a)

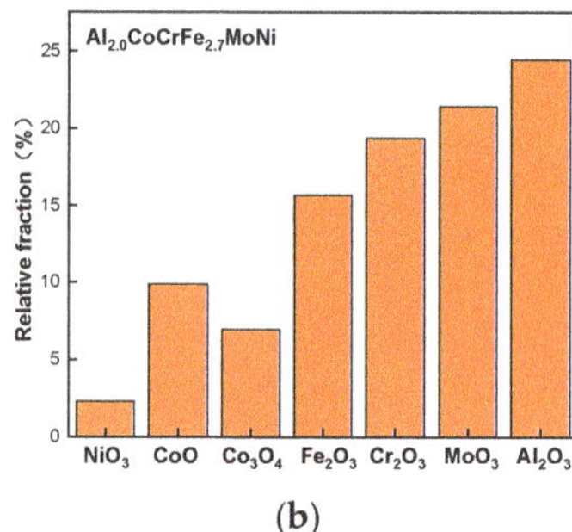

(b)

Figure 7. Composition (in relative at.%) of the surface of $Al_{1.0}$ and $Al_{2.0}$ coatings determined from XPS measurements. (**a**) $x = 1.0$; (**b**) $x = 2.0$.

Table 6. Measured electrical behavior during corrosion of the $Al_xCoCrFe_{2.7}MoNi$ coatings in a 3.5 wt.% NaCl solution.

Alloys	I_{corr} /μA·cm^{-2}	E_{corr} vs. SHE/V
$Al_{0.0}$	2.033×10^{-3}	0.332
$Al_{0.5}$	1.803×10^{-3}	0.581
$Al_{1.0}$	1.355×10^{-3}	0.556
$Al_{1.5}$	2.364×10^{-3}	0.639
$Al_{2.0}$	2.762×10^{-3}	0.586

4. Discussion

4.1. Microstructure and Phase

Results in many studies have shown that the CoCrFeNi alloy consists of a single FCC phase [26,29,30]. Our XRD analysis results show that the $CoCrFe_{2.7}NiMo$ alloy consists of FCC and BCC phases (as shown in Figure 1), which indicates that adding Mo causes formation of the BCC1 phase. This conclusion is consistent with Wu's research [29]. Table 7 shows the enthalpy of mixing between elements. The enthalpy of mixing between Fe and Cr is relatively higher, and it is difficult to form a stable solid solution. Therefore, it is considered that the high melting point of the mixed powders leads to incomplete melting while depositing an Al_0 alloy. The enthalpy of mixing between Al and other elements is lower than that between Mo and other elements, and results from many studies show that adding Al can cause the FCC phase to change to the BCC phase in HEAs [30–32]. Therefore, the FCC phase disappears in the $Al_{1.0}$ alloy and the peak in the XRD pattern becomes more intense due to the increased BCC phase content (BCC1 phase and (Al, Ni)-rich BCC2). Elements such as Co, Ni and Al enrich to form a disordered (Al, Ni)-rich BCC2 phase (random solid solution) as the Al content increases further, as shown in Figure 8a. Meanwhile, Mo dissolves with other elements, e.g., Ni, to form an ordered σ phase (intermetallic), as shown in Figure 8b. A random solid solution tends to have a large configuration entropy due to random mixing of its various elements [1,2]. According to the Gibbs free energy formula $G = H - TS$ (G is the Gibbs free energy, H is the enthalpy of mixing, T is the temperature, and S is the configuration entropy). G is negative when $S > (H/T)$ and the solid solution phase forms easily, and the enthalpy of mixing between Al and other elements is lower than that between Mo and other elements, indicating that other elements dissolve more easily in Al than Mo, forming a solid solution, as shown in Table 7 (the data are derived from the literature [33]). Therefore, the addition of Al generates a large amount of the (Al, Ni)-rich BCC phase, which forces the Mo-containing BCC1 phase to transform into a (Mo, Cr)-rich σ phase. In conclusion, for the alloys studied herein, Mo allows a BCC structure to form more easily when Al is absent in the coating, while both Al and Mo easily form a BCC structure when a small amount of Al is added. Al is the primary driver of BCC formation, and addition of Mo tends to cause formation of the σ phase when more Al is added.

Table 7. Mixing enthalpy values for different element pairs, data from [33].

Element	ΔH_{AB}^{mix}					
	Al	Co	Cr	Fe	Ni	Mo
Al	-					
Co	−19	-				
Cr	−10	−4	-			
Fe	−11	−1	−1	-		
Ni	−22	0	−7	−2	-	
Mo	−5	−5	0	−2	−7	-

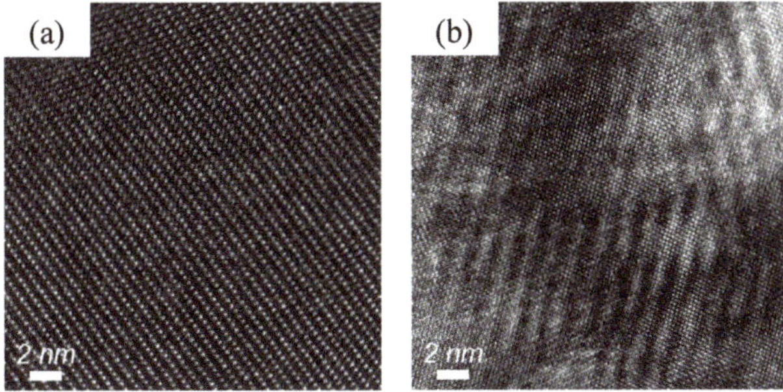

Figure 8. High resolution TEM images of the σ and BCC phases in the $AlCoCrFe_{2.7}NiMo$ coating. (**a**) shows σ phase; (**b**) shows BCC phase.

4.2. Hardness and Wear Resistance

Hardness and wear resistance of alloys are closely related to their microstructures. Figure 9 shows the relationship between the volume fraction of the phases in the coating, obtained by XRD and the hardness of the coating. The presence of the FCC phase minimizes the hardness of the Al_0 alloy, and formation of (Al, Ni)-rich BCC2 causes the hardness of the $Al_{0.5}$ alloy to increase. The volume fraction of BCC2 phase has a greater influence on the hardness and wear resistance of the coatings than does that of σ phase. Greater content of the (Al, Ni)-rich BCC2 phase correlates with higher hardness and wear resistance as the Al content increases. However, the presence of the flaky σ phase causes large flaky exfoliation on the $Al_{1.0}$ coating in wear resistance experiments, which can be attributed to the brittleness of the σ phase. It is worth noting that the σ phase changes from sheet-like to granular as the Al content increases further, which plays a role of dispersion strengthening in the alloy. Furthermore, plastic deformation can be observed in the wear microstructure diagram of the $Al_{1.0}$ coating, while no plastic deformation can be observed in the $Al_{1.5}$ and $Al_{2.0}$ coatings. Oxides appear, which indicates that oxidation wear is the primary wear mechanism, and dispersion strengthening can significantly increase the coating's wear resistance. In summary, introduction of the (Al, Ni)-rich BCC2 phase increases the hardness and wear resistance of the coating. Adding Al reduces the size of the σ phase, which also increases the hardness and wear resistance of the coating.

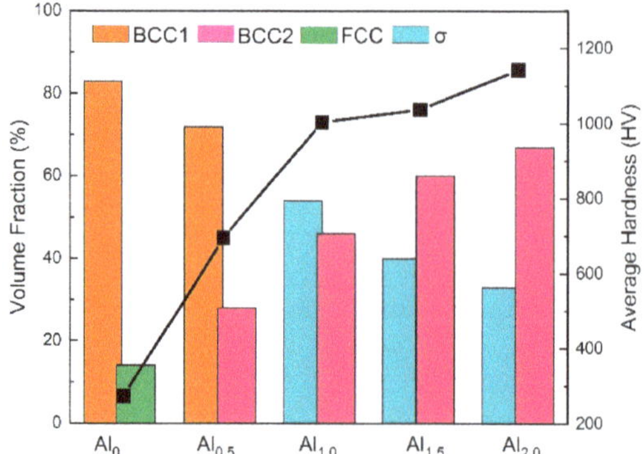

Figure 9. Relationship between the volume fraction of the phases and the average hardness of the coatings.

4.3. Corrosion Resistance

Figure 10 shows a comparison of E_{corr} and I_{corr} in our $Al_xCoCrFe_{2.7}NiMo$ coatings and $Al_xCoCrFeNi$ alloys from the literature [34]. One can see that the addition Mo leads to increased corrosion resistance. This may be attributed to the fact that Mo is prone to produce dense passivation films. In addition, the formation of the σ phase increases the corrosion resistance of the coating, but its dispersion distribution reduces the corrosion resistance of the coating. It is generally believed that Al_2O_3 can effectively resist chloride ion corrosion because of its compact structure. However, increasing the Al content further increases the differences in the content of different metal oxides in the passivation film, which will reduce the coating's corrosion resistance. To date, a definitive understanding of the effect of oxide interaction in passive film on the corrosion resistance of HEAs is yet to emerge. However, the data presented in this paper can provide more important information for other researchers in this emerging field.

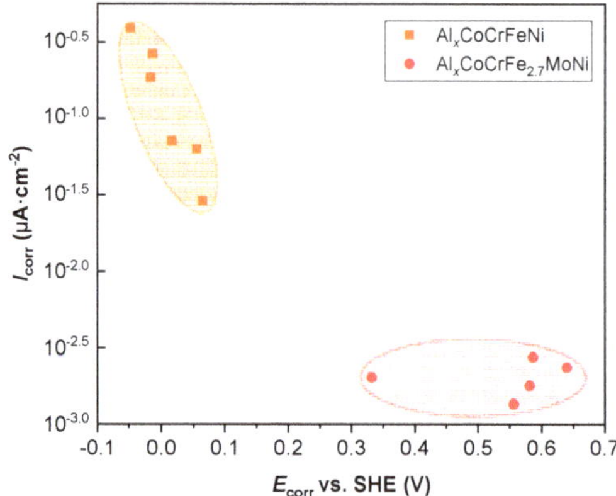

Figure 10. Comparison of the corrosion properties of $Al_xCoCrFe_{2.7}NiMo$ HEAs and $Al_xCoCrFeNi$ HEAs in a 3.5 wt.% NaCl solution.

5. Conclusions

$Al_xCoCrFe_{2.7}MoNi$ coatings were prepared on pure iron via laser cladding, and their microstructure, hardness, wear resistance and corrosion resistance were studied.

The increase of Al content promotes the releasing of Mo from Mo-rich BCC1 phase, and the formation of the (Mo, Cr)-rich σ phase. The increase of Al content causes the increase in volume fraction of BCC2 phase, and correspondingly the increase of hardness and wear resistance. The formation of the strip-shaped σ phase contributes to the improvement of the corrosion resistance of the coating, but the dispersed distribution of the σ phase deteriorates corrosion resistance.

Author Contributions: Conceptualization, M.S. (Minghong Sha), C.J. and S.L.; Funding acquisition, M.S. (Minghong Sha) and S.L.; Formal analysis, C.J.; Investigation, C.J.; Resources, W.F., X.A., Y.-A.J. and M.S. (Minggang Shen); Writing - original draft, C.J.; Writing - review & editing, C.J., M.S. (Minghong Sha) and J.Q.

Funding: This work was financially supported by the National Key Research and Development Program of China (NO.2017YFB0304201), National Natural Science Foundation of China (NO.51774179), Natural Science Foundation of Liaoning Province (NO.20180550546, NO.20170540460), and the Innovation Team Project of Liaoning Education Department (NO.LT2016003).

Conflicts of Interest: The authors declare no conflict of interest.

References

1. Yeh, J.W.; Chen, S.K.; Lin, S.J.; Gan, J.Y.; Chin, T.S.; Shun, T.T.; Tsau, C.H.; Chang, S.Y. Nanostructured high-entropy alloys with multiple principal elements: Novel alloy design concepts and outcomes. *Adv. Eng. Mater.* **2004**, *6*, 299–303. [CrossRef]
2. Senkov, O.N.; Wilks, G.B.; Miracle, D.B.; Chuang, C.P.; Liaw, P.K. Refractory high-entropy alloys. *Intermetallics* **2010**, *18*, 1758–1765. [CrossRef]
3. Miracle, D.B.; Miller, J.D.; Senkov, O.N.; Woodward, C.; Uchic, M.D.; Tiley, J. Exploration and development of high entropy alloys for structural applications. *Entropy* **2014**, *16*, 494–525. [CrossRef]
4. Yeh, J.W. Alloy design strategies and future trends in high-entropy alloys. *JOM* **2013**, *65*, 1759–1771. [CrossRef]
5. Zhang, Y.; Zuo, T.T.; Tang, Z.; Gao, M.C.; Dahmen, K.A.; Liaw, P.K.; Lu, Z.P. Microstructures and properties of high-entropy alloys. *Prog. Mater. Sci.* **2014**, *61*, 1–93. [CrossRef]
6. Gao, M.C.; Liaw, P.K.; Yeh, J.W.; Zhang, Y. *High-Entropy Alloys: Fundamentals and Applications*; Springer: Berlin, Germany, 2016.
7. Lu, Y.; Dong, Y.; Guo, S.; Jiang, L.; Kang, H.; Wang, T.; Wen, B.; Wang, Z.; Jie, J.; Cao, Z. A promising new class of high-temperature alloys: Eutectic high-entropy alloys. *Sci. Rep.* **2014**, *4*, 6200. [CrossRef]
8. Lei, Z.; Liu, X.; Wu, Y.; Wang, H.; Jiang, S.; Wang, S.; Hui, X.; Wu, Y.; Gault, B.; Kontis, P.; et al. Enhanced strength and ductility in a high-entropy alloy via ordered oxygen complexes. *Nature* **2018**, *563*, 546–550. [CrossRef]
9. Huang, H.; Wu, Y.; He, J.; Wang, H.; Liu, X.; An, K.; Wu, W.; Lu, Z. Phase-transformation ductilization of brittle high-entropy alloys via metastability engineering. *Adv. Mater.* **2017**, *29*, 1701678. [CrossRef]
10. He, J.Y.; Wang, H.; Huang, H.L.; Xu, X.D.; Chen, M.W.; Wu, Y.; Liu, X.J.; Nieh, T.G.; An, K.; Lu, Z.P. A precipitation-hardened high-entropy alloy with outstanding tensile properties. *Acta. Mater.* **2016**, *102*, 187–196. [CrossRef]
11. Lu, Y.; Gao, X.; Jiang, L.; Chen, Z.; Wang, T.; Jie, J.; Kang, H.; Zhang, Y.; Guo, S.; Ruan, H.; et al. Directly cast bulk eutectic and near-eutectic high entropy alloys with balanced strength and ductility in a wide temperature range. *Acta Mater.* **2017**, *124*, 143–150. [CrossRef]
12. Lu, Y.; Jiang, H.; Guo, S.; Wang, T.; Cao, Z.; Li, T. A new strategy to design eutectic high-entropy alloys using mixing enthalpy. *Intermetallics* **2017**, *91*, 124–128. [CrossRef]
13. Zhang, S.; Wu, C.L.; Zhang, C.H.; Guan, M.; Tan, J.Z. Laser surface alloying of FeCoCrAlNi high-entropy alloy on 304 stainless steel to enhance corrosion and cavitation erosion resistance. *Opt. Laser. Technol.* **2016**, *84*, 23–31. [CrossRef]
14. Niu, X.; Julius, J.; Dan, Z.; Yang, G.; You, Y. Microstructure and corrosion properties of Al_xFeCoCrNiCu (x=0.25, 0.5, 1.0) thin coatings on steel substrates deposited by electron beam evaporation. *Rare Metal Mater. Eng.* **2017**, *46*, 3621–3625. [CrossRef]
15. Kao, Y.F.; Lee, T.D.; Chen, S.K.; Chang, Y.S. Electrochemical passive properties of Al_xCoCrFeNi (x = 0, 0.25, 0.50, 1.00) alloys in sulfuric acids. *Corros. Sci.* **2010**, *52*, 1026–1034. [CrossRef]

16. Li, Q.H.; Yue, T.M.; Guo, Z.N.; Lin, X. Microstructure and corrosion properties of AlCoCrFeNi high entropy alloy coatings deposited on AISI 1045 steel by the electrospark process. *Metall. Mater. Trans. A* **2013**, *44*, 1767–1778. [CrossRef]
17. Qiao, J.W.; Ma, S.G.; Huang, E.W.; Chuang, C.P.; Liaw, P.K.; Zhang, Y. Microstructural characteristics and mechanical behaviors of AlCoCrFeNi high-entropy alloys at ambient and cryogenic temperatures. *Mater. Sci. Forum.* **2011**, *688*, 419–425. [CrossRef]
18. Sun, R.; Zhang, W.; Fu, H. Effect of solid aluminization on microstructure of AlCoCrFeNi high-entropy alloy. *Heat. Treat. Met.* **2015**, *29*, 250–252. [CrossRef]
19. Manzoni, A.; Daoud, H.; Völkl, R.; Glatzel, U.; Wanderka, N. Phase separation in equiatomic AlCoCrFeNi high-entropy alloy. *Ultramicroscopy* **2013**, *132*, 212–215. [CrossRef]
20. Ma, S.G.; Zhang, Y. Effect of Nb addition on the microstructure and properties of AlCoCrFeNi high-entropy alloy. *Mat. Sci. Eng. A* **2012**, *532*, 480–486. [CrossRef]
21. Na, Y.S.; Lim, K.R.; Chang, H.J.; Kim, J. Effect of trace additions of Ti on the microstructure of AlCoCrFeNi-based high entropy alloy. *Sci. Adv. Mater.* **2016**, *8*, 1984–1988. [CrossRef]
22. Yong, D.; Zhou, K.; Lu, Y.; Gao, X.; Wang, T.; Li, T. Effect of vanadium addition on the microstructure and properties of AlCoCrFeNi high entropy alloy. *Mater. Des.* **2014**, *57*, 67–72. [CrossRef]
23. Cunat, P. *Alloying Elements in Stainless Steel and Other Chromium-Containing Alloys*; ICDA: Paris, France, 2004.
24. Li, X.C.; Dou, D.; Zheng, Z.Y.; Li, J.C. Microstructure and properties of $FeAlCrNiMo_x$ high-entropy alloys. *J. Mater. Eng. Perform.* **2016**, *25*, 2164–2169. [CrossRef]
25. Jiang, L.; Cao, Z.Q.; Jie, J.C.; Zhang, J.J.; Lu, Y.P.; Wang, T.M.; Li, T.J. Effect of Mo and Ni elements on microstructure evolution and mechanical properties of the $CoFeNi_xVMo_y$ high entropy alloys. *J. Alloy. Compd.* **2015**, *689*, 585–590. [CrossRef]
26. Yong, D.; Lu, Y.; Kong, J.; Zhang, J.; Li, T. Microstructure and mechanical properties of multi-component $AlCrFeNiMo_x$ high-entropy alloys. *J. Alloy. Compd.* **2013**, *573*, 96–101. [CrossRef]
27. Liu, W.H.; Lu, Z.P.; He, J.Y.; Luan, J.H.; Wang, Z.J.; Liu, B.; Yong, L.; Chen, M.W.; Liu, C.T. Ductile $CoCrFeNiMo_x$ high entropy alloys strengthened by hard intermetallic phases. *Acta Mater.* **2016**, *116*, 332–342. [CrossRef]
28. Tsai, M.H.; Yuan, H.; Cheng, G.M.; Xu, W.Z.; Jiang, W.W. Significant hardening due to the formation of a sigma phase matrix in a high entropy alloy. *Intermetallics* **2013**, *33*, 81–86. [CrossRef]
29. Wei, W.; Li, J.; Hui, J.; Pan, X.; Cao, Z.; Deng, D.; Wang, T.; Li, T. Phase evolution and properties of $Al_2CrFeNiMo_x$ high-entropy alloys coatings by laser cladding. *J. Therm. Spray. Technol.* **2015**, *24*, 1333–1340. [CrossRef]
30. Chou, H.P.; Chang, Y.S.; Chen, S.K.; Yeh, J.W. Microstructure, thermophysical and electrical properties in $Al_xCoCrFeNi$ ($0 \leq x \leq 2$) high-entropy alloys. *Mat. Sci. Eng. B* **2009**, *163*, 184–189. [CrossRef]
31. Tong, C.J.; Chen, M.R.; Yeh, J.W.; Lin, S.J.; Chen, S.K.; Shun, T.T.; Chang, S.Y. Mechanical performance of the $Al_xCoCrCuFeNi$ high-entropy alloy system with multiprincipal elements. *Metall. Mater. Trans. A* **2005**, *36*, 1263–1271. [CrossRef]
32. Kao, Y.F.; Chen, T.J.; Chen, S.K.; Yeh, J.W. Microstructure and mechanical property of as-cast, -homogenized, and -deformed $Al_xCoCrFeNi$ ($0 \leq x \leq 2$) high-entropy alloys. *J. Alloy. Compd.* **2009**, *488*, 57–64. [CrossRef]
33. Takeuchi, A.; Inoue, A. Classification of bulk metallic glasses by atomic size difference, heat of mixing and period of constituent elements and its application to characterization of the main alloying element. *Mater. Trans.* **2005**, *46*, 2817–2829. [CrossRef]
34. Shi, Y.; Collins, L.; Rui, F.; Zhang, C.; Balke, N.; Liaw, P.K.; Yang, B. Homogenization of $Al_xCoCrFeNi$ high-entropy alloys with improved corrosion resistance. *Corros. Sci.* **2018**, *133*, 120–131. [CrossRef]

Article

Slurry Erosion Behavior of $Al_xCoCrFeNiTi_{0.5}$ High-Entropy Alloy Coatings Fabricated by Laser Cladding

Jianhua Zhao [1,2], Aibin Ma [1,*], Xiulin Ji [2,*], Jinghua Jiang [1] and Yayun Bao [2]

1 College of Mechanics and Materials, Hohai University, Nanjing 210098, China; zhaojh@hhu.edu.com (J.Z.); jinghua-jiang@hhu.edu.cn (J.J.)
2 College of Mechanical and Electrical Engineering, Hohai University, Changzhou 213022, China; paulyayun@163.com
* Correspondence: aibin-ma@hhu.edu.cn (A.M.); xiulinji@gmail.com (X.J.); Tel.: +86-258-378-7239 (A.M.); +86-519-8519-1969 (X.J.)

Received: 7 January 2018; Accepted: 5 February 2018; Published: 11 February 2018

Abstract: High-entropy alloys (HEAs) have gained extensive attention due to their excellent properties and the related scientific value in the last decade. In this work, $Al_xCoCrFeNiTi_{0.5}$ HEA coatings (x: molar ratio, x = 1.0, 1.5, 2.0, and 2.5) were fabricated on Q345 steel substrate by laser-cladding process to develop a practical protection technology for fluid machines. The effect of Al content on their phase evolution, microstructure, and slurry erosion performance of the HEA coatings was studied. The $Al_xCoCrFeNiTi_{0.5}$ HEA coatings are composed of simple face-centered cubic (FCC), body-centered cubic (BCC) and their mixture phase. Slurry erosion tests were conducted on the HEA coatings with a constant velocity of 10.08 m/s and 16–40 meshs and particles at impingement angles of 15, 30, 45, 60 and 90 degrees. The effect of three parameters, namely impingement angle, sand concentration and erosion time, on the slurry erosion behavior of $Al_xCoCrFeNiTi_{0.5}$ HEA coatings was investigated. Experimental results show $AlCoCrFeNiTi_{0.5}$ HEA coating follows a ductile erosion mode and a mixed mode (neither ductile nor brittle) for $Al_{1.5}CoCrFeNiTi_{0.5}$ HEA coating, while $Al_{2.0}CoCrFeNiTi_{0.5}$ and $Al_{2.5}CoCrFeNiTi_{0.5}$ HEA coatings mainly exhibit brittle erosion mode. $AlCoCrFeNiTi_{0.5}$ HEA coating has good erosion resistance at all investigated impingement angles due to its high hardness, good plasticity, and low stacking fault energy (SFE).

Keywords: high-entropy alloy; laser cladding; microstructure; slurry erosion

1. Introduction

Slurry erosion (SE) is a serious concern for hydraulic turbines and other fluid machines due to silt entrained in water flow, especially in the Yellow River regions of China. Slurry erosion results in the surface degradation of flow components of hydroturbine equipment and reduces all efficiency [1,2]. Hydraulic turbine equipment generally made of mild steel, white cast iron or stainless steel. However, these materials are considerably less resistant to erosive wear. It is important to develop new erosion-resistant materials. Recently, high-entropy alloys (HEAs) have attracted extensive attention due to their versatile combinations including high strength and hardness, good thermal stability, excellent corrosion and wear resistance [3–6]. These characteristics make them suitable candidates for structural and functional materials. The main method of preparing high-entropy alloys is vacuum arc melting and then casting [7,8]. As the formation of simple solid solution phase requires high cooling rate, the shape and size of bulk ingots prepared by arc-melting technique are limited. Meanwhile, this preparation method causes high production cost due to many expensive metals such as Ni, Co, and Cr being contained in HEAs. Therefore, some researchers have been turning to explore the effective and economical HEA coating on the low-cost metallic substrate.

Compared to the other processing techniques such as magnetron sputtering, electrochemical deposition, and plasma arc cladding, laser cladding can be used to deposit coatings with thickness more than 1 mm, which is more beneficial for engineering applications. In addition, the coatings can be deposited in a few steps, which eliminates the influence of the substrate and allows gradual composition and property changes through the coating thickness [9]. These favorable advantages have made the laser-cladding alloy attractive among surface modification technologies. Huang et al. [10] prepared TiVCrAlSi HEA coatings on Ti-6Al-4V substrate by laser cladding and investigated the dry sliding wear behavior. The combination of the hard $(Ti,V)_5Si_3$ phase and relatively ductile and tough BCC matrix improved the sliding wear resistance. Yue et al. [11] studied the solidification behavior in laser cladding of AlCoCrCuFeNi high-entropy alloy on magnesium substrates using the Kurz-Giovanola-Trivedi and the Gaümann models. Except for some Cu rejected into the Mg melt, no serious dilution of the HEA composition occurred in the top layer of the coating. This is considered to be important because any dilution of the HEA composition with Mg would likely decrease the corrosion resistance of the HEA. Kunce et al. [12] produced the AlCoCrFeNi high-entropy thin-walled samples using the laser engineered net shaping (LENS) technology. The effect of the cooling rate during solidification on the microstructure of the alloy was studied through different laser-scanning rates. It was found that with an increasing in the laser-scanning rate during the solidification process, the average grain size of the alloy decreased. Vickers microhardness increases with the decrease of the average grain size. AlCoCrFeNiTi alloy system has been investigated in bulk state. Zhou et al. [13] studied the microstructure and strengthening mechanism of the$AlCoCrFeNiTi_{0.5}$ alloy. For $AlCoCrFeNiTi_{0.5}$ alloy, its super-high strength and good plasticity were attributed to its microstructure of intrinsic strong body-centered cubic solid solution, and effective multiple strengthening mechanisms such as solid solution strengthening, precipitation strengthening, and nano-composite strengthening effects, etc. Jiao et al. [14] studied the superior mechanical properties of $AlCoCrFeNiTi_x$ High-Entropy Alloys upon dynamic loading. They found that the ultimate strength and fracture strain of $AlCoCrFeNiTi_x$ alloys are superior to most of bulk metallic glasses and in situ metallic glass matrix composites. However, the slurry erosion properties of $Al_xCoCrFeNiTi_{0.5}$HEA coatings have been rarely studied. In this article, $Al_xCoCrFeNiTi_{0.5}$ HEA coatings with different Al content were fabricated by laser cladding. The effects of Al addition on the microstructure and slurry erosion wear behavior were investigated. It is necessary for practical industrial applications.

2. Experimental Procedure

2.1. Material

As-received Q345 steel plate with dimensions of $25 \times 40 \times 10$ mm^3 was used as the substrate material. The substrate was sandblasted to remove surface contaminants and increased the absorption of laser energy. The $Al_xCoCrFeNiTi_{0.5}$ HEA coatings(x: molar ratio, denoted as $Al_{1.0}$, $Al_{1.5}$, $Al_{2.0}$ and $Al_{2.5}$ alloy, respectively) were prepared in this study by laser cladding. The HEA powder used in the experiment had a high purity (more than 99.5%) with a mesh size of 200–300. The mixed powders with the aid of a high-energy ball milling equipment were pre-placed on the steel specimens with a thickness of approximate 300–400 μm using PVA (Shanghai Zengye Industrial Co., Ltd., Shanghai, China) as a binder. The samples were dried in a vacuum oven (Nanjing Huanke Testing Equipment Co., Ltd., Nanjing, China) at 100 °C for 1h prior to laser cladding. Laser cladding was carried out using an EFW-300 type YAG pulsed laser (Guangda Laser Technology Co., Ltd., Shenzhen, China), which was equipped with a four-axis numerical control working table. With a series of optimization trial runs, the optimized process parameters were obtained: laser power 2.5 kW, laser beam spot diameter 1.2 mm, scanning velocity 1.5 mm·s^{-1}, pulse frequency 20 Hz, pulse width 2.5 ms. High-purity argon gas at a flow rate 5 L·min^{-1} was used as ash ielding gas to prevent oxidation during the cladding experiment. A 50% overlap condition for multi-tracking was employed. Three layers high-entropy alloys were deposited under the same processing parameter.

After laser cladding, metallographic and erosion samples with dimensions of 10 × 10 × 10 mm^3 were cut by electrical sparkle machining. All samples were ground and polished using abrasive papers down to 1200 grit size to obtain a smooth surface. Samples of microstructural observation were etched with alcohol dilute aqua regia. The top surface microstructure was investigated by scanning electron microscopy (SEM, JSM-6360, JEOL Ltd., Tokyo, Japan). The phase composition of HEA coatings was identified by X-ray diffraction (XRD) with a D/max-2550 diffractometer (Rigaku Corporation, Tokyo, Japan) using Cu Ka radiation. The top surface and cross-section morphology of the HEA coatings was examined by SEM. The microhardness of the polished surface of the HEA coatings was performed by a Vickers hardness tester (HXD-1000TC, Shanghai Optical Instrument Factory, Shanghai, China) at a load of 200 g and 15 s loading time. The average of five points was reported for each sample.

2.2. Slurry Erosion Test

Slurry erosion test was performed using man-made jet type rig shown in Figure 1. The test rig provides the flexibility to regulate experiment parameters such as impingement angle, sand concentration, working media and impact velocity. The velocity of the slurry jet is controlled by changing the frequency of the motor converter used for driving the pump. The sand concentration is adjusted by changing the rotation speed of driving motor. The test parameters used for the slurry erosion experiment are shown in Table 1. Irregular sand particles in the size range of 16–40 mesh were used for slurry erosion studies. Slurry with a concentration of 10 kg/m^3 and 30 kg/m^3 was prepared using sand obtained from the Yangtze River Delta. The main composition of river sand is SiO_2. Each sample was tested for 30 min with a cycle time of 5 min. In this study, the distance is 6 cm between the tested specimen and the ejector nozzle. The erosion samples were cleaned thoroughly with industrial acetone solution to remove contaminants and dried. A precision balance to an accuracy level of 0.1 mg was used to measure the mass loss before and after the test at regular intervals. The erosive wear rate is calculated based on the cumulative mass loss of sample with time, i.e., $mg{\cdot}min^{-1}$. The eroded surface characterization was examined by SEM. For comparison, 00Cr16Ni5Mo alloy (denoted as Cr16 alloy), widely used to fabricate various hydraulic turbine components, was tested under the same erosion condition.

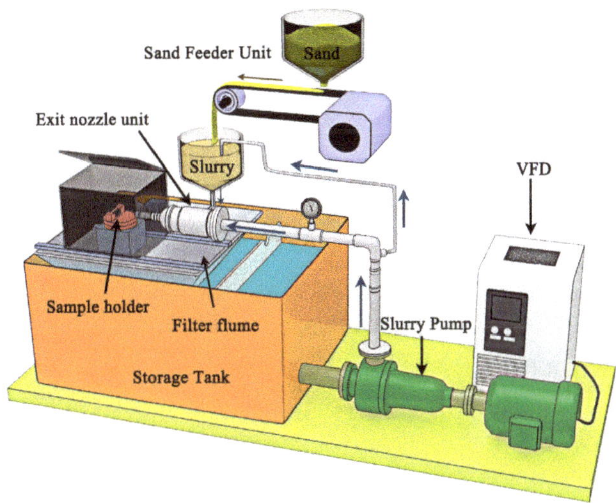

Figure 1. Schematic view of the slurry erosion test rig.

Table 1. Parameters employed for slurry erosion testing.

Parameters (Unit)	Quantitative Value
Impinged angle (°)	15, 30, 45, 60, 90
Impinged velocity (m/s)	10.08
Impinged medium	Tap water mixed with fresh river sand
Nozzle diameter (mm)	8
Erosion time (min)	30

3. Results

3.1. Microstructure and XRD Analysis

The laser-cladding process parameters have great influence on the quality, microstructure, and properties of the HEA coatings. With the aforementioned optimized parameter, $AlCoCrFeNiTi_{0.5}$ HEA coating with few pores could be formed on Q345 substrate as is shown in Figure 2a. It is obvious from Figure 2a that the HEA coating exhibits a typical fish scale lap structure. Figure 2b shows the cross-section SEM image of $AlCoCrFeNiTi_{0.5}$ single-track coating. The bonding line shows a curved shape, rather than a straight line, indicating a good metallurgical bond between the cladding layer and the substrate, which is favorable for the mechanical performance of the coating.

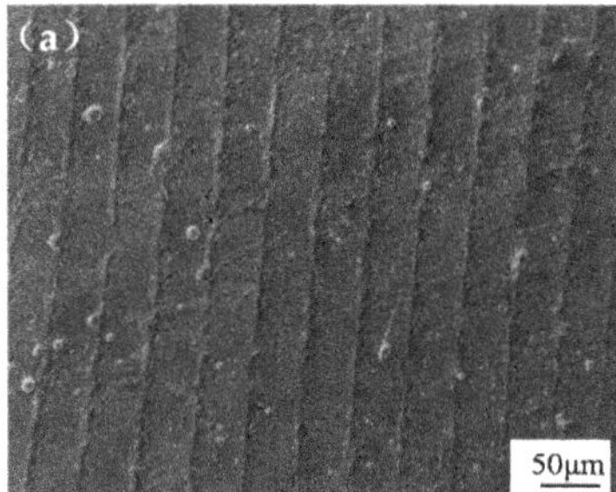

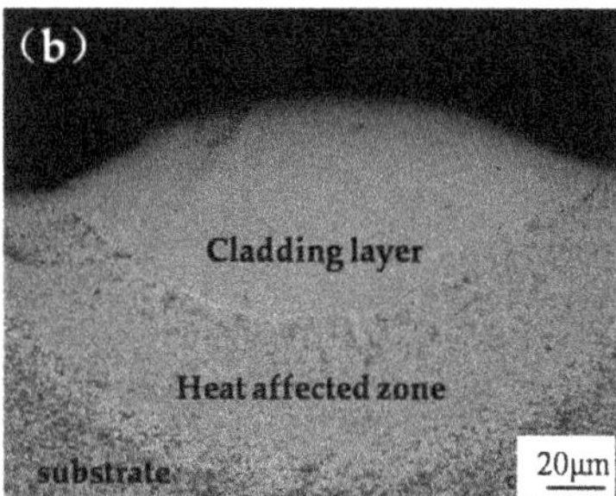

Figure 2. Micro-morphologies of the $AlCoCrFeNiTi_{0.5}$ HEA (High-entropy alloy) coating for (**a**) surface and (**b**) cross-section single-track coating.

The XRD patterns of $Al_xCoCrFeNiTi_{0.5}$ HEA coatings with different Al content are shown in Figure 3. As can be seen, the $Al_xCoCrFeNiTi_{0.5}$ HEA coatings exhibit only simple solid solution structure, specifically face-centered cubic (FCC), body-centered cubic (BCC) and their mixture due to the effect of high mixing entropy [15]. A mixture of FCC + BCC crystal structure is observed in $Al_{1.0}$HEA alloy. The relative intensity of $(110)_B$ peak increases and FCC peak disappears in $Al_{1.5}$HEA alloy. The reflection shift can be partially attributed to the difference of local topologies between FCC and BCC structures [16]. Al has a larger metallic radius, compared with several transition cluster elements such as Co, Cr, Fe, Ni. The increase in the lattice constant with increasing the Al content indicates a corresponding larger lattice-strain effect. To relax the lattice distortion, the metastable FCC phase prefers to transform to a relatively stabilized BCC structure as the Al content in the alloy is increased. Only two BCC phases can be detected in the XRD pattern of the $Al_{2.0}$ and $Al_{2.5}$ HEA alloys. Compared with the XRD pattern of the $Al_{1.5}$ HEA alloy, a minor order BCC peak appears in the $Al_{2.0}$ and $Al_{2.5}$ HEA alloys. The order BCC phase in $Al_xCoCrFeNiTi_{0.5}$ alloy system has been confirmed as NiAl-based intermetallic (IM) phase [17]. The XRD results show that Al addition exhibits a remarkable influence on the phase composition of the $Al_xCoCrFeNiTi_{0.5}$ HEA alloy.

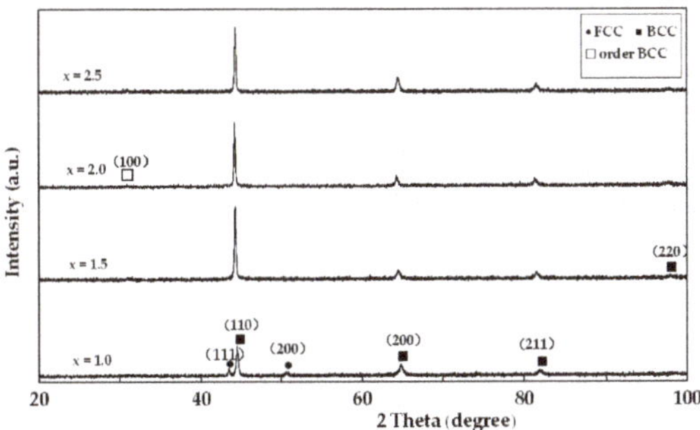

Figure 3. X-ray diffraction patterns of $Al_xCoCrFeNiTi_{0.5}$ HEA coatings.

Figure 4 shows Vickers hardness as a function of Al content. The hardness of the $Al_xCoCrFeNiTi_{0.5}$ HEA coatings exhibited a strong correlation with their aluminum content and phase structure. This suggests that the formation of a BCC type structure is a dominant factor of hardening, and the increase of the relative amount of BCC phase leads to a large increase in hardness. The larger atomic radius, the transformation of the crystal structure and dispersion of nanocrystallite could be responsible for the increased hardness of the alloys [18]. The microhardness of the HEA coatings in this work can reach 667.3 to 801.1 HV, which is at least 1.8 times that of 00Cr16Ni5Mo alloy (370.5 HV).

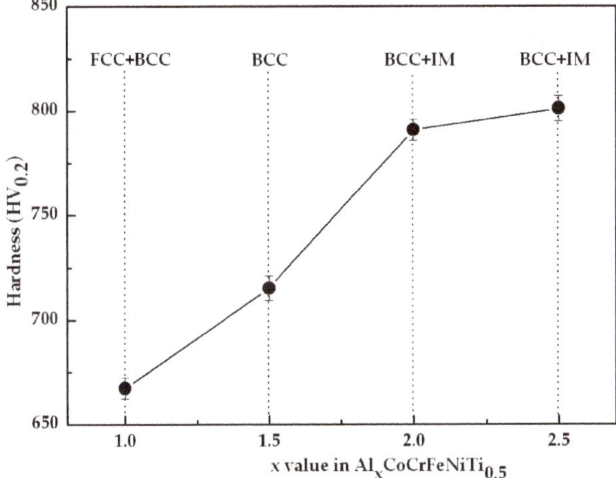

Figure 4. Vickers hardness of $Al_xCoCrFeNiTi_{0.5}$ HEA coatings with different aluminum contents.

Figure 5 presents the SEM images of the $Al_xCoCrFeNiTi_{0.5}$HEA top surface layers. The typical microstructure consists of dendritic (DR) and interdendritic (ID) as a result of faster nucleation and solidification. With the addition of Al content, the solidification structure varies from columnar dendrite to non-equiaxed dendrite grain, and finally to equiaxed dendrite grain. The $Al_{1.0}$ alloy shows the morphology of columnar dendrite and minor non-equiaxed dendrite. When Al content reaches $x = 1.5$, the columnar phase dissolves and single non-equiaxed dendrite appears, which is consistent with the XRD result. The $Al_{2.0}$ and $Al_{2.5}$HEA alloys exhibit the same equiaxed dendrite structure. The morphology features indicate the $Al_{2.0}$ and $Al_{2.5}$HEA alloys should have similar phase composition and solidification behavior.

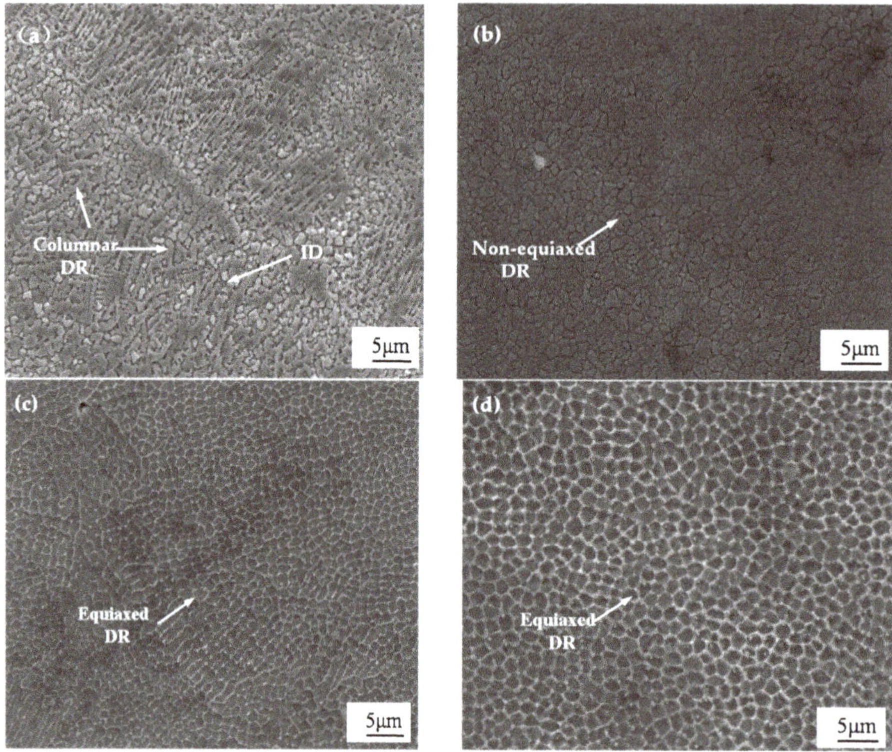

Figure 5. SEM (scanning electron microscopy) images of the as-laser cladding $Al_xCoCrFeNiTi_{0.5}$ HEA coatings: (**a**) $Al_{1.0}$; (**b**) $Al_{1.5}$; (**c**) $Al_{2.0}$; and (**d**) $Al_{2.5}$.

3.2. Results of Slurry Erosion Tests

Figure 6 displays the erosion rate of $Al_xCoCrFeNiTi_{0.5}$ HEA coatings and Cr16 alloy as a function of impingement angle after erosion for 30 min. The $Al_{1.0}$ HEA coating and Cr16 alloy showed the maximum erosion rate at 45°, then the erosion rate decreasing gradually with the impingement angle, which is consistent with the theory of the ductile mode of erosion behavior [19,20]. The erosion rate of the $Al_{1.5}$ HEA coating approached a maximum at 60° and then decreased at 90°. It showed a mixture mode of erosion behavior. In case of the $Al_{2.0}$ and $Al_{2.5}$ HEA coatings, the erosion rate increased monotonically with the impingement angle and arrived at its maximum value at 90° impingement angle, which exhibits the brittle mode of erosion behavior [21]. It is evident from Figure 6 that the $Al_{1.0}$ HEA coating showed smaller erosion rate in comparison with Cr16 alloy at all the investigated impingement angles. The erosion rate of $Al_{1.0}$ HEA coating is 1.78 times lower than Cr16 alloy at 45° impingement angle and 1.68 times lower at 90° impingement angle. The reason behind the high erosion resistances of the $Al_{1.0}$ HEA coating at low impingement angles (from 15° to 45°) is thought to be its higher hardness compared to theCr16 alloy. In slurry erosion wear, the effect of slurry scouring makes the deformation lips or convex bodies easy to be washed away, which results in the more important role of the hardness of target material at a low angle of impingement. At normal impingement angle (90°), the $Al_{1.0}$ and $Al_{1.5}$ HEA coatings still showed significantly lower erosion rate than Cr16 alloy. However, the erosion rate for the $Al_{2.0}$ and $Al_{2.5}$ HEA coatings was similar to theCr16 alloy. Closed to normal impingement, ductility and toughness play a more dominant role [22]. $Al_{1.0}$ HEA alloy, whose yield stress, fracture strength, and plastic strain are as high as 2.26 GPa, 3.14 GPa, and 23.3%, respectively, has the super comprehensive mechanical properties even superior to most of the high-strength alloys such as bulk metallic glasses [23]. It would result in less erosion rate at normal impingement angle. The second possible reason is that $Al_{1.0}$ and $Al_{1.5}$ HEA alloys have lower stacking fault energy (SFE) compared to theCr16 alloy. A decrease of SFE results in an increase of work hardening capability of the material, thereby lowering the material removal rate [24]. Using a combination of discrete Fourier transform (DFT) calculation and XRD analysis, Zaddach et al. [25] reported the SFE of the equiatomic NiFeCoCr alloy to be approximate 20 mJ/m^2, whereas SFE is of the order of 70–80 mJ/m^2 for Cr16 stainless steel [26]. With the Al content increasing, the $Al_xCoCrFeNiTi_{0.5}$HEA coatings showed a transition from the ductile erosion mode to the brittle erosion mode. Limited plasticity likely affects the erosion wear resistance, so the $Al_{2.0}$ and $Al_{2.5}$ HEA coatings with higher hardness as well as bigger brittleness can reduce the erosion wear resistance at normal impingement.

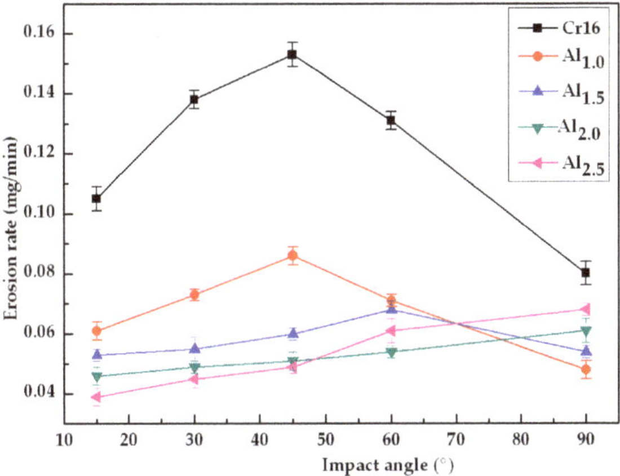

Figure 6. Variation in erosion rate of $Al_xCoCrFeNiTi_{0.5}$ HEA coatings and Cr16 alloy with impact angle at a velocity of 10.08 m/s.

The slurry erosion behavior of $Al_xCoCrFeNiTi_{0.5}$ HEA coatings and Cr16 alloy with sand concentration at 45° and 90° impingement angle is presented in Figure 7. It can be observed that the erosion rates of the test material increased nonlinearly with the increase in the sand concentration, although the percent of the increase is not same for different alloys. Higher sand concentration allows a larger number of sand particles to impact on the surface of the wear specimen, which leads to increase the erosion rate of the material. It is also clear that although sand concentration was increased to 3 times, the erosion rate did not show a similar response. This could be explained by the fact that the shield effect caused by the collisions between the incoming and rebounding particles [27]. Only a portion of particles actually impacts on the target surface while the others lose their way to target surface owing to the interaction between the incoming and rebounding particles. These findings are studied in detail by many researchers [28–30]. For pot-and centrifugal-type impingement experimental rigs, the correlation between erosion rates and sand concentration is highly nonlinear in nature. Some of the investigators have observed the adverse effect of concentration on erosion rate [31,32]. Moreover, as the impingement angle was higher (α = 90°), the shield zone was higher, and therefore the percent increase of erosion rate for the same material at 45° impingement angle is higher than that at 90° impingement angle.

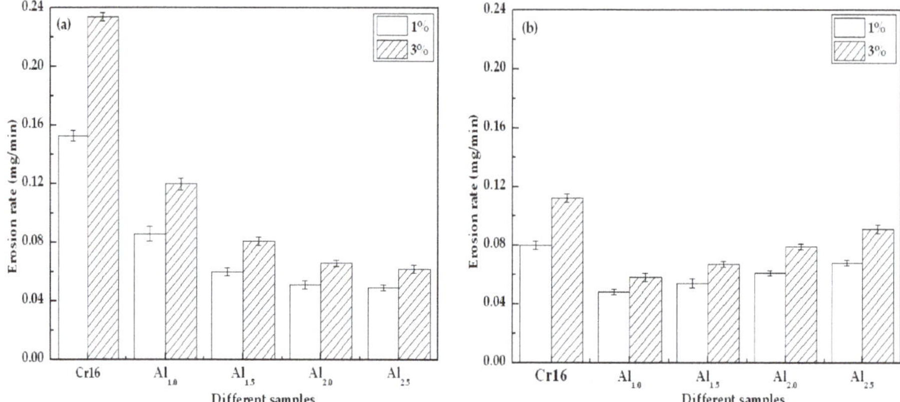

Figure 7. Variation in erosion rate of $Al_xCoCrFeNiTi_{0.5}$ HEA coatings and Cr16 alloy with sand concentration at a velocity of 10.08 m/s and impingement angle of (**a**) 45° and (**b**) 90°.

The slurry erosion behavior of $Al_xCoCrFeNiTi_{0.5}$ HEA coatings and Cr16 alloy with time is shown in Figure 8. The results clearly indicate that the erosion rate of specimens was relatively high during the early cycles with some exceptions, and attained a steady state after 15th or 20th min of testing. Comparison with solid particle erosion, slurry erosion has not obvious incubation period. Similar trends were reported by other researchers [2,33]. The relatively stable erosion rate shows erosion time has no effect on the wear mechanism noticeably during the impingement process. The high erosion rate in the initial stage may be attributed to the initial rough surface of the specimens. The existence of micro-peaks and valleys on the surface of the specimens might have resulted in higher material removal rate during the initial period [34]. The second reason may be work hardening owing to the ductility of sample surface subjected to the repeated impact of sand particles, which results in less material removal and thus reduces the erosion rate.

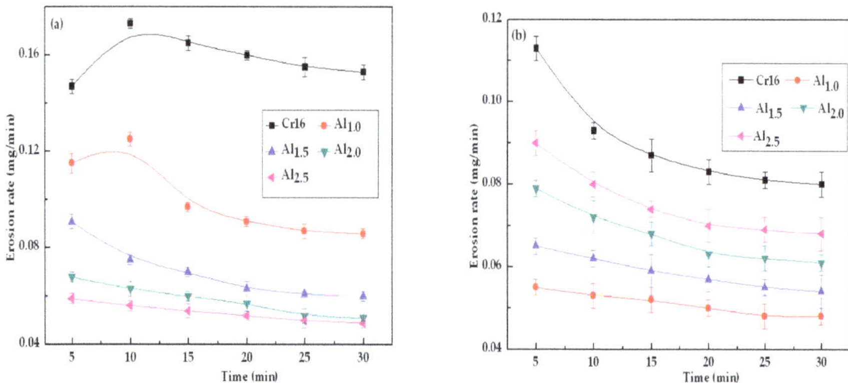

Figure 8. Variation in erosion rate of $Al_xCoCrFeNiTi_{0.5}$ HEA coatings and Cr16 alloy with erosion time at a velocity of 10.08 m/s and impingement angle of (**a**) 45° and (**b**) 90°.

3.3. Observation of Eroded Surfaces

The SEM images showing the eroded surfaces of $Al_xCoCrFeNiTi_{0.5}$ HEA coatings are given in Figures 9 and 10. Detailed examination of the images indicates that microcuting and mixed cutting and ploughing by irregularly shaped erodent particles are responsible for the material removal at 45° impingement angle. Grewal et al. [35] have proposed that with the impact of irregular particles, the primary mode of material removal was a mixture of cutting and ploughing at low impingement angle. Microcutting and mixed cutting-ploughing marks can be seen in Figure 9. At low impingement angle, the normal component of the impact force is too small compared to the tangential one. Material's hardness is a predominant role against these deformation mechanisms. Compared to the hardness of sand particle, which is about 1100 HV, the highest hardness of $Al_xCoCrFeNiTi_{0.5}$ HEA coatings is only 801.3 HV.

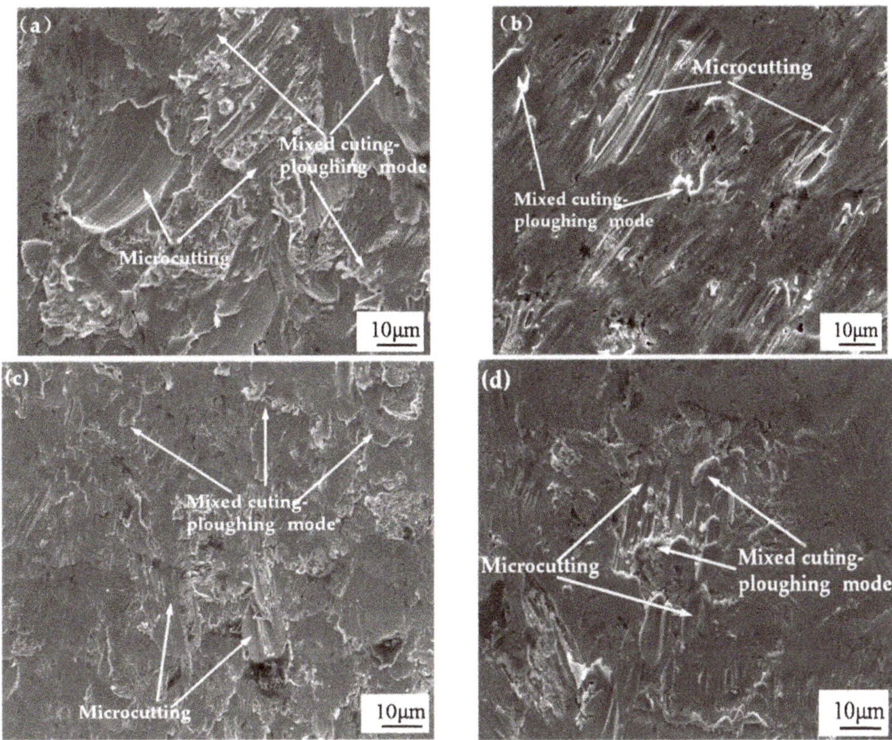

Figure 9. The eroded surface SEM micrographs of $Al_xCoCrFeNiTi_{0.5}$ HEA coatings after 30-min slurry erosion: (**a**) $Al_{1.0}$; (**b**) $Al_{1.5}$; (**c**) $Al_{2.0}$; and (**d**) $Al_{2.5}$ at 10.08 m/s, 1.0 wt. % sand concentration and 45° impingement angle.

Figure 10a,b show the SEM images of eroded surfaces of $Al_{1.0}$ and $Al_{1.5}$HEA coatings. It can be observed that the eroded surfaces showed the presence of many deformed platelets and indentions. The formation of platelets was mainly through indention of impact sand particles. The material extruded from the craters tends to flow outward and accumulates around the periphery in form of platelets, which are removed by a number of subsequent normal impacts of the sand particles. The highly deformed surface of $Al_{1.0}$ and $Al_{1.5}$HEA coatings at 90° impingement indicates significant strain hardening. The major erosion mechanism for $Al_{1.0}$ and $Al_{1.5}$ HEA coatings is the formation and removal of material in the form of platelets at normal impingement angle. Figure 10c,d show the SEM micrographs of the eroded surface of $Al_{2.0}$ and $Al_{2.5}$ HEA coatings at 90° impingement angle. The presence of flattened lips indicates the limited ductility of coatings. Some cracks were also observed, which indicates that the coatings also were removed by brittle fracture. For brittle materials, the energy transfers as a result repeated particle impact results in a fatigue process at or close to normal incidence [36]. Due to high hardness and limited ductility, $Al_{2.0}$ and $Al_{2.5}$HEA coatings were undergone a rapid embrittlement during the continuous impact of erodent particles and fractured easily. In the case, erosion of $Al_{2.0}$ and $Al_{2.5}$ HEA coatings at 90° impingement is carried out by repetitive plastic deformation and brittle fracture. These observations of eroded surfaces appear to be back up the trend in erosion rates as discussed in Figure 6.

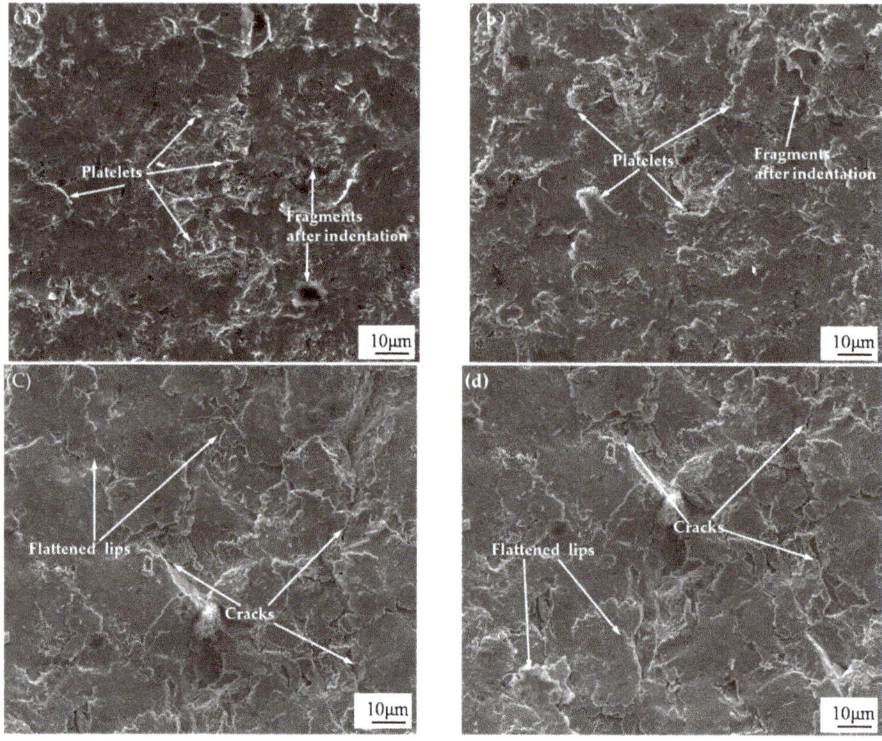

Figure 10. The eroded surface SEM micrographs of $Al_xCoCrFeNiTi_{0.5}$ HEA coatings after 30-min slurry erosion: (**a**) $Al_{1.0}$; (**b**) $Al_{1.5}$; (**c**) $Al_{2.0}$; and (**d**) $Al_{2.5}$ at 10.08 m/s, 1.0 wt. % sand concentration and 90° impingement angle.

4. Conclusions

The phase composition and microstructure of as-laser cladding $Al_xCoCrFeNiTi_{0.5}$ HEA coatings have been studied. The effect of impingement angle, sand concentration and erosion time on the erosion behavior and mechanism of $Al_xCoCrFeNiTi_{0.5}$ HEA coatings were investigated by slurry erosion test. The following conclusions could be made.

(1) $Al_xCoCrFeNiTi_{0.5}$ HEA coatings with few pores and the good metallurgical combination could be fabricated on Q345 substrate by laser cladding with the optimized processing parameters.

(2) The crystal structures of $Al_xCoCrFeNiTi_{0.5}$ HEA coatings evolve from FCC plus BCC mixture phases for x = 1.0 to single BCC phase for x = 1.5, and then to BCC and IM mixture phases for x = 2.0 and 2.5. The microhardness of $Al_xCoCrFeNiTi_{0.5}$ HEA coatings increased obviously with the addition of Al element content. The $Al_{2.5}$ HEA coating has the highest hardness of 801.1 HV.

(3) $AlCoCrFeNiTi_{0.5}$ HEA coating exhibits the ductile erosion mode under slurry erosion, a mixed erosion mode (neither ductile nor brittle) for $Al_{1.5}CoCrFeNiTi_{0.5}$HEA coating, whereas

the $Al_{2.0}CoCrFeNiTi_{0.5}$ and $Al_{2.5}CoCrFeNiTi_{0.5}$HEA coatings exhibit the brittle erosion mode. $AlCoCrFeNiTi_{0.5}$ HEA coating showed good slurry erosion resistance at all the investigated impingement angles due to its high hardness, good plasticity, and low stacking fault energy. The erosion rate of $Al_{1.0}$ HEA coating is 1.78 times lower than Cr16 alloy at 45° impingement angle and 1.68 times lower at 90° impingement angle. The erosion rates of the test materials increase nonlinearly with the increase in the sand concentration at 45° and 90° impingement angles. The erosion time has no effect on the wear mechanism noticeably.

(4) SEM observation confirms the dominant erosion mechanism for all HEA coatings was microcuting and mixed cutting and ploughing at low impingement angle. Platelets were observed to be the primary erosion mechanism for $Al_{1.0}$ and $Al_{1.5}$ HEA coatings at normal impingement angle, compared to repetitive plastic deformation and fatigue fracture being the prevailing material removal phenomenon for $Al_{2.0}$ and $Al_{2.5}$ HEA coatings.

Acknowledgments: The authors thankfully acknowledge to the financial support provided by the National Natural Science Foundation of China (No. 51475140), and the National Natural Science Foundation of China (No. 51774109).

Author Contributions: Aibin Ma and Xiulin Ji conceived and designed the experiments; Jianhua Zhao and Yayun Bao performed the experiments; Jinhua Jiang supervised experimental work and data analysis; Jianhua Zhao wrote the paper.

Conflicts of Interest: The authors declare no conflict of interest.

References

1. Manisekaran, T.; Kamaraj, M.; Sharrif, S.M.; Joshi, S.V. Slurry erosion studies on surface modified 13Cr-4Ni steels: Effect of angle of impingement and particle size. *J. Mater. Eng. Perform.* **2007**, *16*, 567–572. [CrossRef]
2. Nguyen, Q.B.; Lim, C.Y.H.; Nguyen, V.B.; Wan, Y.M.; Nai, B.; Zhang, Y.W.; Gupta, M. Slurry erosion characteristics and erosion mechanisms of stainless steel. *Tribol. Int.* **2014**, *79*, 1–7. [CrossRef]
3. Varalakshmi, S.; Rao, G.A.; Kamaraj, M.; Murty, B.S. Hot consolidation and mechanical properties of nanocrystalline equiatomic AlFeTiCrZnCu high entropy alloy after mechanical alloying. *J. Mater. Sci.* **2010**, *45*, 5158–5163. [CrossRef]
4. Wu, Z.; Bei, H.; Pharr, G.M.; George, E.P. Temperature dependence of the mechanical properties of equiatomic solid solution alloys with face-centered cubic crystal structures. *Acta Mater.* **2014**, *81*, 428–441. [CrossRef]
5. Soare, V.; Mitrica, D.; Constantin, I.; Badilita, V.; Stoiciu, F.; Popescu, A.-M.J.; Carcea, I. Influence of remelting on microstructure, hardness and corrosion behaviour of AlCoCrFeNiTi high entropy alloy. *Mater. Sci. Technol.* **2015**, *31*, 1194–1200. [CrossRef]
6. Shi, Y.Z.; Yang, B.; Liaw, P.K. Corrosion-resistance high-entropy alloys: A review. *Metals* **2017**, *7*, 43. [CrossRef]
7. Chuang, M.H.; Tsai, M.H.; Wang, W.R.; Lin, S.J.; Yeh, J.W. Microstructure and wear behavior of $Al_xCo_{1.5}CrFeNi_{1.5}Ti_y$ high-entropy alloys. *Acta Mater.* **2011**, *59*, 6308–6317. [CrossRef]
8. Senkov, O.N.; Wilks, G.B.; Miracle, D.B.; Chuang, C.P.; Liaw, P.K. Refractory high-entropy alloys. *Intermetallics* **2010**, *18*, 1758–1765. [CrossRef]
9. Ocelik, V.; Janssen, N.; Smith, S.N.; De Hosson, J.T.M. Additive manufacturing of high-entropy alloys by laser processing. *JOM* **2016**, *68*, 1810–1818. [CrossRef]
10. Huang, C.; Zhang, Y.Z.; Vilar, R.; Shen, J.Y. Dry sliding wear behavior of laser clad TiVCrAlSi high entropy alloy coatings on Ti-6A1-4V substrate. *Mater. Des.* **2012**, *41*, 338–343. [CrossRef]
11. Yue, T.M.; Xie, H.; Lin, X.; Yang, H.O.; Meng, G.H. Solidification behaviour in laser cladding of AlCoCrCuFeNi high-entropy alloy on magnesium substrates. *J. Alloys Compd.* **2014**, *587*, 588–593. [CrossRef]
12. Kunce, I.; Polansk, M.; Karczewski, K.; Plocinski, T.; Kurzydlowski, K.J. Microstructural characterisation of high-entropy alloy AlCoCrFeNi fabricated by laser engineered net shaping. *J. Alloys Compd.* **2015**, *648*, 751–758. [CrossRef]
13. Zhou, Y.J.; Zhang, Y.; Kim, T.N.; Chen, G.L. Microstructure characterizations and strengthening mechanism of multi-principal component $AlCoCrFeNiTi_{0.5}$ solid solution alloy with excellent mechanical properties. *Mater. Lett.* **2008**, *62*, 2673–2676. [CrossRef]

14. Jiao, Z.M.; Ma, S.G.; Chu, M.Y.; Yang, H.J.; Wang, Z.H.; Zhang, Y.; Qiao, J.W. Superior mechanical properties of $AlCoCrFeNiTi_x$ high-entropy alloys upon dynamic loading. *J. Mater. Eng. Perform.* **2015**, *25*, 451–456. [CrossRef]
15. Guo, S.; Liu, C.T. Phase stability in high entropy alloys: Formation of solid-solution phase or amorphous phase. *Prog. Nat. Sci. Mater. Int.* **2011**, *21*, 433–446. [CrossRef]
16. Tang, Z.; Gao, M.C.; Diao, H.Y.; Yang, T.F.; Liu, J.P.; Zuo, T.T.; Zhang, Y.; Lu, Z.P.; Cheng, Y.Q.; Zhang, Y.W.; et al. Aluminum alloying effects on lattice types, microstructures, and mechanical behavior of high-entropy alloys systems. *JOM* **2013**, *65*, 1848–1858. [CrossRef]
17. Zhang, K.B.; Fu, Z.Y. Effects of annealing treatment on phase composition and microstructure of $CoCrFeNiTiAl_x$ high-entropy alloys. *Intermetallics* **2012**, *22*, 24–32. [CrossRef]
18. Tong, C.J.; Chen, M.R.; Chen, S.K.; Yeh, Y.W.; Shun, T.T.; Lin, S.J.; Chang, S.Y. Mechanical performance of the $Al_xCoCrCuFeNi$ high-Entropy alloy system with multiprincipal elements. *Metall. Mater. Trans. A* **2005**, *36*, 263–1271.
19. Wang, Y.F.; Yang, Z.G. Finite element model of erosive wear on ductile and brittle materials. *Wear* **2008**, *265*, 871–878. [CrossRef]
20. Okonkwo, P.C.; Shakoor, R.A.; Zagho, M.M.; Mohamed, A.M.A. Erosion behavior of API X100 pipeline steel at various impact angles and particle speeds. *Metals* **2016**, *6*, 232. [CrossRef]
21. Wen, D.C. Erosion and wear behavior of nitrocarburized DC53 tool steel. *Wear* **2010**, *268*, 629–636. [CrossRef]
22. Arora, H.S.; Grewal, H.S.; Singh, H.; Mukherjee, S. Zirconium based bulk metallic glass—Better resistance to slurry erosion compared to hydroturbine steel. *Wear* **2013**, *307*, 28–34. [CrossRef]
23. Zhou, Y.J.; Zhang, Y.; Wang, Y.L.; Chen, G.L. Solid solution alloys of $AlCoCrFeNiTi_x$ with excellent room-temperature mechanical properties. *Appl. Phys. Lett.* **2007**, *90*, 181904. [CrossRef]
24. Nair, R.B.; Selvam, K.; Arora, H.S.; Mukherjee, S.; Singh, H.; Grewal, H.S. Slurry erosion behavior of high entropy alloys. *Wear* **2017**, *386–387*, 230–238. [CrossRef]
25. Zaddach, A.J.; Niu, C.; Koch, C.C.; Irving, D.L. Mechanical properties and stacking fault energies of NiFeCrCoMn high-entropy alloy. *JOM* **2013**, *65*, 1780–1789. [CrossRef]
26. Das, A. Revisiting stacking fault energy of steels. *Metall. Mater. Trans. A* **2016**, *47*, 748–768. [CrossRef]
27. Zhao, J.; Zhang, G.C.; Xu, Y.J.; Wang, R.H.; Zhou, W.D.; Han, L.X.; Zhou, Y. Mechanism and effect of jet parameters on particle water jet rock breaking. *Power Technol.* **2017**, *313*, 231–244. [CrossRef]
28. Burzynski, T.; Papini, M. Analytical models of the interference between incident and rebounding particles within an abrasive jet: Comparison with computer simulation. *Wear* **2007**, *263*, 1593–1601. [CrossRef]
29. Ciampini, D.; Spelt, J.K.; Papini, M. Simulation of interference effects in particle streams following impact with a flat surface: Part II. Parametric study and implications for erosion testing and blast cleaning. *Wear* **2003**, *254*, 237–249. [CrossRef]
30. Goyal, D.K.; Singh, H.; Kumar, H. An overview of slurry erosion control by the application of high velocity oxy fuel sprayed coatings. *J. Eng. Tribol.* **2011**, *225*, 1092–1105. [CrossRef]
31. Padhy, M.K.; Saini, R.P. Effect of size and concentration of silt particles on erosion of Pelton turbine buckets. *Energy* **2009**, *34*, 1477–1483. [CrossRef]
32. Grewal, H.S.; Arora, H.S.; Agrawa, A.; Singha, H.; Mukherjee, S. Slurry erosion of thermal spray coatings: Effect of sand concentration. *Procedia Eng.* **2013**, *68*, 484–490. [CrossRef]
33. Singh, H.; Goyal, K.; Goyal, D.K. Experimental investigations on slurry erosion behaviorof HVOF and HVOLF sprayed coatings on hydraulic turbine steel. *Trans. Indian Inst. Met.* **2017**, *70*, 1585–1592. [CrossRef]
34. Goyal, D.K.; Singh, H.; Kumar, H.; Sahni, V. Slurry erosion behavior of HVOF sprayed WC-10Co-4Cr and Al_2O_3 + $13TiO_2$ coatings on a turbine steel. *Wear* **2012**, *289*, 46–57. [CrossRef]
35. Grewal, H.S.; Agrawal, A.; Singh, H. Slurry erosion mechanism of hydroturbine steel: Effect of operating parameters. *Tribol. Lett.* **2013**, *52*, 287–303. [CrossRef]
36. Wheeler, D.W.; Wood, R.J.K. Erosion of hard surface coatings for use in offshore gate valves. *Wear* **2005**, *258*, 526–536. [CrossRef]

 *metals*

Article

Preparation and Performance Analysis of Nb Matrix Composites Reinforced by Reactants of Nb and SiC

Zhen Lu *, Chaoqi Lan, Shaosong Jiang, Zhenhan Huang and Kaifeng Zhang

National Key Laboratory for Precision Heat Processing of Metals, Key laboratory of Micro-Systems and Micro-Structures Manufacturing Ministry of Education, Department of Materials Science and Engineering, Harbin Institute of Technology, Harbin 150001, China; xiayilcq@163.com (C.L.); jiangshaosong@hit.edu.cn (S.J.); huangfengzhenhan@163.com (Z.H.); kfzhang@hit.edu.cn (K.Z.)

* Correspondence: luzhen-hit@163.com

Received: 5 February 2018; Accepted: 27 March 2018; Published: 3 April 2018

Abstract: In this paper, one kind of new composite material formed with Nb and SiC was prepared by hot pressing sintering. The influence of the addition of SiC particles on the mechanical properties at room and high temperature was analyzed. The composite material consists of three phases: Nb_2C, Nb_3Si, and Nb solid solution (Nbss). The fraction of SiC particles added in the Nb matrix was 3%, 5%, and 7%, respectively. Flexural strength, Vickers hardness, and compressive strength at room temperature were improved with the increasing of SiC content. Among them, compressive strength and fracture toughness were higher than those of Nb/Nb_5Si_3 composites. The compressive strength at high temperature of the new composites was higher than that of Nb-Si alloys, which improved with the increasing of SiC content.

Keywords: Nb/SiC composite material; hot pressing sintering; microstructure; mechanical property

---

1. Introduction

In recent decades, greater demands have been placed on high-temperature structural materials with the rapid development of the aerospace field. At present, there are many high temperature metal materials that can be used under 1100 °C, including superalloy, TiAl intermetallic compound, titanium alloy, and others. Among them, the maximum service temperature of single crystal Ni-based superalloy is 1100 °C. On the other hand, there are some materials (W, Mo, Ta, Zr, Nb alloys) that can be used between 1100 °C and 2000 °C. The superhigh temperature materials still have a high strength at this temperature range. Ni-based alloys are most widely used in high-temperature components, but their operating temperature is close to the limiting temperature due to their melting point [1,2]. Refractory metals have received great attention because of their high melting point. Among the refractory metals, niobium alloy has the lowest density, and has attracted more attention from the public. In addition, Nb has stable physical and chemical properties, good corrosion resistance, and great ductility and machinability. Niobium alloy is expected to become a promising candidate for high-temperature applications [3].

Niobium alloy of Nb matrix composites are widely researched by scientists. The laminated Nb/Nb_5Si_3 composites had been synthesized in situ by Yu et al. [4]. Nb_5Si_3 phase was the reinforcement and Nbss had great plasticity, which could enhance the fracture toughness. In addition, Yu et al. [5] synthesized in situ Nb-silicide composite by powder metallurgy. Its nominal composition was Nb-16Si-10Ti-10Mo-5Hf. Solid solution strengthening and silicide hardening were the important strengthening mechanisms at high temperature. The reinforcements were Nb_5Si_3 and a small amount of Nb_3Si. Zhang et al. [6] studied the microstructure and properties of the Nb-Ti-C-Si system. The results indicated that borides and carbides effectively increased the strength and hardness of the composites, while Nbss toughened the composites. Higher boron content increased the hardness and strength, but decreased the toughness. These studies show that the addition of reinforcements can significantly enhance the Nbss. The hardness of SiC particles is much higher than that of Si particles. Comparing to Si particles, the SiC particles are more easily surrounded by Nb particles. So, the mixed powders after ball milling are more uniform, the particles are finer and the structure of sintered material is more homogeneous. In addition, a study by Sha et al. [7] showed that Nb-based composites containing silicide and carbide phases had higher high-temperature yield strength and Vickers hardness than the composites with a single-silicide reinforced phase. There is little research on Nb-based materials reinforced by a mixture of carbide and silicide. Therefore, it is expected to produce a new type of composite with higher strength and better properties by adding SiC to the Nb solid solution. The vacuum melting method is the most common method to produce Nb-based composites [8,9]. However, composite materials produced by the vacuum melting method have shrinkage, poor density, and other defects. Beyond that, the reinforcements are prone to uneven distribution, which reduces the service performance of these materials. Hot pressing sintered materials have higher relative density and uniform structure [10]. However, research on preparing Nb-based composites with reinforcements of silicide and carbide phases is rare.

In the present study, Nb matrix composites reinforced by reactants of Nb and SiC (with SiC fractions of 3%, 5%, and 7%) were produced by hot pressing sintering technology. The effects of the amount of additional SiC on the microstructure and mechanical properties of composites were studied.

2. Experimental Procedure

Figure 1 shows the powder morphology of the original Nb and SiC powders, which was detected by scanning electron microscopy (SEM, FEI corporation, Hillsboro, OR, USA). It can be seen that the original Nb powders had irregular shape. The average particle size of Nb was 60 μm, while the particle size of SiC powders was 2 μm. The Nb-xSiC powders (x = 3%, 5%, 7% molar ratio) were ball milled in a QM-BP planetary ball mill under Ar protective environment. The powders, weighing 30 g, were put together with steel balls. The ball-to-powder ratio was 15:1. In order to prevent the powder from welding during the ball milling, 20 mL ethanol was added to the pot. The ball milling time was 25 h and the speed was 250 r/min. The powders, after milling, were sintered at 1550 °C and 30 MPa for 1 h in the vacuum hot press sintering furnace. The sintered materials were marked as Nb-3SiC, Nb-5SiC, and Nb-7SiC.

Figure 1. Morphology of original powders: (**a**) original Nb powders; (**b**) original SiC powders.

Samples were taken from the sintered materials for property testing and microstructure observation. Actual density was measured according to the Archimedean drainage method. Flexural strength and fracture toughness at room temperature were determined on the Instron-1186 universal testing machine. Flexural strength at room temperature was defined by a three-point bending test on specimens with 16 mm span length (total length = 20 mm), 4 mm width, and 3 mm thickness. Fracture toughness was determined by the single-edge notched beam three-point bending test on specimens with 16 mm span length (total length = 20 mm), 4 mm width, 2 mm thickness, and a notch (2 mm depth). In the above two experiments, the loading speed of the pressing head was 0.5 mm/min, and the final result was the average value from five testing samples. Vickers hardness was measured on a MICRO-586 Vickers hardness tester with a load of 4 kg and a pressure of 15 s. The average hardness value was obtained from five indents on each sample.

High-temperature compression tests were performed on a Gleeble-1500D thermal simulation tester (DSI corporation, Poestenkill, NY, USA). The sample for testing was a cylinder with 4 mm diameter and 6 mm height, and the axis direction was the same as the hot pressing direction. Compression experiments were carried out at 1050 °C, 1100 °C, and 1150 °C. The strain rate was $1 \times 10^{-3} \cdot s^{-1}$. The deformation was 20% (true strain was 0.223), the heating rate was 20 °C/s, and the holding time was 20 s.

Scanning electron microscopy was used to observe the morphology of the powders during the different ball milling stages. Backscattered electron (BSE) images were produced to reveal the microstructure of the hot-pressed materials and high-temperature compressed samples. The samples were wire-cut from the material after hot-pressing and high-temperature compressing, then sanded with sandpaper. The final specimens were obtained by ion beam polishing. X-ray diffraction (XRD, Panalytical corporation, Almelo, Holland) was used to analyze the phase composition of the composite powders after milling and the sintered materials. Different stages of mechanically alloyed powders were placed on glass slides with square grooves, flatting these powders and determining the phase compositions. For sintered materials, after grinding and polishing the sintered block samples, the phase compositions were measured on the polished surface.

3. Results and Discussion

3.1. Preparation of Nb/SiC Composite Powders

Ball milling has the benefit of obtaining a dense and uniform structure. By studying the micromorphology of composite powders, we can observe the refinement of powders and obtain the optimum milling time. Figure 2 shows the morphologies of composite powders with different ball milling times at a speed of 250 r/min. The morphology of the pure Nb powders after 20 h milling is shown in Figure 2a. The Nb particles were obviously refined into slices and sticks in comparison with the initial powders (10–20 μm). As shown by red arrows in Figure 2a, cracks formed in the Nb powders. After adding SiC and ball milling for 25 h, the powders were obviously refined. Most of the powders were in the form of crumbs (3–4 μm), and the particles were evenly distributed. Continuing milling to 30 h, Nb powders were not further refined.

Figure 2. Scanning electron microscope micrographs of Nb/SiC composite powders ball milled for (**a**) 20 h; (**b**) 25 h; (**c**) 30 h.

The results of XRD analysis of Nb and SiC powders milled for different times are shown in Figure 3. Compared with the XRD patterns of the original Nb powders, it can be seen that the diffraction peaks of Nb powders became broader and the intensities decreased when the Nb powders were milled for 25 h. This is because Nb powders undergo strong plastic deformation during the milling process, which causes lattice distortions and increased dislocation, resulting in an increased amount of subgrains and an area of subgrain boundaries [11]. When the number of subgrains increases to a certain extent, the subgrains will transform into grains so that the grains will be refined. Grain refinement and lattice distortion will lead to broadening diffraction peaks. Meanwhile, grain refinement can also lead to reducing diffraction peak intensity [12].

When powders were milled for 30 h, the secondary peak showed in the left of the main peak of Nb powders, indicating that the main diffraction peaks of Nb powders began to move toward the direction of low angle and the lattice constant of Nb powders increased [13]. This is because the iron element inevitably enters the composite powders during ball milling because of the stainless-steel balls and cans used. The atomic radius of Fe is 1.72 Å and the atomic radius of Nb is 2.08 Å. If an Fe atom enters the lattice interstices of Nb to form an interstitial solid solution, the lattice constant of Nb will increase and the diffraction peak will shift toward the lower angle. As shown in Figure 3, the diffraction peak width of Nb powders does not increase significantly from 25 to 30 h, indicating that the particle size of Nb powders does not change significantly during this process. In addition, the longer the milling time, the more impurities will be introduced. Based on the above analysis, 25 h is the optimal milling time.

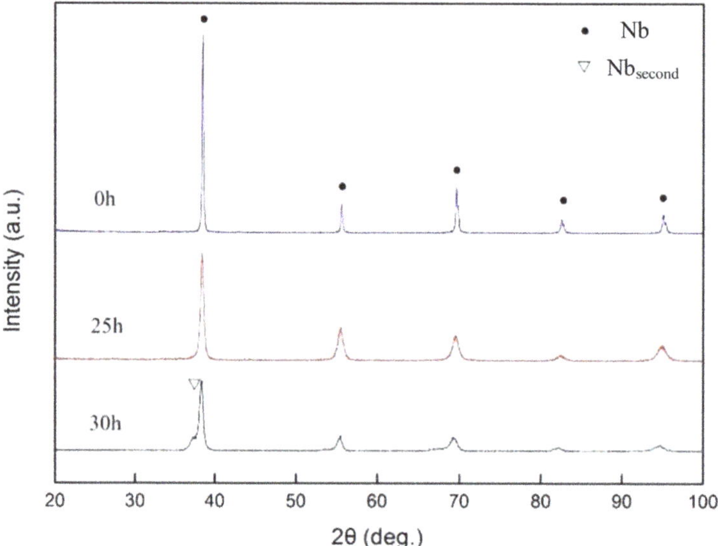

Figure 3. X-ray diffraction (XRD) patterns of composite powders after different milling times.

Figure 4 shows the XRD patterns of three composite powders after ball milling for 25 h. It can be seen from the figure that no diffraction peak of SiC was detected. This is because at the initial stage of ball milling, Nb particles turn into sheets due to plastic deformation. The brittle SiC powders not only have little content, but also have small particle size (generally no more than 2 μm). The SiC powder particles are easily broken into particles that are less than 1 μm under the high-speed impact of the ball. Only a few SiC powders embed into Nb powders. Pressing composite powders and making samples for the XRD test at this time, the flaky Nb particles will cover small SiC particles, which makes it difficult to detect the diffraction peak of SiC.

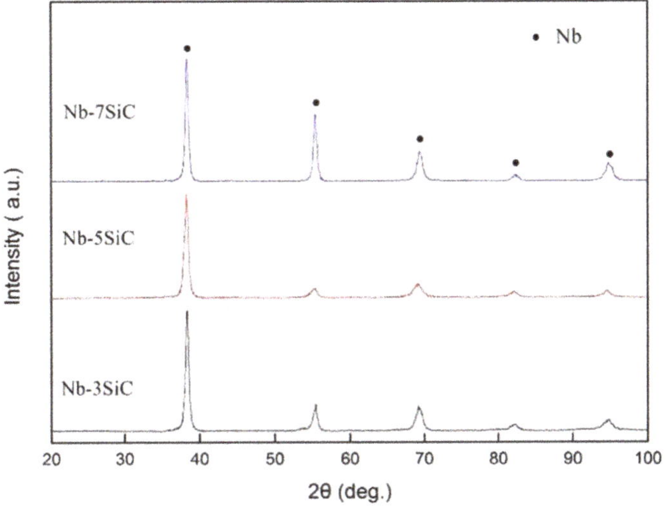

Figure 4. XRD patterns of composite powders after milling for 25 h.

3.2. Microstructure and Mechanical Properties at Room Temperature

3.2.1. XRD Patterns and Microstructure of Sintered Material

XRD patterns of the sintered materials are shown in Figure 5. Compared with the XRD pattern of the milled composite powders, a new phase was formed in the sintered materials. The diffraction peaks of Nb_3Si, Nbss, and Nb_2C can be detected in the XRD patterns of the three kinds of sintered materials. The diffraction peaks of Nb-Si intermetallic compound were not detected in the XRD pattern after ball milling for 25 h (Figure 3). Therefore, the Nb-Si intermetallic compound found in the sintered materials was generated by the reaction of Nb and SiC in the hot pressing process. Sintered materials consisted of three phases: Nbss, Nb_3Si, and Nb_2C. Comparing the XRD patterns of the three composites, the diffraction peak of Nbss was strongest and the area the peak covered was the largest, indicating that there was the most content of Nbss in three phases, and the three composite materials were based on Nbss.

Figure 6 shows BSE images of the three kinds of composite. The material was dense, and the grains were more refined with increasing SiC content. The average grain size of Nb-3SiC was 10 μm, the average crystal size of Nb-5SiC was 8 μm, and the average grain size of Nb-7SiC was 4 μm. The composite materials consisted of three contrasting phases of black, gray, and white. It was known from the XRD patterns that the material consisted of Nb_2C, Nb_3Si, and Nbss phases. Generally it is well known that phases containing heavier elements exhibit brighter contrast in BSE images, thus the black grains are Nb_2C, the white grains are Nbss, and the gray grains are Nb_3Si. From the image it can be observed that the black grains were always adjacent to the gray grains, indicating that SiC particles react with the surrounding Nbss during hot press sintering. C and Si atoms diffuse locally: C atoms diffuse into the black grains, and Si atoms diffuse into the gray grains. The original Si and C atoms are adjacent, so the Nb_2C and Nb_3Si grains produced by the reaction are also adjacent, which proves that the black grains are also Nb_2C. In addition, certain areas in Nb_2C were darker in color, indicating that C was locally enriched in these regions and was unevenly distributed in the Nb_2C grains.

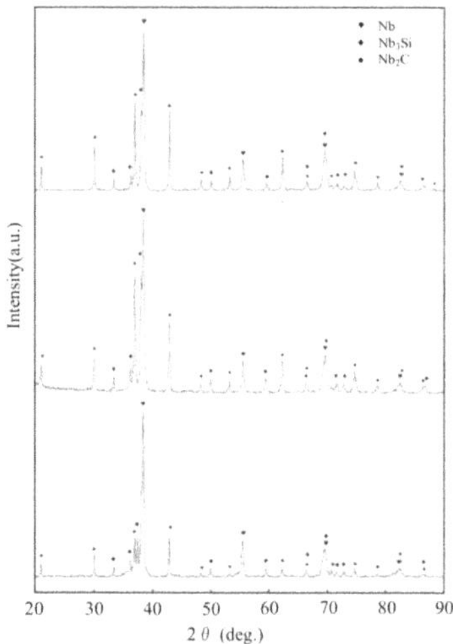

Figure 5. XRD patterns of Nb-xSiC composites.

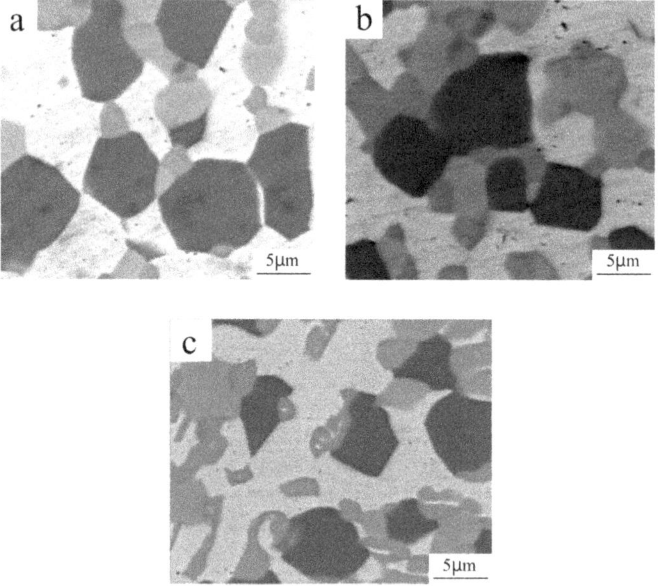

Figure 6. Backscattered electron images of the three composites: (**a**) Nb-3SiC; (**b**) Nb-5SiC; (**c**) Nb-7SiC.

3.2.2. Mechanical Properties of Nb Matrix Composites at Room Temperature

Adding SiC to Nbss, mixed reinforcement consisting of carbide and silicide would be produced, which made it possible to give the Nbss better mechanical properties at room temperature. Figure 7a shows the room temperature flexural strength and fracture toughness of the new Nb matrix composites obtained with different amount of SiC. It can be seen from the figure that flexural strength increased with the addition of SiC. Flexural strength of the three components was 754.52 MPa, 793.63 Mpa, and 814.38 Mpa, respectively. The reason is that the content of Nb_2C and Nb_3Si formed by the reaction increased with the addition of SiC. Nb_2C and Nb_3Si, which are both hard and brittle phases, embed into Nbss, hindering the movement of dislocation during deformation and having the function of dispersion hardening. The fracture toughness of the new Nb matrix composites decreased with the addition of SiC. The fracture toughness of the three materials was 13.82 $MPa^{1/2}$, 12.74 $MPa^{1/2}$, and 11.37 $MPa^{1/2}$, respectively. Nb/Nb_5Si_3 composites have been widely studied by researchers. The room temperature fracture toughness of Nb was 3$MPa^{1/2}$ [14], and it could go up to 8 $MPa^{1/2}$ of Nb/Nb_5Si_3 composites. The room temperature fracture toughness of Nb and Nb/Nb_5Si_3 composites were both lower than that of the new Nb matrix composites. Figure 7b shows the room temperature compressive strength and Vickers hardness of composites with different amounts of SiC. Compressive strength and Vickers hardness both increased with the amount of SiC. The compressive strength of the three materials was 1659 MPa, 1780 MPa, and 1915 MPa, respectively. The Vickers hardness was 567 HV, 601 HV, and 623 HV, respectively. The compressive strength of Nb/Nb_5Si_3 composites was 1400 MPa. It can be seen that the compressive strength of the new Nb matrix composites improve greatly compared with Nb/Nb_5Si_3 composites.

Figure 8a shows the Vickers indentations of the new Nb matrix composites with different amounts of SiC. It can be seen from the figure that cracks existed only in the Nb_2C and Nb_3Si grains and ended at the grain boundary of Nbss, also known as crack arrest (CA). Cracks passed through the hard and brittle phases in the manner of transgranular fractures (TF) [15] and deflected when the cracks propagated. The Nbss phase is ductile. When a crack propagates in the Nbss, it undergoes plastic deformation, which consumes the energy required for crack propagation during deformation and applies compressive stress to the tip of a crack to close it. Finally, it offsets tensile stress at the tip and slows the propagation of the crack. With the increasing of SiC content, the content of the Nbss phase reduces relatively and the fracture toughness decreases. Besides, there are a greater number of grains (smaller grain size) in the same volume with the increasing of SiC content. So during the deformation process, the deformation that disperses to each grain is smaller. Thus, it is difficult to form cracks in grains. Macroscopically, the material can withstand large deformation without breaking quickly. In addition, with the increasing of SiC content, the grain boundary surface area increases. It hinders the movement of dislocations during deformation. Meanwhile, the amounts of Nb_2C and Nb_3Si increase and strengthen the Nbss, so the compressive strength increases with the increasing of SiC content. Furthermore, the increase of Nb_2C and Nb_3Si also augments the number of hard and brittle phases in the indentation area, resulting in increasing hardness.

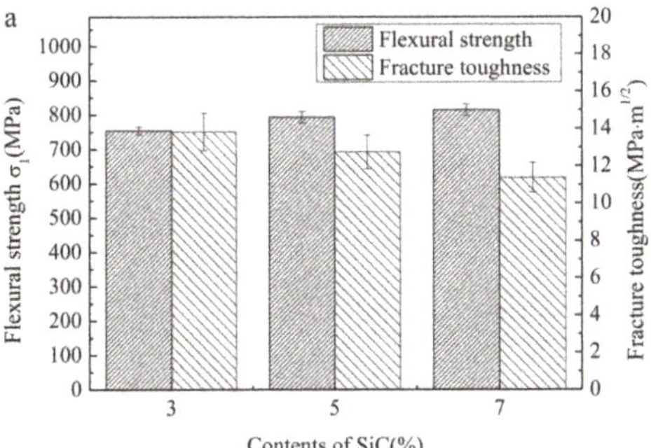

Figure 7. *Cont.*

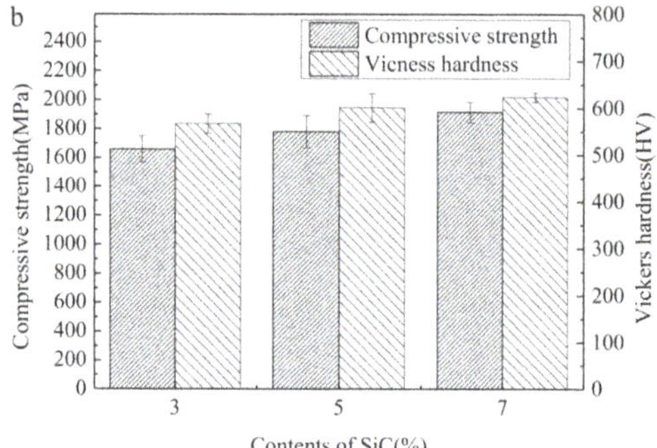

Figure 7. Room temperature mechanical properties of the new Nb matrix composites: (**a**) flexural strength and fracture toughness; (**b**) compressive strength and Vickers hardness.

Figure 8b shows the fracture morphology of an Nb-5SiC specimen after the three-point bending test. As the figure shows, Nb_2C exhibited a transgranular cleavage fracture mode, inlaid into the Nbss like skeletons. In the process of flexural fracture, Nb_2C gradually bends when it suffers bending load. When deformation reaches a certain extent, the cracks begin to germinate, and they propagate inside the N_2C grains, which is known as transgranular fracture. The more SiC is added, the more Nb_2C forms, and the greater the load suffered during the deformation process. Besides, in the process of propagation, cracks continually change direction, which virtually slows the crack growth rate, delaying fracture. Therefore, the room temperature flexural strength of composites increased with the amount of SiC.

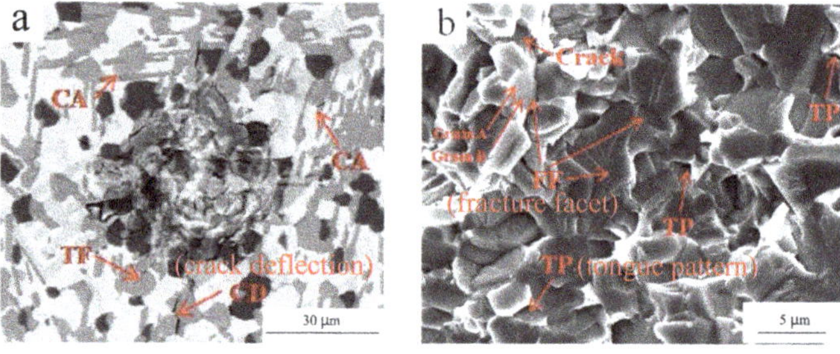

Figure 8. Vickers hardness indentation and fracture morphology of composites: (**a**) Vickers hardness indentation; (**b**) fracture morphology.

3.3. Behavior of Nb Matrix Composites at High Temperature

3.3.1. Stress-Strain Curve at High Temperature

Figure 9 shows the true stress-strain curves of the three materials at 1050 °C, 1100 °C, and 1150 °C. When compression started, there was an increase in flow stress as dislocations interacted and multiplied, and the true stress grew extremely fast. When it reached the peak, the true stress declined gradually. When the dislocation density increased to a certain degree, dynamic restoration occurred. The climb of edge dislocations, cross-slip of screw dislocations and counteraction of unlike dislocations are basic to the softening mechanism of the dynamic recovery, which contributes to lowering the dislocation density and opening dislocation tangle. Meanwhile, the rate of dislocation migration accelerates with increasing temperature, which decreases the deformation resistance [16]. Thus, the slope of the flow stress curve slowed down gradually.

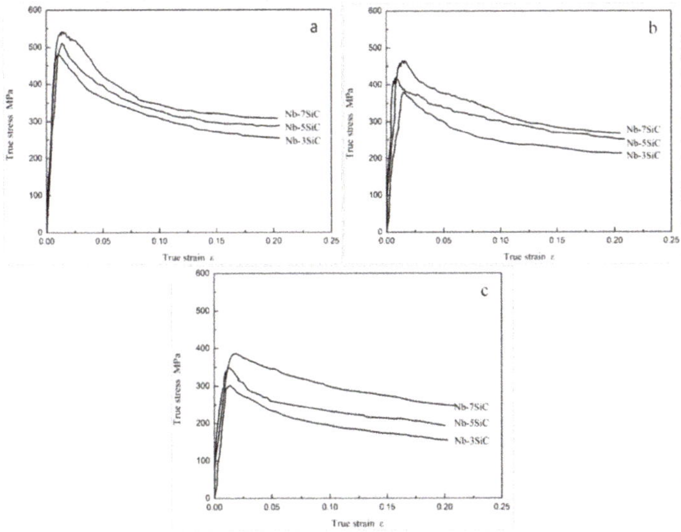

Figure 9. True stress-strain curves of the three materials at different temperatures: (**a**) 1050 °C; (**b**) 1100 °C; (**c**) 1150 °C.

Figure 10 shows the compressive strength of the three materials at 1050 °C, 1100 °C, and 1150 °C. It can be observed that with the increasing of SiC, compressive strength gradually increases. Compressive strength of the three materials at 1050 °C was 480 MPa, 513 Mpa, and 548 MPa. Compressive strength at 1100 °C was 381 MPa, 417 Mpa, and 470 MPa, respectively. When the temperature was 1150 °C, compressive strength was 301 MPa, 348 Mpa, and 387 MPa for the three materials, respectively. Introducing Hf, Cr, and other elements into the Nb-Si alloys could improve the mechanical properties of the Nb-Si alloys. Compressive strength of the 0Hf-2B-3Cr-54Nb-22Si–based alloy and the 2Hf-2B-3Cr-52Nb-22Si-based alloy at 1250 °C was 194 MPa and 291 Mpa [17], respectively; compressive strength of Nb-8Si-20Ti-6Hf-(6,10,14)Cr at 1150 °C was 268 MPa, 278 Mpa, and 313 MPa [18]. The compressive strength of ordinary Nb-Si alloys was lower than that of the new Nb matrix composites, which shows that the addition of SiC could significantly improve the high-temperature strength of the material.

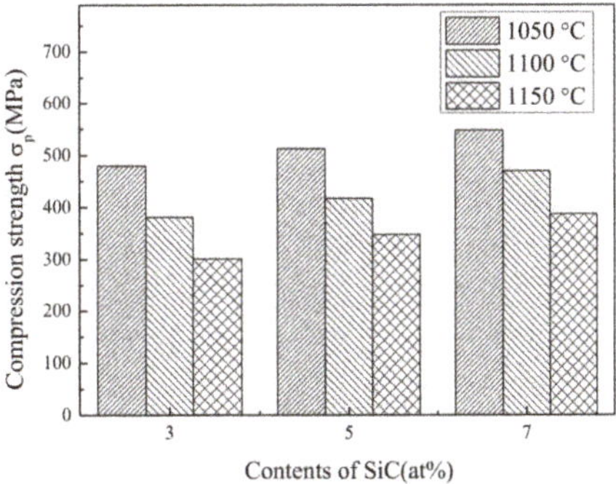

Figure 10. Compressive strength and peak stress at different temperatures.

With the increase of added SiC, deformation resistance and compressive strength of the material increased and the Nbss was strengthened. This is because, as the content of SiC increases, Nb_2C and Nb_3Si increase, and the bonding areas with Nbss also increase, hindering the movement of dislocation during deformation. In addition, the hard and brittle phases are of great help for second-phase strengthening, which also improves the high-temperature strength of the material.

3.3.2. Microstructure of Material after High-Temperature Deformation

During the compression process, the specimen was divided into three parts. As shown in Figure 11a, zone 1 is the dead zone, with the least deformation, zone 2 is the pressure zone, with the largest deformation, and zone 3 is the tension stress zone. The corresponding areas on the actual compression specimen are shown in Figure 11b, a longitudinal cross-sectional view of the compressed specimen. The three regions indicated by the arrows labeled A, B, and C in Figure 11b correspond to the regions labeled 1, 2, and 3 shown in Figure 11a.

Figure 12a–c show the BSE images corresponding to the three regions A, B, and C, respectively, in Nb-5SiC (Figure 11b). The area shown in Figure 12a is close to the pressing head and belongs to the undeformed area, retaining the equiaxed grain structure of original tissue. Figure 12b is the deformation area of the compression sample. The middle part of the equiaxed grain structure was squashed, showing a strip-shaped structure. Compared with the A region, the morphology of the C region did not have significant changes, because the total compression deformation was small and the deformation in the C region was smaller, resulting in insignificant changes of the tissue.

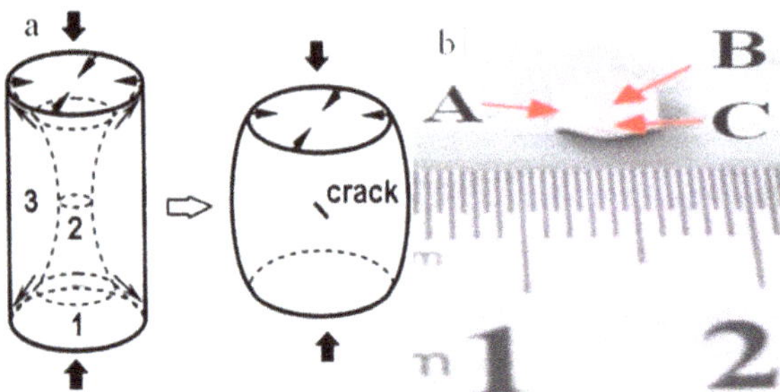

Figure 11. Stress states of different parts in the compressive sample: (**a**) schematic diagram; (**b**) actual sample.

Figure 13 shows BSE images of the deformed area B in the middle of Nb-5Si and area C after deformation at different temperatures. There was a common feature among these diagrams: cracks existed in grain boundaries between Nb_2C and Nb_3Si, but there were no cracks in the grain boundaries between hard, brittle phases and Nbss. This is because with increased deformation, there is inconsistent deformation between Nb_2C and Nb_3Si, so that cracks occur in the grain boundaries. However, the deformation between the hard brittle phases and the ductile Nbss is coordinate, and thus it is not easy to produce cracks. This indicates that the material is about to undergo intergranular fracture, while the fracture mode of the material is transgranular cleavage fracture at room temperature. The maximum load of material at room temperature was greater than at high temperature. This is because at room temperature, the grain boundary strength is high and the intragranular strength is low, so the fracture mode is mainly transgranular fracture. Meanwhile, because of the small grain size of the material, the grain boundary not only impedes dislocation, but also withstands large deformation, so that it is difficult to crack. Therefore, the material has good plasticity and high strength at room temperature. As the temperature increases, the grain boundary strength and intragranular strength both decrease, but the strength of the grain boundary decreases more, resulting in the transformation of transgranular fracture at room temperature to intergranular fracture at high temperature. Because of the low grain boundary strength at high temperature, the material will crack with small deformation. However, the ductile Nbss can absorb the energy of crack growth and delay crack propagation, playing a significant role in improving deformation of the material and preventing premature fracture.

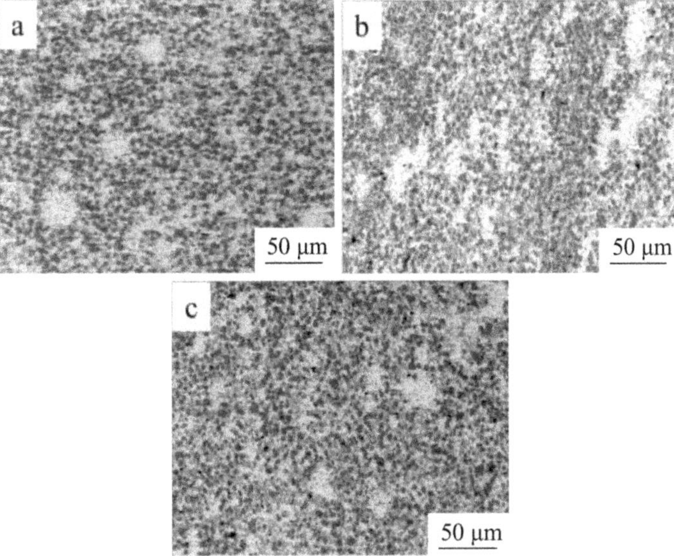

Figure 12. Backscattered electron images of different parts of Nb-5SiC at 1050 °C: (**a**) zone A; (**b**) zone B; (**c**) zone C.

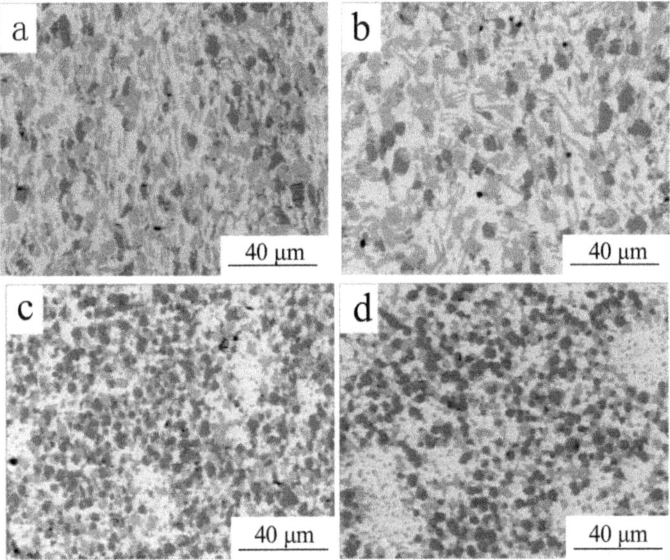

Figure 13. *Cont.*

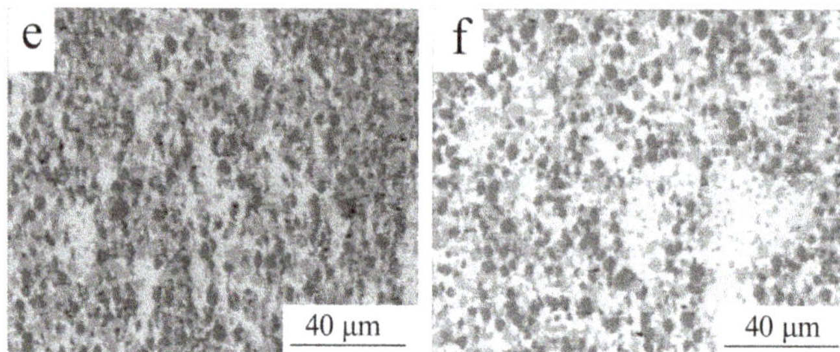

Figure 13. Backscattered electron images of Nb-5Si at different temperatures: (**a**) zone B at 1050 °C; (**b**) zone C at 1050 °C; (**c**) zone B at 1100 °C; (**d**) zone C at 1100 °C; (**e**) zone B at 1150 °C; (**f**) zone C at 1150 °C.

4. Conclusions

1. The new Nb matrix composites were prepared by hot pressing sintering process. The reinforcements consisted of Nb_2C, Nb_3Si, and Nbss. The relative density of material sintered was 97%, indicating that the material had good compactness.
2. At room temperature, the flexural strength of Nb-3SiC, Nb-5SiC, and Nb-7SiC composites was 754.52 MPa, 793.63 Mpa, and 814.38 MPa, respectively. The fracture toughness was 13.82 $MPa^{1/2}$, 12.74 $MPa^{1/2}$, and 11.37 $MPa^{1/2}$, respectively. Compressive strength was 1659 MPa, 1780 MPa, and 1915 MPa, respectively. Vickers hardness was 567 HV, 601 HV, and 623 HV, respectively. The performance was better than that of Nb/Nb_5Si_3 composites (fracture toughness = 8 $MPa^{1/2}$, compressive strength = 1400 MPa).
3. The compressive strength of Nb-3Si, Nb-5Si, and Nb-7Si was 480 MPa, 513 Mpa, and 548 Mpa, respectively, at 1050 °C. The compressive strength of Nb-7Si at 1050 °C, 1100 °C, and 1150 °C was 584 MPa, 470 MPa, and 387 MPa, respectively. In addition, the high-temperature compressive strength of composite materials was higher than that of ordinary Nb-Si alloys.

Acknowledgments: The authors gratefully acknowledge financial support from the National Natural Science Foundation of China (grant No. 51675126) and the Natural Science Foundation of Heilongjiang Province (grant No. DC2013C048).

Author Contributions: Zhen Lu designed this experiment and revised the manuscript. Chaoqi Lan analyzed the experimental data and completed this paper. Zhenhan Huang performed the nanoindentation experiments. Shaosong Jiang and Kaifeng Zhang helped to complete the microstructure analysis.

Conflicts of Interest: The authors declare no conflict of interest.

References

1. Xu, Y.; Sun, W.; Dai, W.; Hu, C.; Liu, X.; Zhang, W. Experimental and numerical modeling of the stress rupture behavior of nickel-based single crystal superalloys subject to multi-row film cooling holes. *Metals* **2017**, *7*, 340. [CrossRef]
2. Rodríguez-Millán, M.; Díaz-Álvarez, J.; Bernier, R.; Cantero, J.L.; Rusinek, A.; Miguelez, M.H. Thermo-viscoplastic behavior of ni-based superalloy haynes 282 and its application to machining simulation. *Metals* **2017**, *7*, 561. [CrossRef]
3. Lei, R.; Wang, M.; Xu, S.; Wang, H.; Chen, G. Microstructure, hardness evolution, and thermal stability mechanism of mechanical alloyed Cu-Nb alloy during heat treatment. *Metals* **2016**, *6*, 194. [CrossRef]
4. Yu, C.; Zhao, X.; Xiao, L.; Cai, Z.; Zhang, B.; Guo, L. Microstructure and mechanical properties of in-situ laminated Nb/Nb_5Si_3 composites. *Mater. Lett.* **2017**, *209*, 606–608. [CrossRef]
5. Yu, J.; Weng, X.D.; Zhu, N.; Liu, H.; Wang, F.; Li, Y.; Cai, X.; Hu, Z. Mechanical properties and fracture behavior of an nb-silicide in situ composite. *Intermetallics* **2017**, *90*, 135–139. [CrossRef]

6. Zhang, X.; He, X.; Fan, C.; Li, Y.; Song, G.; Sun, Y.; Huang, J. Microstructural and mechanical characterization of multiphase nb-based composites from Nb-Ti-C-B system. *Int. J. Refract. Met. Hard Mater.* **2013**, *41*, 185–190. [CrossRef]
7. Sha, J.; Hirai, H.; Tabaru, T.; Kitahara, A.; Ueno, H.; Hanada, S. Effect of carbon on microstructure and high-temperature strength of Nb Mo Ti Si in situ composites prepared by arc-melting and directional solidification. *Mater. Sci. Eng. A* **2003**, *343*, 282–289. [CrossRef]
8. Geng, J.; Tsakiropoulos, P. A study of the microstructures and oxidation of Nb-Si-Cr-Al-Mo in situ composites alloyed with ti, hf and sn. *Intermetallics* **2007**, *15*, 382–395. [CrossRef]
9. Geng, J.; Tsakiropoulos, P.; Shao, G. A study of the effects of hf and sn additions on the microstructure of Nbss/Nb_5Si_3 based in situ composites. *Intermetallics* **2007**, *15*, 69–76. [CrossRef]
10. Nassef, A.; El-Garaihy, W.H.; El-Hadek, M. Characteristics of cold and hot pressed iron aluminum powder metallurgical alloys. *Metals* **2017**, *7*, 170. [CrossRef]
11. Wang, X.; Wang, G.; Zhang, K. Effect of mechanical alloying on microstructure and mechanical properties of hot-pressed Nb-16Si alloys. *Mater. Sci. Eng. A* **2010**, *527*, 3253–3258. [CrossRef]
12. Jayasankar, K.; Pandey, A.; Mishra, B.; Das, S. Evaluation of microstructural parameters of nanocrystalline Y_2O_3 by X-ray diffraction peak broadening analysis. *Mater. Chem. Phys.* **2016**, *171*, 195–200. [CrossRef]
13. Wang, X.; Zhang, K. Mechanical alloying, microstructure and properties of Nb-16Si alloy. *J. Alloy. Compd.* **2010**, *490*, 677–683. [CrossRef]
14. Zhang, L.; Wu, J. Ti5Si3 and Ti5Si3-based alloys: Alloyingbehavior, microstructure and mechanical property evaluation. *Acta Mater.* **1998**, *46*, 3535–3546. [CrossRef]
15. Gulizzi, V.; Rycroft, C.; Benedetti, I. Modelling intergranular and transgranular micro-cracking in polycrystalline materials. *Comput. Methods Appl. Mech. Eng.* **2018**, *329*, 168–194. [CrossRef]
16. Jiang, S.Y.; Zhang, Y.Q.; Zhao, Y.N. Dynamic recovery and dynamic recrystallization of NiTi shape memory alloy under hot compression deformation. *Trans. Nonferrous Metals Soc. China* **2013**, *23*, 140–147. [CrossRef]
17. Zhang, S.; Guo, X. Effects of B addition on the microstructure and properties of Nb silicide based ultrahigh temperature alloys. *Intermetallics* **2015**, *57*, 83–92. [CrossRef]
18. Sha, J.; Yang, C.; Liu, J. Toughening and strengthening behavior of an Nb-8Si-20Ti-6Hf alloy with addition of Cr. *Scr. Mater.* **2010**, *62*, 859–862. [CrossRef]

Review

Corrosion, Erosion and Wear Behavior of Complex Concentrated Alloys: A Review

Aditya Ayyagari [1], Vahid Hasannaeimi [1], Harpreet Singh Grewal [2], Harpreet Arora [2] and Sundeep Mukherjee [1,*]

1 Department of Materials Science and Engineering, University of North Texas, Denton, TX 76203, USA; aa0715@unt.edu (A.A.); vahidHasannaeimi@my.unt.edu (V.H.)

2 Surface Science and Tribology Lab Department of Mechanical Engineering, School of Engineering, Shiv Nadar University, Uttar Pradesh 201314, India; harpreet.grewal@snu.edu.in (H.S.G.); harpreet.Arora@snu.edu.in (H.A.)

* Correspondence: sundeep.mukherjee@unt.edu; Tel.: +1-940-565-4170; Fax: +1-940-565-2944

Received: 18 June 2018; Accepted: 24 July 2018; Published: 3 August 2018

Abstract: There has been tremendous interest in recent years in a new class of multi-component metallic alloys that are referred to as high entropy alloys, or more generally, as complex concentrated alloys. These multi-principal element alloys represent a new paradigm in structural material design, where numerous desirable attributes are achieved simultaneously from multiple elements in equimolar (or near equimolar) proportions. While there are several review articles on alloy development, microstructure, mechanical behavior, and other bulk properties of these alloys, then there is a pressing need for an overview that is focused on their surface properties and surface degradation mechanisms. In this paper, we present a comprehensive view on corrosion, erosion and wear behavior of complex concentrated alloys. The effect of alloying elements, microstructure, and processing methods on the surface degradation behavior are analyzed and discussed in detail. We identify critical knowledge gaps in individual reports and highlight the underlying mechanisms and synergy between the different degradation routes.

Keywords: corrosion; surface degradation; wear; high entropy alloys; complex concentrated alloys; potentiodynamic polarization; erosion-corrosion; slurry-erosion; oxidation wear; highly wear resistant coatings

---

1. Introduction

Development of materials having superior surface degradation resistance has been a major thrust area of research in modern metallurgy. Loss of material in the form of corrosion, erosion, and wear results in economic impact in the range of billions of dollars worldwide by some estimates. Several technologies have not realized their full potential due to the lack of materials that can withstand surface degradation in critical applications. Specific examples include core walls, diverters, and reactor vessels in nuclear reactors that can withstand hot corrosion, contact with molten metals, high-pressure water, and exposure to super critical temperatures. Similarly, there is high demand for developing materials with improved wear resistance for enhancing the energy efficiency of turbines, windmill rotors, and automobiles. These call for structural components that can withstand high torque and resist metallurgical changes that are caused by frictional heat and high pressure. Problems, such as white matter in bearings, spalling, and deterioration in mechanical properties under operating conditions remain as a major challenge. In addition, synergistic combination of different surface degradation mechanisms leads to accelerated material loss. Examples include simultaneous wear and corrosion seen in food processing and chemical handling industries. Erosion is another significant source of material loss, where particulate materials, such as sand and debris entrapped in a moving/impinging liquid, degrade the surface integrity of materials.

Traditionally, development of materials that simultaneously meet multiple application requirements has been done by adding minor proportions of alloying elements to the base material and tailoring the heat treatment. Examples include aluminum alloys, where tempering treatments are used to obtain a balance in mechanical properties and corrosion resistance. In that regard, multi-principal element alloys represent a new paradigm in structural material design, where numerous desirable attributes are achieved simultaneously from multiple elements in equimolar (or near equimolar) proportions [1–7]. These alloys are typically referred to as high entropy alloys (HEAs) or more generally as complex concentrated alloys (CCAs). High configurational entropy leads to single-phase solid solutions in a certain subset of these multi-component systems. It was initially believed that the core effects, such as high configurational entropy [8], lattice distortion [9], and sluggish diffusion [10] may have resulted in a gamut of attractive properties including high strength-ductility combination [6,10,11], resistance to oxidation, corrosion and wear properties [12,13]. However, recent reports suggest that these may not be the only structure-property determining parameters, thus leaving a large scope for understanding the physical metallurgy of complex concentrated alloys [14–16]. Another advantage of the complex concentrated approach is the vast number of alloy systems that can be developed from a small palette of elements by focusing on the central region of the multi-component phase space, rather than the edges [16].

With exponentially growing interest in complex concentrated (or high entropy) alloys, there are several reports in literature on the surface degradation behavior of these multi-component systems. The corrosion behavior of high entropy alloys has been discussed in a recent review [17]. However, a clear understanding of the underlying mechanisms and synergy between the different surface degradation routes is lacking. Here, we provide a comprehensive overview of corrosion, erosion, and wear behavior of complex concentrated alloys to elucidate the similarities and in certain cases the unique differences in response to different environments. The effect of alloying elements, microstructure, and processing methods on the different surface degradation routes are analyzed and discussed in detail.

2. Evaluation of Surface Degradation Mechanisms

In this section, the methods used in literature for evaluation of corrosion, wear and erosion behavior of complex concentrated (or high entropy) alloys are summarized along with the pertinent metrics for quantifying the extent of damage.

2.1. Corrosion Characterization

Corrosion behavior of complex concentrated alloys has been evaluated by immersion (or mass loss/gain) test, open circuit potential measurement with time, potentiodynamic polarization, and anodic polarization. Immersion test is the simplest, where the change in mass of the sample is measured by assessing the damage that is caused by the environment in which it is immersed (ASTM G31). The corrosion rate is calculated as:

$$\text{Corrosion rate} = \frac{(K \times W)}{(A \times T \times D)} \quad (1)$$

where, K is a constant, T is time of exposure in hours, A is area in cm^2, W is mass loss/gain in g, and D is the density in gm/cm^3.

Accelerated assessment of corrosion performance can be made using electrochemical corrosion tests. When no external current or potential is applied to a metal immersed in an electrolyte, the system eventually reaches equilibrium and the net current measured is zero. The potential developed on the surface of the electrode when the metal is immersed into the electrolyte is called the open circuit potential (OCP). For potentiodynamic polarization, three types of reference electrodes are typically used, namely saturated calomel electrode (SCE), Ag/AgCl electrode, and standard hydrogen electrode (SHE). In a three-electrode set up, one of the aforementioned electrodes is connected as reference electrode, the sample as working electrode, and platinum or graphite as counter electrode. In the potentiodynamic polarization test, the sample is subjected to a potential sweep typically from −250 mV with respect to OCP to at least +250 mV at a scan rate of 0.16 mV/s. Scanning beyond 250 mV above OCP may cause further anodic reactions, such as breakdown of the protective surface oxides and pitting. Potentiodynamic polarization tests are extensively used to identify critical corrosion parameters such as pitting potential, passivation range, corrosion current, and re-passivation potentials in an accelerated way. Corrosion rate is calculated as:

$$\text{Corrosion Rate} = \frac{K \times i_{\text{corr}} \times EW}{\text{Density}} \tag{2}$$

where, K is 3.27×10^{-3} mm g/(μA·cm·year), i_{corr} is the corrosion current density, and EW is the equivalent weight. Equivalent weight is calculated from the expression:

$$\text{Equivalent weight} = \left\{ \sum \frac{f_i \times n_i}{w_i} \right\}^{-1} \tag{3}$$

where, f_i is the mass fraction, w_i is the atomic weight, and n_i is the valence of the ith element in the alloy [18].

2.2. Wear Testing

The wear behavior of complex concentrated alloys has been evaluated using three techniques, namely sliding reciprocating wear test, pin on disc test, and modified pin on disc test (pin-on-belt test). The fundamental working principle is the same in all three tests—a normal load is applied on to a sample while it is in contact with a reference material. Depending on the test type, either the reference material or the sample are moved to cause a relative motion between the surfaces. In pin-on-disc and pin-on-belt tests, the sample is made into the form of a stationary cylinder called "pin", which is brought in contact with a rotating disc made of hardened steel. A test load is applied normal to the pin, producing wear at the interface of the two materials as shown in Figure 1a. The rotating steel disc may be replaced with a moving belt, typically coated with Alumina or Silica abrading media, as shown in Figure 1b. In the sliding reciprocating test, the sample is made in the form of a flat plate and loaded under a hard counterface, such as WC or Si_3N_4 ball indenter or steel pin, as shown in Figure 1c. The stage slides at a set frequency and stroke length. The wear volume loss is quantified while using weight loss, contact profilometry, or interferometry.

Quantification of loss during wear test is done from the volume of worn material removed (V_w) and relating it to total sliding distance (L) and load (F). Wear volume loss for most engineering materials increases with decreasing hardness, as given by Archard's relation:

$$V_w = K \frac{L \times F}{H} \tag{4}$$

where, K is the dimensionless wear coefficient and H is the hardness. Certain high entropy alloys followed the Archard's wear relation in sliding wear test. The engineering unit of wear resistance is measured as wear volume loss per unit distance of sliding and expressed in the dimensions of $[L]/[L]^3$.

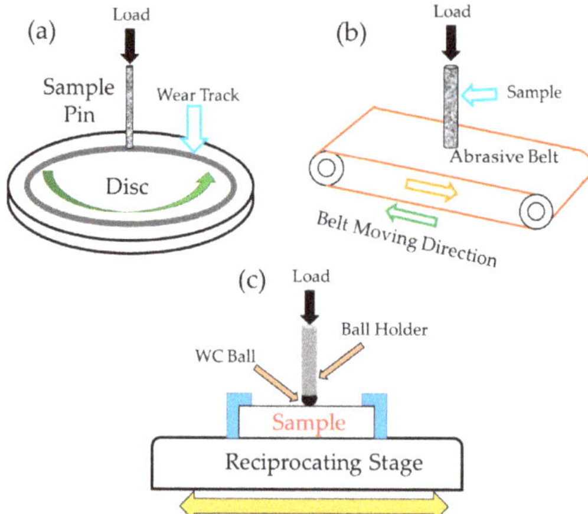

Figure 1. Illustration of (**a**) pin-on-disc test setup; (**b**) pin-on-belt setup; and (**c**) sliding reciprocating wear stage.

2.3. Erosion and Erosion Corrosion Characterization

Erosion is a form of material degradation characterized by the progressive loss of material from a solid surface due to mechanical interaction between the surface and a fluid or impinging liquid containing solid particles. There are very limited number of reports on erosion behavior of complex concentrated (or high entropy) alloys. Erosion that is caused by the impact of solid particles entrained in gaseous medium is termed as solid particle erosion. On the other hand, if a liquid is used as a carrier medium, the process is termed as slurry erosion [19,20]. The impact of the entrained abrasive particles results in micro-cutting and severe plastic deformation of the target surface depending on the operating parameters (Table 1). Other forms of erosion, which result from the interaction between a solid surface and fluid alone, are cavitation erosion and liquid droplet erosion [21–23]. In the case of cavitation erosion, degradation takes place due to implosion of cavities/bubbles in the liquids. Implosion of such cavities results in the formation of high velocity micro jets or shockwaves affecting the solid surfaces. Contact pressures at the point of impact can reach several hundred Giga-Pascals, which is sufficient for the deformation and removal of material. Several parameters influence the erosion processes, which may be classified into flow related, erodent related, and materials related parameters (Table 1) [24–28]. Chemical/electrochemical interactions (corrosion) are also possible along with erosion depending on the working environment. The synergy between erosion and corrosion can further aggravate the material degradation. The synergistic effect in tribo-corrosion process due to interaction between erosion and corrosion is given as [25]:

$$S = W - (E + C) \tag{5}$$

where, S is the synergy, W is the material removal rate by combined erosion and corrosion process, E is the material removal rate by pure erosion process, and C is the material removal rate by pure corrosion process. The synergy, S, is further composed of: (1) erosion induced corrosion (ΔC_E) and (2) corrosion induced erosion (ΔE_C). The factors that are responsible for erosion-induced corrosion are increase in surface area due to roughening effect, increased strain hardening and dislocation density, mechanical damage of the passive layer, and increased local temperatures. The factors contributing towards corrosion-induced erosion may be dislodgement of hard particles due to corrosion of the matrix, weakened grain boundaries, inter-granular pitting, and accelerated cracking due to crevice corrosion.

Table 1. Process parameters that affect the erosion process [24].

Flow Related Parameters	Erodent Related Parameters	Materials Related Parameters
• Velocity • Impact angle • Concentration • Viscosity of fluid • Temperature • Flow type • Surface tension • Density • Amplitude and frequency of vibrating probe (in case of cavitation erosion)	• Size • Distribution • Shape • Hardness • Defects • Density	• Yield and ultimate strength • Fatigue strength • Fracture resistance • Toughness • Hardness • Work hardenability • Microstructure • Composition • Porosity • Binder content and composition • Inter-splat bonding • Adhesion and cohesion

3. Corrosion Behavior of Complex Concentrated Alloys

Majority of the complex concentrated alloys studied so far for their corrosion behavior are based on the CoCrFeNi equimolar system. The observed corrosion behavior in these alloys may be broadly classified based on their composition and the resulting surface passivation layers, microstructural heterogeneity, phase segregation and associated galvanic corrosion, and finally, the test environment. CoCrFeNi-Cu_x (where "x" indicates varying proportions) was one of the earliest developed alloys, where the effect of increasing copper content on the microstructure and corrosion properties was reported [29]. Immersion and potentiodynamic polarization tests were conducted in 3.5 wt% NaCl solution. The as-cast CoCrFeNi-Cu_x alloys showed face centered cubic (FCC) phase mixture having distinct dendritic (copper lean) and inter-dendritic (copper rich) phases. In this alloy, the bright inter-dendritic regions were Cu rich as shown in Figure 2a. The X-ray diffraction (XRD) results show a single set of FCC peaks for several of these compositions, although the microstructure shows segregation between dendrites. This may be due to the very close *d*-spacing of the two phases that could not be resolved in XRD. The corrosion behavior of these alloys was comparable to SS304 stainless steel, with the copper free composition showing highest resistance to pitting (Figure 2b). The $x = 0.5$ alloy showed higher corrosion current density and pitting density, as shown in Figure 2c. This was attributed to higher galvanic action prompted from the Cu segregated in the inter-dendritic regions. Galvanic coupling results in initiation and propagation of localized corrosion pits causing rapid dissolution of the more anodic phase (in this case the bright Cu rich phase).

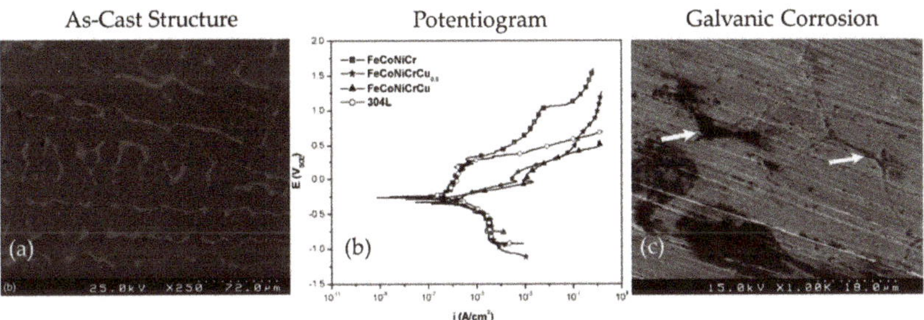

Figure 2. (**a**) As-cast microstructure of CoCrFeNi-Cu alloy showing dendritic microstructure. The copper rich interdendritic regions appear white, while the dendrites are darker; and, (**b**) Potentiodynamic polarization plots of the alloys in 3.5% NaCl. The Cu free CoCrFeNi alloy showed highest resistance compared to other two complex concentrated alloys (CCAs) and SS304L; (**c**) Microstructure after corrosion tests showing that Cu rich inter-dendritic regions corroded faster as compared to the dendritic regions lean in Cu. This may be due to the galvanic effect arising from the difference in composition [29] (reprinted with permission from Elsevier).

The effect of Al addition to CoCrFeNi has also been systematically studied and the resulting microstructure and corrosion behavior has been reported [30]. The $Al_{0.1}CoCrFeNi$ alloy shows a single-phase FCC structure with good microstructural stability. The XRD patterns for the alloy in recrystallized and as-cast state are shown in Figure 3a,b, respectively. The corresponding SEM microstructures are shown in Figure 3c,d.

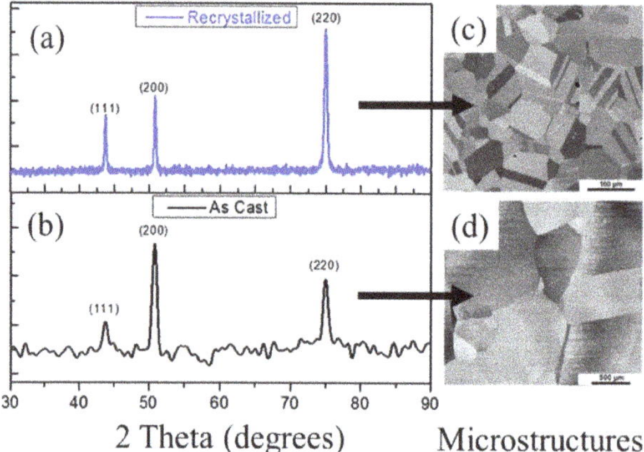

Figure 3. X-ray diffraction curves for $Al_{0.1}CoCrFeNi$ alloy in (**a**) recrystallized state heat-treated at 900 °C for 20 h, as compared to its (**b**) as-cast state. Corresponding back scattered electron scanning electron microscopy (SEM) microstructures of the alloys in (**c**) recrystallized and (**d**) as-cast states [31] (reprinted with permission from Elsevier).

The corrosion behavior of $Al_{0.1}CoCrFeNi$ has been reported in as-cast and recrystallized states in 3.5 wt% NaCl solution [31,32]. The corrosion behavior of the alloy was superior to SS304 steel as seen in Figure 4a. The corrosion potential, corrosion current density, and pitting resistance (referred to as E_{BD} in [32]) were comparable between as-cast and recrystallized states for $Al_{0.1}CoCrFeNi$ alloy. Minor variations between reported values in literature may be explained based on the microstructural differences between as-cast and recrystallized samples. Surface finish also plays an important role in determining the corrosion behavior. Potentiodynamic polarization for $Al_{0.1}CoCrFeNi$ alloy was compared with another single phase HEA, CoCrFeMnNi, as shown in Figure 4b. Both alloys showed wide passivation region and transient pitting, which may be an indication of local corrosion and the re-passivation on the surface.

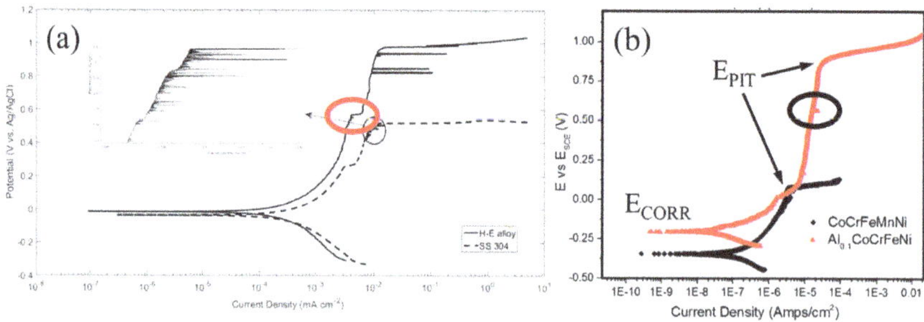

Figure 4. Potentiodynamic polarization curves of (**a**) as-cast $Al_{0.1}CoCrFeNi$ CCA versus SS304 steel in 3.5 wt% NaCl solution [32] (**b**) rolled and recrystallized $Al_{0.1}CoCrFeNi$ CCA as compared to CoCrFeMnNi CCA [31] (reprinted with permission from Elsevier).

The pitting resistance (ΔE) or breakdown resistance (E_{BD}) of $Al_{0.1}$CoCrFeNi alloy measured as the difference of pitting potential (E_{PIT}) and corrosion potential (E_{CORR}) in both conditions was ~1 V. Transient pitting was observed in both conditions around 0.6 V (highlighted with circle on the polarization curve), which may be an indicator of localized surface instability. More events of transient pitting were seen for the as-cast condition as compared to the recrystallized sample, which may be explained from the microstructural heterogeneity. Both as-cast and wrought alloys showed a high passivation resistance, 193 kΩ [31] and 115.5 kΩ [32], respectively, when tested for their EIS response. The superior corrosion resistance of $Al_{0.1}$CoCrFeNi alloy has been explained based on the relatively high content of Cr and Ni that form a strong passivating surface layer. Pitting resistance is typically quantified based on the wt% of passivating elements that are present in the alloy. Particularly, Cr, Mo, and Ni enhance pitting resistance of most engineering alloys. Since CCAs have passivating elements as high as 20%, are reported to have excellent pitting and corrosion resistance, provided that there are no extraneous corrosion promoters, such as galvanic phases or physical surface aberrations. $Al_{0.1}$CoCrFeNi high entropy alloy showed unique corrosion microstructures, as shown in Figure 5. Corrosion was initially observed to occur in the form of tiny pits, as shown in Figure 5a. Unique hierarchical features developed as a result of extensive grain boundary corrosion as well as micro/nano porosity formation within the grains, as shown in Figure 5c,d.

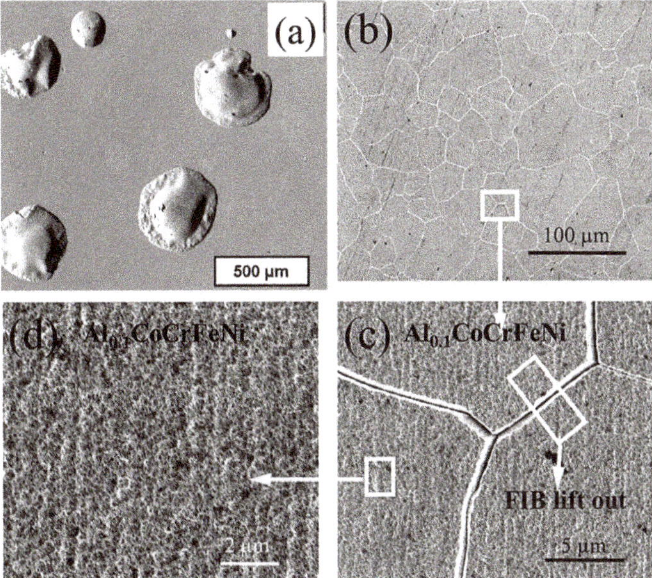

Figure 5. Pitting morphologies in $Al_{0.1}$CoCrFeNi alloy after polarization test in 3.5% NaCl [32]. (**a**) large pitting on the sample when tested to current density of 10 mA/cm^2; (**b**) low magnification image showing grain boundary corrosion; (**c**) high magnification image showing grain boundary corrosion and micro-porosity formation; (**d**) high magnification image showing small micro-porous structures on the surface after corrosion [31] (reprinted with permission from Elsevier).

The effect of increasing Al content on the corrosion behavior of Al_xCoCrFeNi alloy system was investigated in sulfuric acid [33] as well as in NaCl solution [34]. Besides the effect of alloying elements, the effect of experimental variable, i.e., scanning rate and temperature of the alloys in the corrosive media was also studied. This is an important metric to be systematically studied since scan rate can significantly alter the values of corrosion rate measured [33,34]. Increasing Al content induced microstructural changes to Figure 7. Pitting morphology hat significantly affected the corrosion behavior. The single phase FCC alloys for $x = 0$, $x = 0.25$, and $x = 0.3$ were more corrosion resistant when compared to the alloys containing higher fraction of Al. This was primarily attributed to phase separation induced by increasing Al content. Figure 6a shows secondary passive region for alloys with $x = 0.5$ and $x = 1.0$, which was attributed to the selective corrosion of dual phase face centered cubic (FCC)–body centered cubic (BCC) alloy (for $x = 0.5$) and BCC-ordered structure (for $x = 1.0$). In comparison, Figure 6b shows the potentiodynamic polarization charts for alloys with $x = 0.5$ and $x = 0.7$, displaying multiple transient pitting sites and continuous corrosion, both of which indicate the corrosion of secondary phases and partial passivation behavior of the matrix.

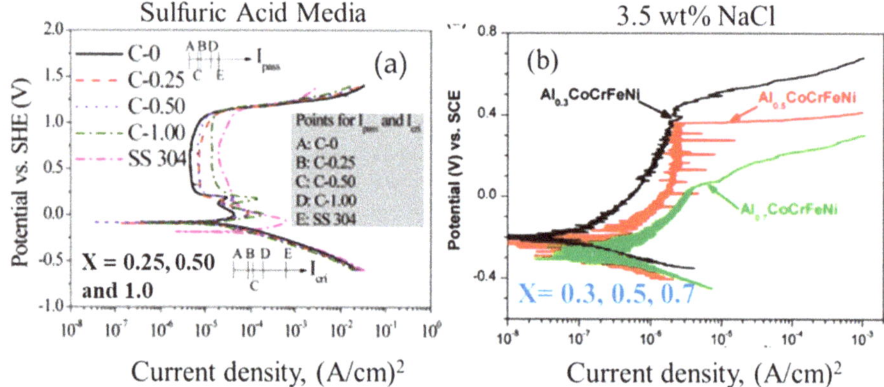

Figure 6. Potentiodynamic polarization curves of $Al_xCoCrFeNi$ alloys in (**a**) 0.5 M H_2SO_4 solution at Al content 0, 0.25, 0.5, 1.0; as compared with SS304, test performed at 25 °C; (**b**) 3.5 wt% NaCl solution for Al = 0.30, 0.50, 0.70 alloys. Both tests show gradual deterioration in corrosion in corrosion behavior with increasing Al content. Tests in NaCl solution resulted in extensive unstable pit formation on the sample, seen as short current spikes on the anodic branch [33,34] (reprinted with permission from Elsevier).

The single phase FCC alloys for $x = 0$ to 0.3 in $Al_xCoCrFeNi$ system showed significantly lower pitting density and pit depth when compared to the two phase alloys resulting from higher Al content. The $x = 0.5$ alloy showed pits on the inter-phase boundary between the FCC and BCC phases, indicating the formation of a galvanic couple. For $x = 0.7$ and $x = 1.0$, the BCC phase was observed to significantly/completely dissolved in H_2SO_4 solution and NaCl solution. Increasing Al content promotes the formation of BCC phase in $Al_xCoCrFeNi$ alloys, which undergoes selective dissolution. Increasing Al content likely results in the formation of porous Al oxide on the surface at the expense of more compact and passivating Cr oxide. A clear pattern of evolution of pitting morphology is seen as the Al content in gradually increased. Al = {0.1–0.3}: uniform pitting → Al = {0.5}: interphase galvanic corrosion → Al = 0.70–1.0 complete dissolution of BCC phase. Therefore, increasing the Al content beyond a threshold value resulted in higher pitting susceptibility, as seen in Figure 7.

Corrosion morphologies on $Al_xCoCrFeNi$ alloy system

Figure 7. Pitting morphology on $Al_xCoCrFeNi$ alloys after polarization experiments. The microstructure of the alloys with (**a**) Al = 0; (**b**) Al = 0.25; and (**c**) Al = 0.3 was reported to be of single phase, and consequent absence of galvanic corrosion sites. The microstructures with Al = 0.5–1.0 showed two-phase microstructure. This promoted accelerated corrosion at the interface between the two-phases (**d**) pits on face centered cubic-face centered cubic (FCC-FCC) interphase boundary; (**e**) BCC-BCC interphase boundary and (**f**) corrosion on ordered BCC phases [34] (reprinted with permission from Elsevier).

Heat treatment of $Al_{0.5}CoCrFeNi$ alloy resulted in phase separation and formation of BCC + FCC phases from a single-phase parent FCC cast alloy [35]. The overall corrosion resistance of the alloy was lower when compared to SS304 steel. Corrosion morphology on the single-phase FCC phase alloy comprised mostly of hemispherical pits nucleating randomly on the surface. This is an indicator of no preferred pit initiation site, while the hemispherical morphology indicates an equal propensity for pit to propagate into the material (Figure 8a). No dendritic coring or secondary pitting was seen. However, in contrast, the alloy with FCC + BCC phases showed preferred pitting along the interphase boundary. This is a clear indication of galvanic coupling between the two phases, governed by composition difference and the partitioning of elements between the two phases. Figure 8b shows the random pitting morphology preferentially occurring along the grain boundaries of the two phases.

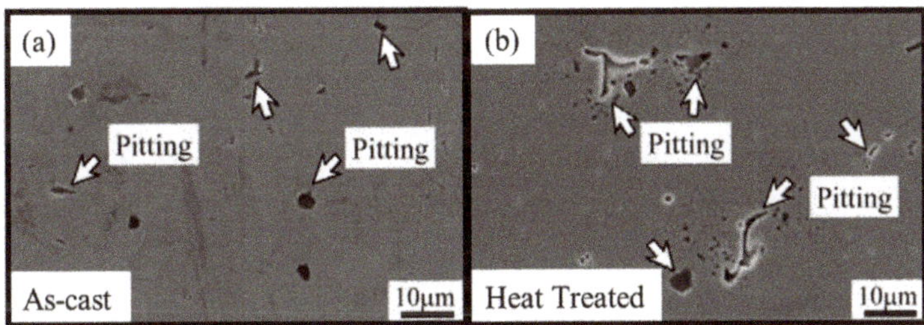

Figure 8. Pitting morphology for $Al_{0.5}CoCrFeNi$ alloy in (**a**) as-cast condition and (**b**) after heat treated at 800 °C. Heat treatment resulted in phase separation forming BCC phase. Corrosion was observed to preferentially nucleate along the interphase boundaries [35] (reprinted with permission from Elsevier).

In addition to intrinsic chemistry and crystal structure, melt-solidification history was found to affect corrosion resistance of CCAs [36]. Understanding the effect of re-melting on the microstructure and consequent corrosion properties can help in casting alloys with superior chemical homogeneity and properties. A multicomponent AlCoCrFeNiTi alloy was prepared by induction melting in Ar atmosphere. The alloy was subsequently re-melted several times in order to homogenize the distribution of elements. The re-melting may have eliminated macro-segregation arising from incomplete melting of elemental metal chunks used in alloy making. However, this may not have completely eliminated micro-segregation in the form of coring. This was evident in the form of a dendritic microstructure with micro-segregation in the inter-dendritic areas. The dendrites were rich in Al, Co, Ni, and Ti, while the inter-dendritic regions contained a higher fraction of Fe and Cr. The microstructure consisted of BCC phases along with complex intermetallics such as $AlFe_3$. Despite the complex microstructure, the corrosion performance of the alloy was better than SS410 alloy. The addition of Ti improved the corrosion rate of the alloy (0.0216 mm/year) by almost a factor of four compared to the Ti-free alloy (0.08 mm/year). This improvement may have resulted from the complex surface oxides that promote strong passivation. Re-melting the alloy homogenized the microstructure by removing macro-segregation, which contributed to improved corrosion resistance.

With increasing interest in the additive manufacture of these complex alloys, the first step is to be able to process the alloys using power-technology. The corrosion behavior of AlCoCrFeNi alloy system was studied as a function of Cu addition via the powder metallurgy route. The corrosion properties were evaluated in 1 mol/L NaCl [37]. The alloy produced using powder metallurgy route showed a microstructure that is similar to the conventional casting route. Microstructure of the AlCoCrFeNi–Cu alloy after sintering is shown in Figure 9a. The microstructure was complex and showed a two phase-mixture of FCC + BCC phases. The potentiodynamic polarization curve for this alloy is shown in Figure 9b. The corrosion potential was −0.012 V and the corrosion current density was 3.23 nA/cm^2. The alloy did not show any pitting up to potentials as high as 1.5 V versus saturated calomel electrode (SCE). This observation is insightful since powder-technology route was observed to possess improved corrosion resistance compared to conventional melt route. This may have resulted from the fact that powder particles individually possess oxide on the surface that are retained when the compacted is sintered. The larger surface oxide present on the bulk of the material may have imparted a nobler corrosion resistance as compared to its fused counter parts. The highly symmetric pattern seen in Figure 9a may have its origins in the parent (oxide covered) powder particles that explains the improved corrosion resistance that is seen in Figure 9b.

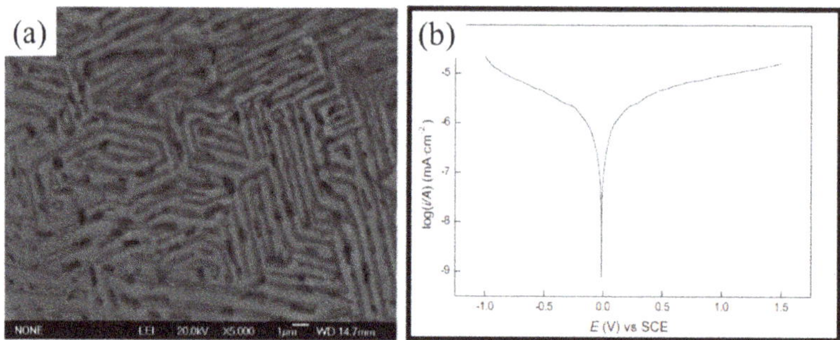

Figure 9. (**a**) Microstructure of AlCoCrFeNi-Cu alloy obtained by powder-metallurgy route; (**b**) Potentiodynamic polarization curve in 1 mol/L NaCl showing corrosion potential close to 0 V with respect to saturated calomel electrode [37] (reprinted with permission from Elsevier).

Increasing B content led to the precipitation of boride containing phases as seen in Figure 10a–d [38]. The corrosion behavior of the alloys was tested in 1 N H_2SO_4 [39]. The corrosion current density of the alloy increased from 787 $\mu Amps/cm^2$ to 2848 $\mu Amps/cm^2$ with the increase in boron content and boride phase fraction. The increasing boride phase fraction promoted the formation of "stringy precipitates" rich in Cr, Fe, and Co borides. This difference in composition led to the formation of local micro galvanic couples, making them susceptible to corrosion, as seen in Figure 10e–h. The phases rich in strongly passivating elements (Cr, Co) showed high pitting resistance while the matrix and inter-dendritic regions preferentially corroded. The morphology of the secondary phase changed with progressively increasing B content. Increasing B promoted stronger phase separation and formation of anodic regions that corroded more aggressively. The precipitation of hard boride phases may improve other surface properties, such as hardness and wear resistance (as discussed in subsequent sections), but certainly deteriorated the corrosion resistance due to galvanic corrosion. A balance of mechanical degradation resistance and galvanic corrosion resistance must be achieved by properly tailoring the composition to suite the application requirements.

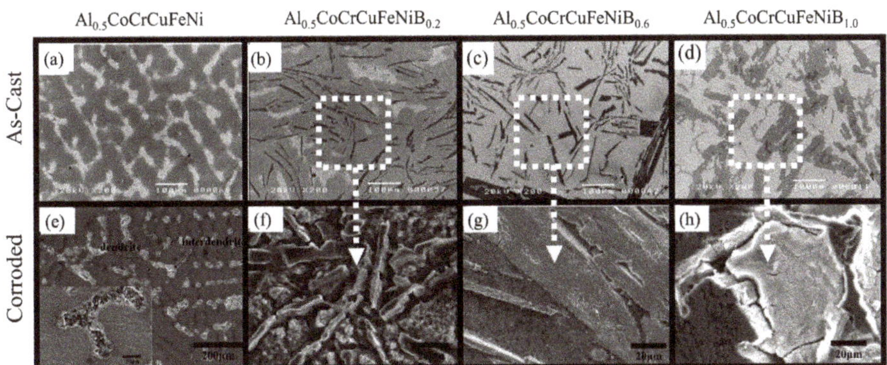

Figure 10. Microstructure of the as-cast alloys with increasing Boron content: (**a**) boron free alloy; (**b**) Boron = 0.2; (**c**) Boron = 0.6; and (**d**) Boron = 1.0. Corroded features on respective microstructures. Preferred corrosion of (**e**) inter-dendritic phase (**f**) of stringer features (**g**) and (**h**) borides [38,39] (reprinted with permission Journal of the Electrochemical Society and Springer).

The corrosion behavior resulting from addition of Cr and Ti to the base composition of AlCoCuFeNi has been reported and corresponding microstructures are shown in Figure 11a–d [40]. The bright phase in the images is Cu-rich FCC, whereas the BCC phases are rich in Al and Ni and they show a darker contrast. Adding Cr resulted in the formation of dendrites, whereas BCC formed into lamellar Widmanstatten type structures. Ti caused Al-, Co-, Ni-, and Ti-rich BCC phases (A2/B2), whereas adding both Cr and Ti refined the grain structure and led to copper segregation. The corrosion behavior of the alloys studied in 0.5 mol/L H_2SO_4 solution showed very low corrosion current densities—in the range of 5–8 $\mu A/cm^2$ for the Cr containing alloy. In contrast, the Ti and Ti-Cr containing alloys showed much higher corrosion activity. Cr and Ti typically improve corrosion resistance. The anomalous finding in this study may be due to the heterogeneous microstructure of the alloy that resulted in the formation of micro-anode and micro-cathode regions that accelerated corrosion. The corroded microstructures are shown in Figure 11e–g, indicating that the Cu-rich FCC phase was highly susceptible to corrosion. Corrosion of Cu-rich phases may be due to the higher galvanic character of the alloy, thus making it susceptible to dissolution.

In contrast to adding passivating elements, such as Cr and Ti, in the aforementioned study, the effect of changing corroding species, such as Cu and Al content on mechanical and corrosion behavior was studied for $Al_xCoCrCu_{0.5}FeNi$ system [41]. The corrosion resistance of this alloy was studied in 0.5 M H_2SO_4 and 0.5 M NaCl solution. The $x = 0.5$ alloy showed single-phase FCC structure, while alloys with $x = 1.0$ and $x = 1.5$ showed a mixture of FCC and BCC phases. The microstructure of the two-phase alloys was a BCC-rich dark matrix and light inter-dendritic FCC phase, as shown in Figure 12a. Potentiodynamic polarization curves showed that the alloys with two-phase microstructure had lower corrosion resistance due to the formation of micro-galvanic couples. NaCl environment caused pitting, while the alloy showed passivation behavior in H_2SO_4 solution. Solutionizing heat treatment improved the corrosion resistance of the $x = 1.5$ alloy as the FCC phase dissolved leaving behind a largely BCC alloy. The potentiograms in Figure 12b,c show the alloys' behavior in 0.5 M NaCl and 0.5 M H_2SO_4 solution.

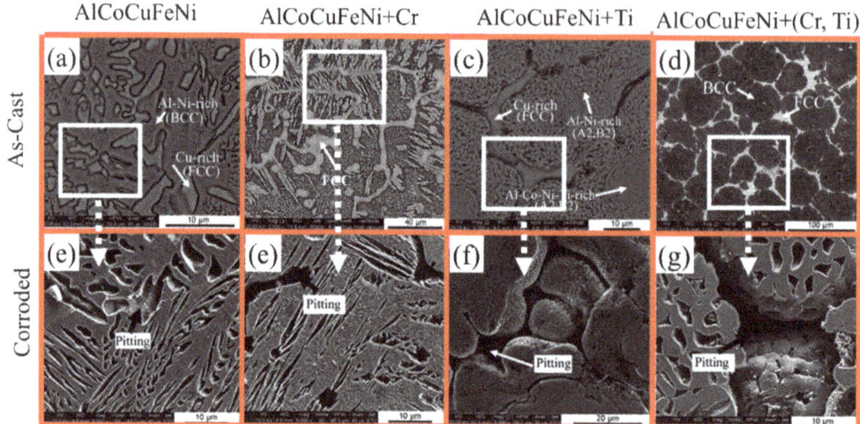

Figure 11. As-cast (**a–d**) and corroded (**e–g**) microstructures of AlCoCuFeNi-Cr/Ti alloys. Inter-dendritic phases containing Cu were observed to corrode and dissolve rapidly. In the case of alloys containing both Cu and Ti, the lighter FCC phase dissolved leaving behind rounded dendrite features [40] (reprinted with permission from Elsevier).

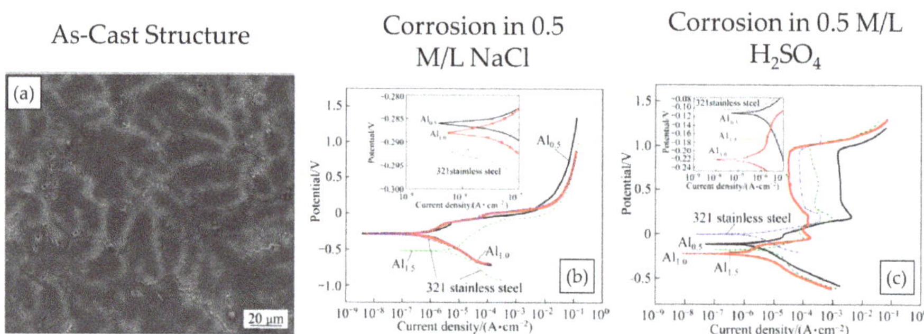

Figure 12. (**a**) As-cast microstructure of $Al_{1.5}CoCrCu_{0.5}FeNi$ CCA alloy, potentiodynamic polarization plots of $Al_xCoCrCu_{0.5}FeNi$ (x = 0.5, 1.0, and 1.5) in (**b**) 0.5 M/L NaCl, and (**c**) 0.5 M/L H_2SO_4 [41]. The potentiograms show better corrosion resistance of the alloys as compared to SS321 alloy (reprinted with permission from Elsevier).

The pitting corrosion resistance of $CoCrFeNiTiMo_x$ was evaluated as a function of Mo content, varying from $x = 0$ to $x = 0.8$ [42,43]. The corrosion behavior was tested in acidic, basic, and saline solution. The as-cast microstructures of $CoCrFeNiTiMo_x$ are shown in Figure 13a–d. Increasing Mo content altered the microstructure to result in phase-partitioning—a dark dendritic phase and a bright interdendritic (ID) phase. Mo was observed be uniformly distributed between the two phases at 0.1 at%, however, increasing the Mo content partitioned in to the interdendritic (ID) regions. This may be due to the strong single-phase forming tendency of Co-Cr-Fe-Ni system, as established by various studies. While the average composition of Co, Cr, Fe, and Ti varied by a mere 2–4% between the two regions, Ni content variation between the dendritic and interdendritic phase was as high 50%, as measured by EDS. A broad observation is that the interdendritic region is rich in Mo and lean in Ni. This information in conjunction with individual binary phase diagrams of Mo and Co, Cr, Fe, Ti, (elements in the ID region), and enthalpy of mixing values suggests that the σ phase that is formed in the alloy might not be a simple Mo-Cr phase, akin to SS316L. The changing phase composition and partitioning of elements between dendritic and ID regions may have resulted in galvanic coupling and the consequent increased corrosion of Mo containing alloys as compared to Mo free/lean alloys.

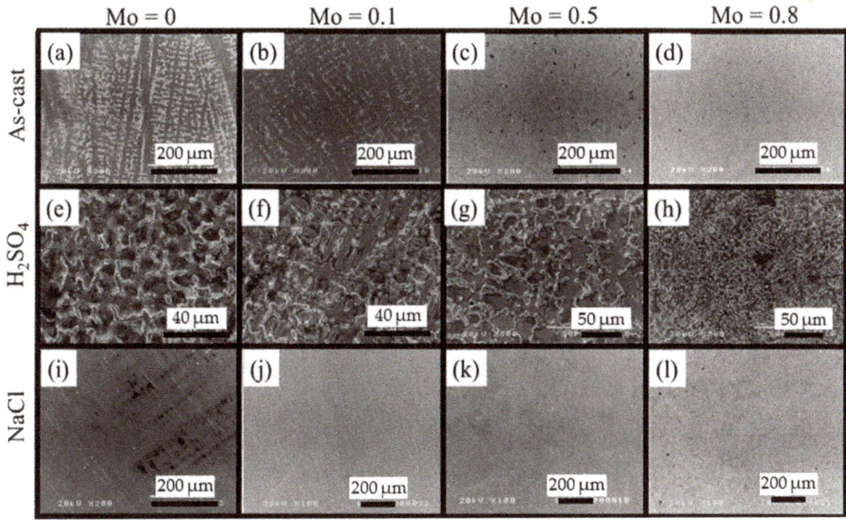

Figure 13. Corrosion and pitting of $CoCrFeNiTiMo_x$ (x = 0, 0.1, 0.5, and 0.8) alloys (**a**–**d**) As-cast, (**e**–**h**) after corrosion in H_2SO_4, (**i**–**l**) after corrosion in NaCl [42] (reprinted with permission from Elsevier).

The corrosion behavior by immersion test was studied for two CCAs, namely, $CrCu_{0.5}FeMnNi$ and $Cr_{0.5}CuFeMnNi$ [44]. The as-cast microstructure of the two alloys are shown in Figure 14a,d. Both of the alloys showed dendritic microstructure with FCC or FCC + BCC solid solution phases. The corrosion behavior of the alloys was characterized by an immersion test and potentiodynamic polarization in 1 M H_2SO_4. The microstructures after immersion test are shown in Figure 14b,e, while that after accelerated corrosion are shown in Figure 14c,f. Both tests showed preferred corrosion of the inter-dendritic phase due to galvanic coupling from the partitioning of alloying elements. Superior corrosion resistance of the dendritic phase was explained by the passive layers of NiO, $Ni(OH)_2$, $NiSO_4$, and Cr_2O_3. The corrosion resistance of both the alloys was superior to stainless steel. Between the two alloys, the one with lower Cu content showed lesser elemental segregation and higher corrosion resistance.

Complex concentrated (or high entropy) alloys have been synthesized in the form of coatings by several processing routes, including melt cladding and deposition, chemical vapor deposition (CVD), physical vapor deposition (PVD), electro spark processing [45], direct current magnetron sputtering [46], and laser cladding techniques [47]. Surface clad CCAs showed a desirable microstructure because of rapid solidification, good metallurgical bonding to substrate, and lesser compositional segregation [48,49]. In the form of coating, CoCrFeMnNi CCA showed spontaneous passivation in NaCl solution. Although this CCA coating showed i_{corr} value similar to 304SS, the passivation potential window for 304SS being wider. The CCA showed better corrosion resistance than 304SS in H_2SO_4, with a stable passive film formation. The initiation of corrosion for CoCrFeMnNi coating started with the depletion of chromium between the dendrites, and the subsequent weakening of the microstructure.

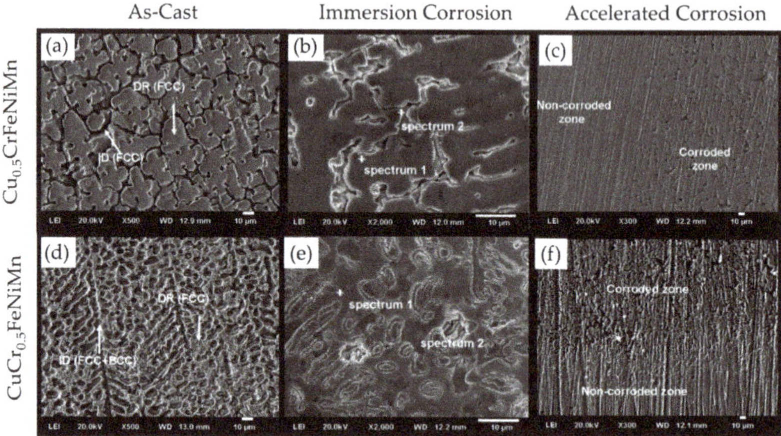

Figure 14. (**a**,**d**) Typical microstructures of $Cr_{0.5}CuFeMnNi$ and $CrCu_{0.5}FeMnNi$ system, respectively; (**b**,**e**) after immersion test in 1 M H_2SO_4 and (**c**,**f**) after polarization test in 1 M H_2SO_4 [44] (reprinted with permission from Wiley).

Addition of Ti up to a certain percentage to AlCoCrFeNi CCA coating resulted in better corrosion and cavitation erosion performance in NaCl [28]. Further increase in Ti content resulted in the formation of Ti_2-Ni and NiAl intermetallic compounds and decreased the passivation resistance. The alloy with the highest Ti content showed the worst corrosion resistance. Cavitation erosion behavior is primarily dictated by mechanical strength of a material in a non-corrosive medium. Therefore, the coating with the highest Ti content showed improved cavitation erosion resistance in distilled water, because the intermetallic compounds acted as a deformation barrier on the surface. In contrast, the same alloy (highest Ti content) showed the worst cavitation resistance in NaCl solution due to the synergistic effect of cavitation and corrosion in NaCl medium. Interestingly, AlCoCrFeNi coating without Ti showed remarkable cavitation erosion resistance, better than 304 stainless steel in NaCl with a lower i_{corr} value [50]. Ti addition up to a certain percentage to $Al_2CrCoCuFeNiTi_x$ coating fabricated by laser cladding resulted in good corrosion performance in HNO_3 [51]. Ti promoted the formation of a BCC phase in this CCA coating and affected both the corrosion and wear properties. Due to rapid cooling rates that were achieved during laser cladding, lesser segregation and uniformly refined grains (down to nanoscale) were reported [52]. The homogenous microstructure resulted in better corrosion resistance. Addition of Ti to AlCoCuFeNi CCA produced by arc melting resulted in a two-phase heterogeneous microstructure and micro-anode/cathode regions in the electrolyte [53]. The corrosion resistance was found to decrease for this alloy at both 298 K and 366 K in H_2SO_4. Laser processed AlCoCuFeNi CCA coating showed similar corrosion current densities to the coatings containing Ti [40,54]. The improvement in corrosion performance was attributed to the reduced dilution rate and formation of a compact CCA phase by controlling the laser parameters. Niobium also showed a similar effect as Ti in CCA coatings. The addition of Nb prevented less noble elements from dissolving in corrosive environments [54]. Corrosion performance of CoCrCuFeNi CCA coating increased considerably with the addition of Nb due to modification of microstructure and the formation of a very stable passive film [55]. Addition of Nb reduced the Cu segregation in interdendrite regions, resulted in the formation of a finer FCC phase and very stable surface oxide films.

Al addition up to a certain percentage showed improved corrosion current density for $Al_xCoCuFeNi$ CCA coating made by laser cladding [56]. A monotonic increase in corrosion resistance was reported for $Al_xCoCrFeNiTi$ [57]. A similar trend was observed for addition of Ni to $Al_2CoCrCuFeNi_xTi$ laser cladded CCA coating [58]. Increasing Ni content in this CCA coating, with x values up to 2, resulted in better corrosion performance in NaOH and NaCl solutions. In summary, the presence of corrosion resistant elements, such as Cr, Ti, Al, and Ni in limited quantity improved corrosion resistance in CCA coatings. However, beyond a certain mole fraction, microstructural segregation and lattice distortion led to a worsening of corrosion resistance.

The presence of Co in CCA coatings typically resulted in improved corrosion performance. In some CCA coatings, Co formed a passive film of CoO, which after exposure to corrosive medium, formed $Co(OH)_2$ [59]. $Co(OH)_2$ acted as a protective passivation layer and prevented corrosive species, such as Cl^- and O^{2-}, from diffusing into the coating. AlCoCrCuFe-$X_{0.5}$ CCA coating in which X was Si, Mo, and Ti, showed no passivation in NaCl [60]. In addition, extensive pitting was observed for the alloy containing Mo and Ti on the Cr-depleted Fe_2Mo and Fe_2Ti phases, as shown in Figure 15. However, the dendritic regions enriched with Cr remained passivated after polarization tests.

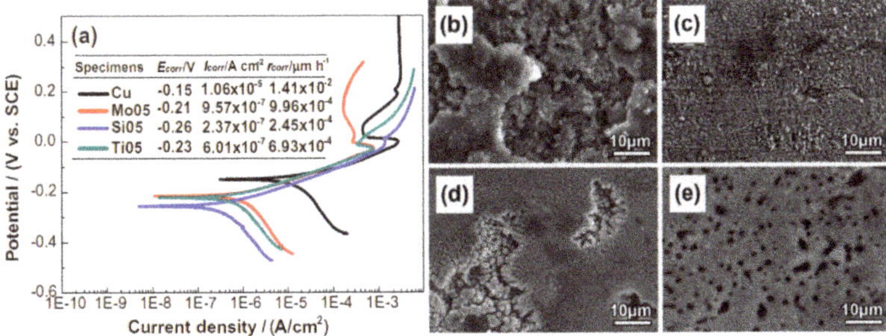

Figure 15. (**a**) Potentiodynamic polarization plots of AlCoCrCuFe-$X_{0.5}$ CCA coatings, SEM micrograrphs of surface morphologies after polarization tests for (**b**) X = Cu, (**c**) X = Si05, (**d**) X = Mo05, and (**e**) X = Ti05 [60] (reprinted with permission from Elsevier).

The electrochemical behavior of CCA coatings has been reported to be different from their bulk counterparts with identical chemical composition [18,61,62]. Due to rapid cooling rates, CCA coatings possess more homogenous microstructure with lesser elemental segregation. In contrast, bulk as-cast HEAs typically consist of dendrites and inter-dendritic regions with different chemical compositions, resulting in micro-galvanic cells that accelerate the corrosion process. This effect was clearly demonstrated for AlCoCrFeNi CCA coating fabricated through electro-spark method. Relatively uniform corrosion was seen for the coating (Figure 16a), while a non-uniform attack was seen for the as-cast alloy (Figure 16b). The inhomogeneous corrosion of the cast CCA was attributed to the micro-galvanic coupling between the matrix precipitates and the matrix itself. However, the CCA coating was free from intercellular segregation and precipitates.

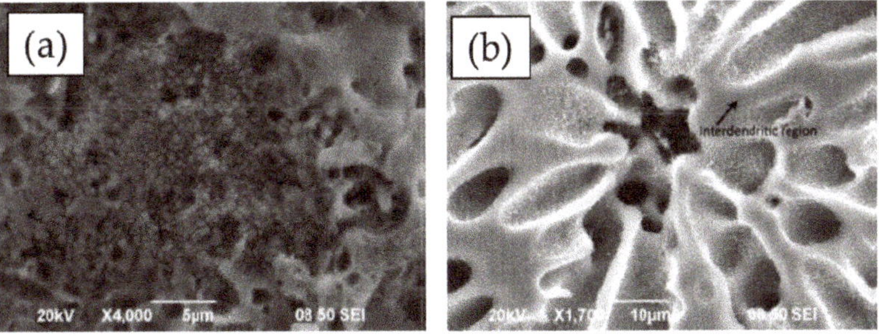

Figure 16. (**a**) Uniform corroded surface of AlCoCrFeNi CCA coating processed by electrospark after polarization test in NaCl, (**b**) Non-uniform corroded surface of a cast AlCoCrFeNi CCA in the same solution [62] (reprinted with permission from Springer).

Direct current magnetron sputtering has also been used for the fabrication of CCA coating with a uniform and homogenous microstructure consisting of very fine grains and low levels of segregation. Coatings fabricated via this method typically showed amorphous microstructure at initial stages of deposition, which crystallized with the increase in deposition time. AlCoCrCuFeMn CCA coating fabricated by magnetron sputtering with a thickness of 1–2 μm showed better corrosion resistance than 201 stainless steel in NaCl, NaOH, and H_2SO_4, with a wide passive region due to fine grains and limited segregation in the microstructure.

Overall, the corrosion resistance of several CCAs are reported to be comparable or better than stainless steels. This may be attributed to the larger fraction of constituent elements, such as Co, Cr, and Ni in the alloy that improve the pitting resistance and improve passivation. Addition of copper was found to induce phase separation and formation of galvanic couples. Vast majority of CCAs reported so far have corrosion potential between −200 to −400 mV and corrosion current density less than 2 μAmps/cm^2 as summarized in Figure 17. The corrosion current density is lower than stainless steels although corrosion potentials are comparable. Another metric for evaluating the corrosion behavior is the pitting resistance (ΔE), measured as the difference between corrosion potential and pitting potential. The pitting resistance of several CCAs are compared with stainless steels in Figure 18. Some CCAs show two times higher pitting resistance when compared to stainless steel. Table 2 is a summary of reported CCAs, their microstructure, corrosion environment, and type of polarization along with the major finding in each case.

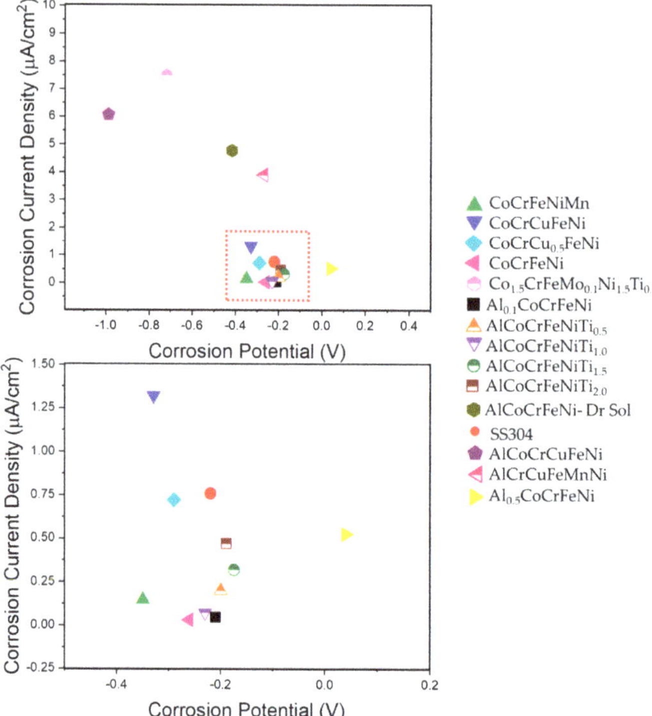

Figure 17. Corrosion current density versus corrosion potential for CCAs/high entropy alloy (HEAs) in 3.5 wt% NaCl solution [18,28,29,31,32,34,51,63–65].

The overall corrosion behavior of CCAs was observed to be dependent on three major factors. First, the composition of the alloys—this in turn affects the nature of the passivation layers, and the relative galvanic characteristic of the constituent phases; second, the environment in which corrosion is being evaluated; and third, the processing parameters. Most of the CCAs investigated showed better corrosion resistance as compared to stainless steels. This is primarily due to the high content of elements that form a passivating oxide layer. For example, SS316 has ~18% Cr, ~12% Ni, and ~2% Mo. In contrast, most of the CCAs that are made of equimolar proportions have at least 20% Cr, 20% Co, and 20% Ni, all of which provide strong passivating effect that translates into better corrosion resistance. Further, the HEA subset showed high resistance to uniform corrosion since these alloys form a single phase structure devoid of galvanic coupling. The general observation of lowering of corrosion resistance with multi-phase CCAs is in line with the galvanic series of alloys. Cu was observed to be particularly detrimental in several of these alloys since it is not only anodic with respect to the passivating elements, but also precipitated in the form of secondary phases that acted as the preferred corrosion sites. No particular relationship between the crystal structure (FCC or BCC) and corrosion resistance was observed. However, phase mixtures had lower corrosion resistance when compared to isomorphous systems. Intermetallic phases, such as borides, aluminides, and Nickel Titanates acted as corrosion initiation sites. The matrix region around the intermetallics dissolve due to anodic character that cause the particles to dislodge. Rupture of passive layer promotes rapid dissolution of the underlying alloy and associated material degradation. Similar effects were observed in Al, B, Mo, and Ti; however, the extent of deterioration varied significantly. Phase morphology was also found to play an important role. Secondary phases with needle and plate-like features dissolved more rapidly when compared to uniformly distributed equiaxed phases, likely because of unfavorable anode to cathode ratio at the tips.

The test environment and corroding species determined the electrochemical kinetics. In general, Cl^- containing solutions caused more corrosion damage as compared to acidic or alkaline solutions. There are limited reports on in vitro and in vivo corrosion studies for bio-medical applications. Processing parameters affect the microstructure, which in turn affects the corrosion behavior of CCAs. Powder processing route was observed to produce more corrosion resistant alloys due to homogeneous elemental distribution, whereas lower corrosion performance was seen in alloys that were produced via melt-casting routes due to coring and segregation. There are significant knowledge gaps on the response of CCAs to welding and joining treatments and associated weld-induced sensitization.

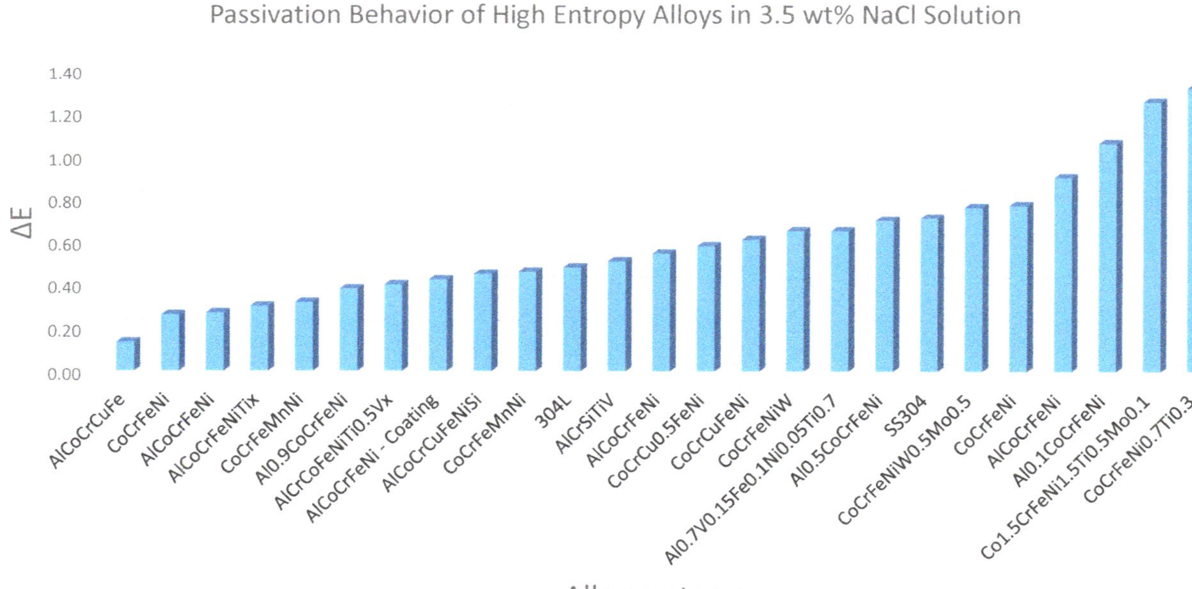

Figure 18. Pitting resistance of CCAs/HEAs in 3.5 wt% NaCl solution. Pitting resistance is calculated as the difference of corrosion potential and breakdown potential [18,28,29,31,32,34,51,63–65].

Table 2. Summary of potentiodynamic/anodic polarization tests of various complex concentrated (high entropy) alloys in aqueous, acidic, and sulfate solutions.

Complex Concentrated Alloy	Microstructure	Corrosion Environment	Test Procedure/Analysis	Major Finding
$AlCoCrCu_{0.5}FeNiSi$ [18]	Two Phase: Dendritic phase: mixture of amorphous and BCC; Inter-dendritic phases: amorphous and nano-scale precipitates	3.5 wt% NaCl, H_2SO_4 at 30–70 °C	Anodic polarization	CCA showed overall better corrosion resistance than SS304, but had poor pitting resistance; Corrosion resistance was lower than SS304 at higher temperatures
$Al_{0.1}CoCrFeNi$ [31]	Single phase FCC	3.5 wt% NaCl	Potentiodynamic Polarization	Very high corrosion resistance, passive region as wide as 1 V; Grain boundary corrosion, very low corrosion *current density*
$Al_{0.3}CrFe_{1.5}MnNi_{0.5}Ti_x$ [61]	BCC with increasing intermetallic with increasing Ti	3.5 wt% NaCl	Potentiodynamic Polarization	Adding Ti lowered corrosion resistance
$Al_{0.3}CrFe_{1.5}MnNi_{0.5}Si_x$ [61]	BCC with increasing intermetallic with increasing Si	3.5 wt% NaCl	Potentiodynamic Polarization	Adding Si lowered corrosion resistance
$Al_{0.5}CoCrCuFeNiB_x$ [39]	FCC + BCC crystal structure	1 N H_2SO_4	Anodic polarization	CCA was nobler than SS304 in terms of corrosion current density, and corrosion potentials; not susceptible to localized corrosion in sulfate solutions
$Al_{0.5}CoCrFeNi$ [35]	FCC solid solution matrix with secondary phases rich in Al–Ni	3.5 wt% NaCl	Potentiodynamic Polarization	Secondary phases rich in Al–Ni were more susceptible to corrosion
$Al_2CrFeNiCoCuTi_x$ [51]	x = 0.0 FCC + BCC_1 x = 0.5 FCC + BCC_2 x = 1.0 BCC_1 + BCC_2 x = 1.5 BCC_1 + BCC_2 x = 2.0 FCC + BCC_2	0.5 M HNO_3	Potentiodynamic Polarization	Increasing Ti content increased corrosion resistance in terms of corrosion current density
AlCoCrCuFeNi [37]	FCC + BCC two phase mixture	1 mol/L NaCl	Potentiodynamic Polarization	Good corrosion resistance despite two phase structure
AlCoCrFeNi [62]	BCC	3.5 wt% NaCl	Potentiodynamic Polarization	Corrosion resistance after processing was three orders of magnitude better than steel and one order of magnitude better than unprocessed HEA
AlCoCrFeNiTi [36]	Complex Microstructure: Al, Co, Ni and Ti rich dendritic phase. Fe and Cr rich inter-dendritic phase. Ti and Ni rich third phase. Ordered phase—A2, B2, $D0_3$ and A12	3.5 wt% NaCl	Polarization	Ti addition improved corrosion resistance of the alloy; Through the re-melting process, the distribution of elements in the alloy improved, improving the corrosion resistance
AlCrCuFeMnNi [63]	Complex Microstructure: BCC dendritic phase, interdendritic area with two phases—a eutectic type and FCC solid solution phase	3.5 wt% NaCl	Potentiodynamic Polarization	CCA was easier to passivate; higher corrosion resistance than SS304L; Galvanic coupling reduced by dissolving Cu during re-melting
$Al_xCoCrCu_{0.5}FeNi$ [41]	x = 0.5 is FCC x = 1.0 is BCC x = 1.5 FCC + BCC	0.5 mol/L H_2SO_4 + 0.5 mol/L NaCl solution	Potentiodynamic Polarization	Single-phase alloys had better corrosion resistance that phase mixtures; BCC alloy was comparable with 321 stainless steel
AlCoCrCuFe [49]	FCC and BCC phase mixture	1 mol/L NaCl and 0.5 mol/L H_2SO_4	Potentiodynamic Polarization	Segregation of Cu was seen in the microstructure; CCA performed better in NaCl solution than in acidic solution
$Al_xCoCrFeNi$ (x = 0, 0.25, 0.50, 1.00) [33]	x = 0 x = 0.25 x = 0.50 x = 1.00	0.5 mol/L H_2SO_4	Potentiodynamic Polarization	Corrosion current density decreased with Al content at 23 °C; Overall superior corrosion resistance compared to steels
$Al_xCoCrFeNi$ [34]	x = 0.3 x = 0.5 x = 0.7	3.5 wt% NaCl	Potentiodynamic Polarization	Increasing Al content decreased corrosion resistance by formation of intermetallic phases

Table 2. *Cont.*

Complex Concentrated Alloy	Microstructure	Corrosion Environment	Test Procedure/Analysis	Major Finding
$Al_xCrFe_{1.5}MnNi_{0.5}$ [66]	x = 0.0 FCC x = 0.3 BCC + FCC x = 0.5 BCC	1 mol/L NaCl + 0.5 mol/L H_2SO_4	Potentiodynamic Polarization	Alloys showed extended passive region, greater than 1 V; Increasing Al lowered corrosion resistance in terms of pitting behavior
$B_xCoCrFeNi$ [67]	x = 0.5 FCC x = 0.75 FCC x = 1.0 FCC x = 1.25 FCC + M_2B	3.5 wt% NaCl	Potentiodynamic Polarization	Corrosion resistance improved with increasing B content up to 1%., beyond which corrosion resistance decreased; The CCAs showed superior corrosion resistance than SS304
$Co_{1.5}CrFeNi_{1.5}Ti_{0.5}Mo_{0.1}$ [43]	FCC solid-solution structure	0.001 to 1 M NaCl and sulfate doped 1 M NaCl	Potentiodynamic Polarization	Sulfate ions increased the pitting potential and critical pitting potential of the alloys
$Co_{1.5}CrFeNi_{1.5}Ti_{0.5}Mo_x$ [42]	FCC solid-solution	0.5 M H_2SO_4 1 M NaCl and NaOH	Potentiodynamic Polarization	Mo addition lowered the overall corrosion resistance
$CoCrCu_{0.5}FeNi$ [68]	Dendritic Structure: Copper lean dendritic phase, copper rich interdendritic phase, aged at different temperatures	3.5 wt% NaCl	Potentiodynamic Polarization	Corrosion current density lowered while corrosion potential decreased with aging temperature; Corrosion properties worsened when heat treated at 1100–1350 °C. Pitting increased with aging temperature
CoCrCuFeNiNb [55]	FCC and Laves phases	6 M HCl	Potentiodynamic Polarization	Alloying with Nb lowered corrosion current density
CoCrFeMnNi [31]	Simple single phase FCC	3.5 wt% NaCl	Potentiodynamic Polarization	~500 mV wide passivation region; Corrosion rate as low as one micron per year
$CoCrCu_xFeNi$ [29]	FCC and Cu rich FCC	3.5 wt% NaCl	Potentiodynamic Polarization	Addition of Cu deteriorated the corrosion resistance
			Immersion Test	Galvanic corrosion between inter-dendritic region and dendrite resulting in localized corrosion
$CoCuFeNiSn_x$ [69]	Single phase FCC solid solution when Sn < 0.09, small BCC phase beyond that	3.5 wt% NaCl and 5% NaOH	Potentiodynamic Polarization	CCAs showed wide passivation range in NaOH and relatively smaller region in NaCl; Better resistance than SS304; $FeCoNiCuSn_{0.04}$ showed improved corrosion resistance
$CoCuFeNiSn_x$ [69] x = 0–0.09	FCC when $x < 0.09$, small BCC for $x > 0.09$	3.5 wt% NaCl and 5% NaOH	Potentiodynamic Polarization	Better corrosion resistance than SS304 alloy when tested in NaCl, while lower corrosion resistance when tested in NaOH
$Cr_{0.5}NbTiZr_{0.5}$, $Cr_{0.5}NbTiVZr_{0.5}$ $Cr_{0.5}MoNbTiZr_{0.5}$ [70]	Dendritic Structure: BCC Disordered Solid Solution phase and Cr_2Zr phase	3.5 wt% NaCl and 0.5 M H_2SO_4	Potentiodynamic Polarization	Superior corrosion resistance, with passive region more than 1400 mV; Mo and V addition decreased corrosion resistance but improved pitting resistance in NaCl and H_2SO_4
$CoCrCu_xFeNi$ [29]	FCC crystal structure, Cu rich interdendritic phase	3.5 wt% NaCl	Potentiodynamic Polarization	Increasing Cu content caused segregation into inter-dendritic phases, and consequent deterioration of corrosion resistance; General corrosion trend was seen as FeCoNiCrCu > $FeCoNiCrCu_{0.5}$ > FeCoNiCr
$CuCr_2Fe_2Ni_2Mn_2$ $Cu_2CrFe_2NiMn_2$ [44]	FCC FCC + BCC	1 M H_2SO_4	Potentiodynamic Polarization	Cr_2 alloy showed better corrosion resistance while Cu_2 alloy promoted segregation and had lower corrosion resistance
AlCoCuFeNi AlCoCuFeNiCr AlCoCuFeNiTi AlCoCrCuFeNiTi [40]	FCC + A2 + B2	0.5 mol/L H_2SO_4	Potentiodynamic Polarization	Adding Ti decreased the corrosion resistance of the AlCoCuFeNi alloys, whereas adding Cr improved corrosion resistance

4. Erosion and Erosion Corrosion of CCAs

There are very limited number of studies on the erosion behavior of CCAs/HEAs. The two alloy systems that have been studied the most are Al_xCoCrFeNi and Al_xCoCrCuFeNi. Slurry-erosion behavior of Al_3CrCoFeNi laser cladded CCA was compared with conventionally used 17-7 precipitation hardened (PH) stainless steel. The Al_3CrCoFeNi CCA coating showed excellent erosion resistance when compared to 17-7 PH stainless steel with seven times higher resistance at 15° impingement angle [71]. Maximum erosion rate for both CCA and 17-7 PH steel were observed at 45° impingement angle. Thereafter, the erosion rates were more or less constant with a further increase in impingement angle. Higher erosion resistance of the Al_3CrCoFeNi CCA was due to high hardness (~750 HV) of the BCC phase and severe lattice strains. Significant lattice distortion was attributed to the high mole fraction of large sized Al atom. The effect of heat treatment on erosion behavior of Al_3CrCoFeNi CCA coating was also investigated. Increased erosion resistance (~15% compared to untreated HEA) was seen with increase in annealing temperature with maximum corresponding to the 950 °C heat-treated sample. Authors attributed the enhanced erosion resistance of the CCA coating treated at 950 °C to increased hardness (765 HV), resulting from Cr_3Ni_2 precipitation and reduced roughness.

In Al_xCoCrFeNi alloy system, decrease in Al content results in transition from pure BCC to BCC + FCC and finally to pure FCC structure [72]. $Al_{0.1}$CoCrFeNi alloy shows a single-phase solid solution of face centered cubic (FCC) structure with good thermal stability. Slurry erosion behavior of $Al_{0.1}$CrCoFeNi CCA was evaluated at different impingement angles (30° to 90°) and a constant impact velocity (20 m/s). Despite the low hardness of 150 HV, the cast $Al_{0.1}$CrCoFeNi alloy displayed erosion resistance that is comparable to or better than mild steel (of hardness 205 HV) at acute angles (Figure 19a). At normal impingement, $Al_{0.1}$CrCoFeNi showed much better erosion resistance when compared to mild steel, which was attributed to the significant work hardening ability of the alloy. Continuous impact of abrasive particles during the erosion test results in significant work hardening of the HEA due to its low stacking fault energy and nano-twin formation [73]. The stacking fault energy (SFE) for $Al_{0.1}$CrCoFeNi high entropy alloy is reported to be about 30 mJ/m^2 [74]. However, $Al_{0.1}$CoCrFeNi alloy showed lower erosion resistance when compared to stainless steel SS316L due to higher hardness and strength of the later. Correlation with different mechanical properties showed that the ultimate strength and ultimate resilience significantly affected the erosion behavior in these multi-component metallic systems.

The slurry erosion-corrosion behavior of AlCrCoCuFeNi CCA after annealing at different temperatures (600 °C and 1000 °C) was studied [27]. Both untreated and heat-treated AlCrCoCuFeNi alloy showed high erosion resistance compared to SS304 stainless steel. Untreated AlCrCoCuFeNi alloy also showed higher corrosion resistance compared to SS304 stainless steel. However, sample annealed at 600 °C showed significantly reduced corrosion resistance, which was attributed to precipitation of intermediate phases. Further increase in annealing temperature improved corrosion resistance from the resulting microstructural homogeneity. In contrast to corrosion studies, the combined erosion-corrosion test showed distinctly different behavior for the sample annealed at 600 °C, exhibiting the lowest mass loss. The improvement in erosion-corrosion resistance was predominantly due to increased hardness (~500 HV) due to the formation of ordered B_2 or disordered A_2 structures from the annealing process. Addition of higher Al fraction in the AlCrCoCuFeNi system was observed to increase hardness due to the formation of BCC/B_2 structure and improved erosion-corrosion resistance. However, lowering the Al content improved the corrosion behavior [34]. The $Al_{0.1}$CrCoFeNi HEA showed high slurry erosion-corrosion resistance (~1.8 times higher) as compared to SS316L stainless steel with significant negative synergy, as shown in Figure 19b. The negative synergy for $Al_{0.1}$CrCoFeNi CCA indicates the positive contribution of corrosion in lowering the mass loss during the erosion-corrosion test. The $Al_{0.1}$CrCoFeNi CCA also showed high pitting and protection potentials as compared to SS316L steel, indicating the formation of stable passive layer. Stability of the passive layer was partly attributed to the high mixing entropy resulting in high activation energy for diffusion.

Ti addition was found to enhance the corrosion and cavitation erosion resistance of complex concentrated alloys [28]. Cavitation erosion-corrosion behavior of laser cladded AlCrCoFeNiTi$_x$ CCA (x = 0.5 to 2) was evaluated. Maximum cavitation erosion resistance was observed for AlCrCoFeNiTi$_2$ HEA. However, the trend reversed completely for the cavitation erosion-corrosion test, with AlCrCoFeNiTi$_2$ CCA showing the least resistance. This trend reversal was explained by the formation of Ti$_2$Ni and NiAl intermetallic compounds. These intermetallics significantly enhanced the hardness and reduced erosion rates due to increased resistance to plastic deformation. At the same time, formation of intermetallic compounds degraded the corrosion resistance due to the formation of localized galvanic cells and unstable passive layer.

The cavitation erosion-corrosion performance of AlCrCoFeNi laser cladded CCA was compared with 304 stainless steel [50]. The AlCrCoFeNi coating showed 7.6 times better cavitation erosion-corrosion resistance compared to 304 stainless steel. The better performance of the CCA was attributed to combined effect of high hardness and corrosion resistance. The high hardness resulted from the BCC solid solution and corrosion resistance from the homogeneous microstructure without any intermetallic phases. When compared to E_{pit} of 96 mV observed for SS304 steel, AlCrCoFeNi laser cladded CCA showed significantly high pitting potential of 257 mV indicating higher passive layer stability of the later.

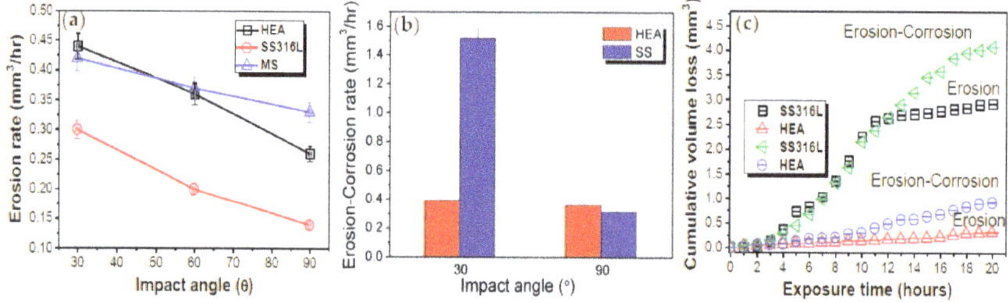

Figure 19. (**a**) Slurry erosion rate [26]; (**b**) slurry erosion-corrosion rate and [75]; (**c**) cumulative volume loss under cavitation erosion and erosion-corrosion of Al$_{0.1}$CrCoFeNi high entropy alloy compared to stainless steel SS316L [76].

Al$_{0.1}$CrCoFeNi alloy showed remarkable resistance to cavitation erosion and erosion-corrosion compared to 316L stainless steel as shown in Figure 19c [76]. Additionally, the alloy showed a much longer incubation period of 6.5 h as compared to 2.5 h for 316L stainless steel. This was attributed to comparatively greater degree of work hardening and superior corrosion resistance of Al$_{0.1}$CrCoFeNi alloy. The strain hardening exponent for Al$_{0.1}$CrCoFeNi (n = 0.77) was more the two times that of the 316L stainless steel (n = 0.3). Higher strain hardening increased the incubation period and lowered the erosion rates by effectively increasing the flow stresses. Superior erosion-corrosion resistance of the alloy may also be explained by its higher pitting potential (E_{pit} = 490 mV) and protection potential (E_{pp} = 184 mV) when compared to SS316L (E_{pit} = 359.8 mV; E_{pp} = −10.58 mV).

Scanning electron microscopy (SEM) images of Al$_{0.1}$CrCoFeNi alloy and 316L stainless steel after slurry erosion and cavitation tests under identical conditions are shown in Figure 20 [26,76]. The SS316L steel and Al$_{0.1}$CrCoFeNi CCA both showed ductile mode of erosion in slurry and cavitation erosion. In the case of slurry erosion, micro-cutting and ploughing were the prominent material removal mechanisms observed at oblique angles. For normal impingement, material removal was mainly through formation and removal of platelets (platelet mechanism). When compared to micro-cutting, the material removal for Al$_{0.1}$CrCoFeNi alloy at an oblique angle was mainly through ploughing mechanism due to higher ductility as compared to SS316L steel. In addition, few micro-indents were also observed for samples that were tested at 90°. Micro-indentation resulted in severe plastic deformation and the removal of the strained material once accumulated strain reached a critical value [77]. From the cavitation erosion-corrosion test, the formation of craters and pits were observed as the primary damage mechanism (Figure 21). The size of the craters were significantly larger for SS316L steel, while they were virtually absent for the CCA as seen in Figure 21. High ductility and strain hardening for $Al_{0.1}$CrCoFeNi alloy played an important role in limiting crack formation/propagation and material loss.

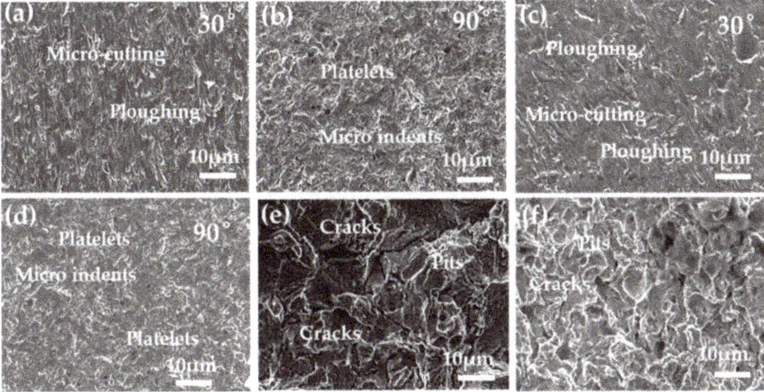

Figure 20. Scanning electron microscope images showing the damage mechanism due to slurry erosion of (**a**,**b**) stainless steel 316L and (**c**,**d**) $Al_{0.1}$CrCoFeNi high entropy alloy at different impingement angles. SEM images of cavitation eroded (**e**) SS316L steel and (**f**) $Al_{0.1}$CrCoFeNi CCA samples tested for 20 h [26,76] (reprinted with permission from Elsevier and Wiley).

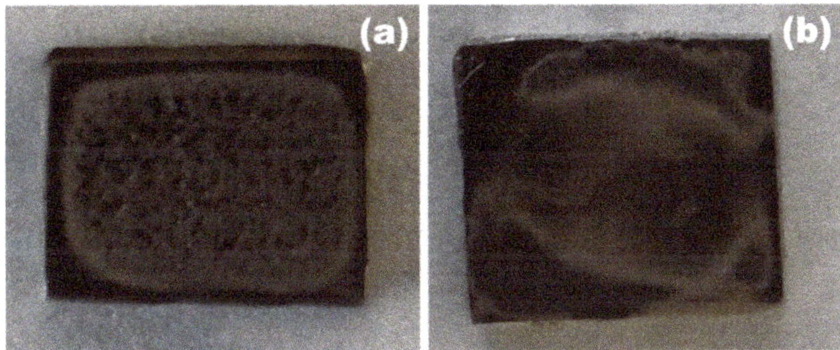

Figure 21. Macrographs showing the (**a**) SS316L steel and (**b**) $Al_{0.1}$CrCoFeNi high entropy alloy (HEA) samples after cavitation erosion-corrosion testing for 20 h.

Complex concentrated alloy coatings have also been studied for their erosion behavior [28,45,78–81]. Al_xCrCoFeNi (x = 0.1 to 3) CCA coatings were developed using microwave processing on SS316L steel substrate, with microstructure consisting of intermetallic phases, as shown in Figure 22 [82]. The matrix in these coatings was composed of either FCC or BCC phases depending on the Al fraction. The average micro-hardness showed a direct correlation with Al fraction in the coatings. The average hardness for $Al_{0.1}$CrCoFeNi coating was 438 HV, which was significantly higher when compared to the bulk counterpart, which is mainly due to the difference in microstructure [26]. Maximum hardness of 624 HV was obtained for Al_3CrCoFeNi CCA coating. Slurry erosion studies of these coating showed a significantly higher erosion resistance at an oblique impingement angle (30°) when compared to SS316L steel. The maximum erosion resistance was observed for the equimolar composition. However, for normal impingement all of the coatings showed an increase in erosion rate indicating brittle behavior. The increase in erosion rate at normal impingement was attributed to the brittle intermetallic σ and B2 phases. Lowest erosion rate observed for the equimolar AlCrCoFeNi

CCA coating at both acute angles and normal impingement was attributed to the combination of higher hardness resulting from the secondary phases and fracture toughness. SEM image of the slurry eroded CCA coating that was tested at 90°shows the presence of large craters for the non-equimolar compositions as shown in Figure 22. The cracks resulted in disintegration of the secondary phase, which is most prominent for $Al_3CrCoFeNi$. In contrast, the equimolar composition showed lesser tendency for brittle fracture.

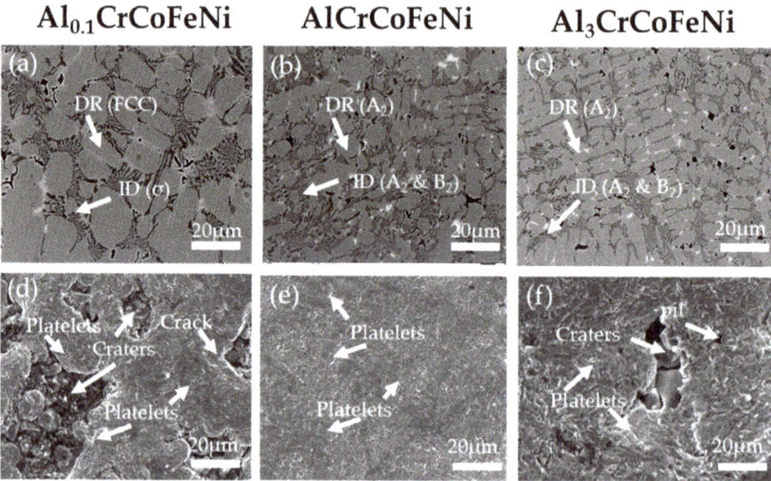

Figure 22. Microwave processed coatings of $Al_xCrCoFeNi$ high entropy alloys (x= 0.1 to 3). (**a–c**) cellular microstructure of the synthesized coatings with dendrite region (DR) and inter dendrite region (ID); and, (**d–f**) microstructures after slurry erosion test at normal impingement angle (90°) for $Al_{0.1}CrCoFeNi$, AlCrCoFeNi and $Al_3CrCoFeNi$ high entropy alloys [82].

A comparative analysis of cavitation erosion and erosion-corrosion behavior of CCAs with respect to conventional structural materials [28,50,83–87] is shown in Figure 23 in terms of the mean depth erosion rates (MDER). For both of the test conditions, CCAs show much lower MDER compared to conventional alloys, such as stainless steels. Therefore, the use of CCAs in applications demanding high erosion and erosion-corrosion resistance can effectively improve the durability and service life of the susceptible components.

The existing literature provides a fair understanding on the erosion and erosion-corrosion behavior of $Al_xCrCoFeNi$-X high-entropy alloy systems. There is a large scope for investigation of erosion-corrosion behavior of alloys for several other compositions with high hardness and corrosion resistance. The alloys in the $Al_xCrCoFeNi$-X system tend to have higher erosion-corrosion resistance as compared to stainless steels, which may be due to the higher content of passivating elements. The studies reported so far have investigated the room-temperature properties of the alloys. However, critical knowledge gaps exist for high-temperature erosion-corrosion behavior. For example, boiler tubes are susceptible to extreme conditions and high temperature erosion-corrosion. In addition, the synergistic effects in cavitation erosion and slurry erosion due to the presence of corrosive media need to be investigated for better understanding of degradation in marine environments. Biofouling behavior of HEAs is another area for future research with high impact.

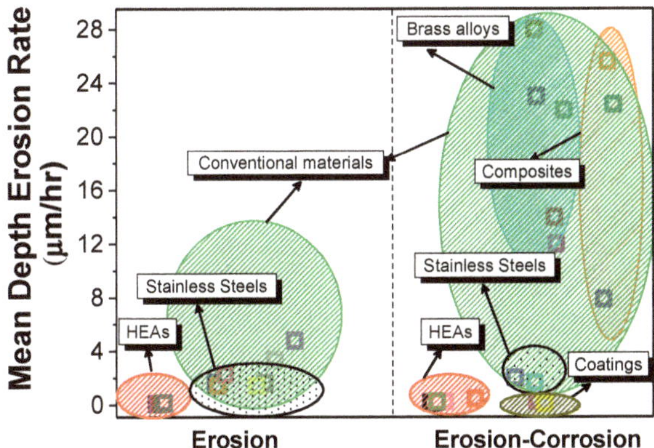

Figure 23. Comparison of high entropy alloys with conventional structural materials for erosion and erosion-corrosion resistance. "Conventional materials" in the figure refers to a broad range of materials developed for erosion/corrosion applications such as SS304 [28,50], SS304L, mild steel, Bainitic Steel [86] and Copper Alloys [87]. "Coatings" refer to coatings on AA6061 [88] for erosion mitigation.

5. Wear Behavior of CCAs

All of the complex concentrated (high entropy) alloy systems that have been studied so far for their wear behavior are summarized in Figure 24. They are broadly classified based on the alloy chemistry and processing. In addition to wear behavior, the hardening response, phase stability, and hot hardness was reported for some alloys. The copper containing alloys were typically single phase or mixture of two simple phases. The copper free alloys were based on AlCoCrFeNi system and modified with Ti or Mo. The copper free alloys showed higher wear resistance when compared to the copper containing alloys. In addition, wear behavior of alloys that are composed of purely refractory elements have also been reported. Surface modification was done for some conventional steels and Ti alloys by LASER cladding, tungsten inert gas (TIG), and sputtering to enhance the wear properties.

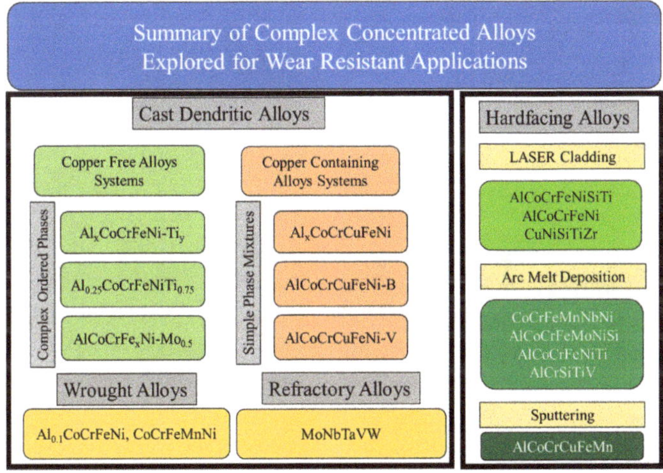

Figure 24. Summary of complex concentrated alloys studied for their tribological behavior.

The wear behavior of $Al_xCoCrCuFeNi$ [89] alloy system has been systematically investigated by varying the Al content from $x = 0.5$ to $x = 2$. In the $Al_xCoCrCuFeNi$ cast alloys, there was Cu segregation in the inter-dendritic region, while other elements enriched the dendrites. At $x = 0.5$, the dendritic and inter-dendritic phases were both FCC, which changed with increasing Al content. Increasing Al content stabilized BCC phases and resulted in hardness increase by several times The Archard's wear relation was found to hold true in the case of $Al_xCoCrCuFeNi$ alloy system with improving wear resistance from the increased hardness.

The softer composition in $Al_xCoCrCuFeNi$ system with $x = 0.5$ showed ductile deformation, grooving, and disc-like wear debris, as shown in Figure 25, all of which are in line with ductile character of the FCC-rich alloy. On the other hand, the BCC-rich alloys showed smoother surface deformation, finer wear debris that were enriched with oxygen, indicating oxidative wear. The steady state friction value was the lowest for $Al_{0.2}CoCrCuFeNi$, which is in line with the surface oxidation and high hardness that protected the alloy from tribo-degradation. Based on variation in hardness, ductile deformation, and steady state friction values in $Al_xCoCrCuFeNi$ alloy system, the $Al_{0.5}CoCrCuFeNi$ composition was chosen for further modification. Addition of V to $Al_{0.5}CoCrCuFeNi$ alloy resulted in a phase mixture consisting of FCC, BCC, and sigma phase [90]. Fixing aluminum content at 0.5 and gradually increasing V content increased the BCC phase fraction. The change in microstructure with increasing V mole fraction is shown in Figure 26. Increasing the vanadium concentration increased BCC phase fraction and hardness of the alloy, but this did not translate into improved wear resistance. There was marginal increase in wear resistance even though the hardness increased from 200 HV to a peak hardness of ~650 HV. The optimum composition range for enhanced mechanical and tribological properties was in the range of 1.0 to 1.2 mole fraction of Vanadium. Therefore, multiple competing factors affected the wear behavior, including complex microstructure, elemental segregation, BCC phase fraction, and morphology.

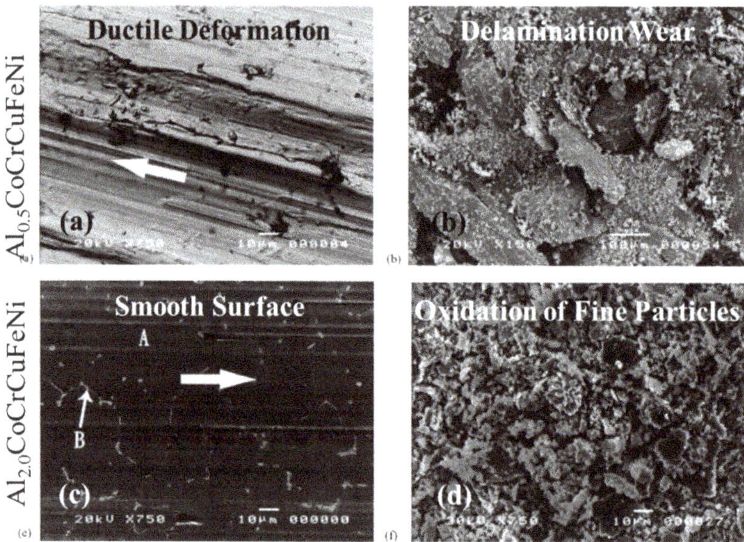

Figure 25. Deformation and wear mechanisms for $Al_{0.5}CoCrCuFeNi$ alloys: (**a**) Ductile deformation showing long grooves with minimal lateral cracks (**b**) wear particles showing large flakes indicating delamination (**c**) smooth surface characteristic of bcc and high hardness alloys, (**d**) smaller wear particles showing significantly higher oxidation seen from lighter shades from oxide charging [89] (reprinted with permission from Elsevier).

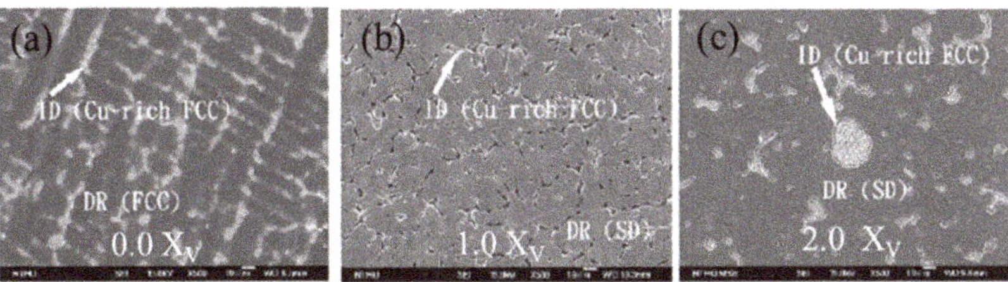

Figure 26. Microstructure of the $Al_{0.5}CoCrCuFeNi$ alloy at (**a**) 0.0, (**b**) 1.0 and (**c**) 2.0 mole fraction V. The initial microstructure had the lowest hardness value, which was observed to increase with increasing vanadium content. All three microstructures showed segregation of copper into inter-dendritic regions [90] (reprinted with permission from Springer).

Boron is known to be a BCC phase stabilizer in ferrous alloys. In contrast, this effect was not seen with boron addition to $Al_{0.5}CoCrCuFeNi$. The alloy retained its FCC structure when B was less than 10%, while small quantities of ordered FCC phases evolved when the content was increased to ~15% [38]. In contrast to Vanadium addition, changing B content led to a significant strengthening effect. Addition of boron to the AlCoCrCuFeNi system was observed to increase the hardness and wear resistance of the alloy to 736 HV and 1.76 m/mm^3, respectively. The wear resistance of the alloy was superior when compared to wear resistant SUJ bearing steels. Hardness increased from ~300 HV to 750 HV along with nearly doubling of wear resistance when boron content was increased from 0% to 15%. The changes in hardness and wear resistance with the addition of Vanadium, Boron, and Aluminum are summarized in Figure 27.

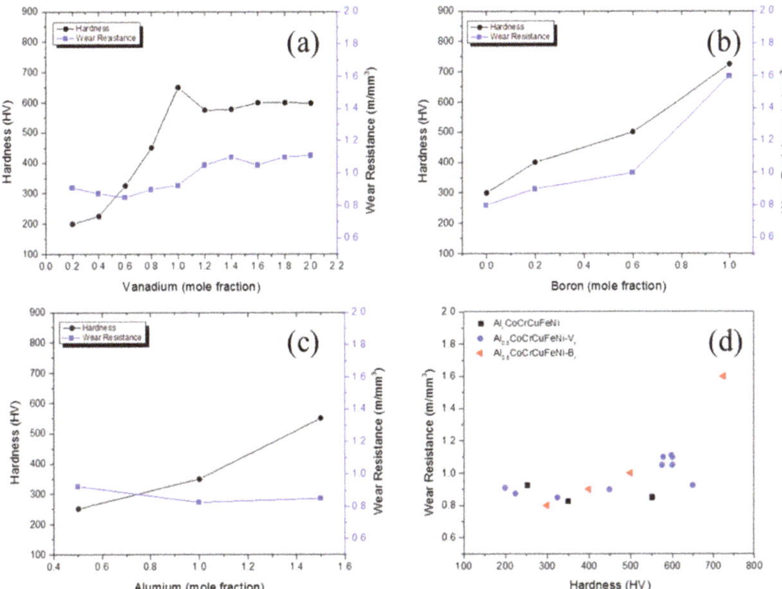

Figure 27. Summary of mechanical, tribological and metallurgical properties of $Al_{0.5}CoCrCuFeNi$-X alloy system. Improved wear resistance of the alloys with increasing proportions of (**a**) Vanadium (**b**) Boron and (**c**) Aluminum; (**d**) relative comparison and summary of wear resistance of $Al_{0.5}CoCrCuFeNi$-X alloys where X is V, B and $Al_yCoCrCuFeNi$ where y is 0.3, 1.0 and 1.5 [38,89,90].

The CoCrFeNi alloy system has been shown to be very versatile in terms of compositional and microstructural modifications. Addition of Ti results in the formation of Ni_3Ti intermetallic compounds that impart exceptional high temperature strength. Therefore, high entropy forming CoCrFeNi composition may be significantly strengthened by these intermetallic phases. Extensive studies have been done to understand the competing effects of Al and Ti in the $Al_xCo_{1.5}CrFeNi_{1.5}Ti_y$ system by creating a series of alloys [91]. Al-free $Co_{1.5}CrFeNi_{1.5}Ti_{0.5}$ and $Co_{1.5}CrFeNi_{1.5}Ti_{1.0}$ alloys were developed for isolating the effect of Ti, while $Al_{0.2}Co_{1.5}CrFeNi_{1.5}Ti_{0.5}$ and $Al_{0.2}Co_{1.5}CrFeNi_{1.5}Ti_{1.0}$ were developed to identify the effect of Al.

Addition of $Ti_{1.0}$ improved the hardness by about 100 HV when compared to the $Ti_{0.5}$ bearing alloys. Addition of Al caused slight drop in hardness due to suppression of $(Ni,Co)_3Ti$ formation. The overall hardness of $Al_{0.2}Co_{1.5}CrFeNi_{1.5}Ti_{1.0}$ improved by over 200 HV as compared to Al free counterparts. Among the four alloy systems $Al_{0.2}Co_{1.5}CrFeNi_{1.5}Ti_{1.0}$ showed highest wear resistance, being as high as 5500 m/mm^3, which is nearly double the values on the other alloys. The tribological properties of the alloys can be compared to commercial wear resistant steels, such as SUJ2 and SKH51. As far as the mechanisms are concerned, the softer alloys displayed characteristic features of wear on ductile materials, such as delamination, grooving, and plastic deformation, as shown in Figure 28. The degradation mechanism for the high Ti content alloys was predominantly oxidative wear in the tribo-system, shallow wear tracks, and marks on the surface. These oxide layers typically protect the underlying alloy from meta-metal contact, thereby reducing adhesive wear. Although the tests were conducted at room temperature, the contact temperature during wear may rise to temperatures where softening and oxidation may be of concern. Therefore, high temperature hardness is important in these applications. Unlike other alloys derived from the CoCrFeNi base system, the AlCoCrFeNiTi alloy forms high temperature deformation resistant intermetallic particles. These particles effectively improve the alloy performance and contribute to improved tribological properties.

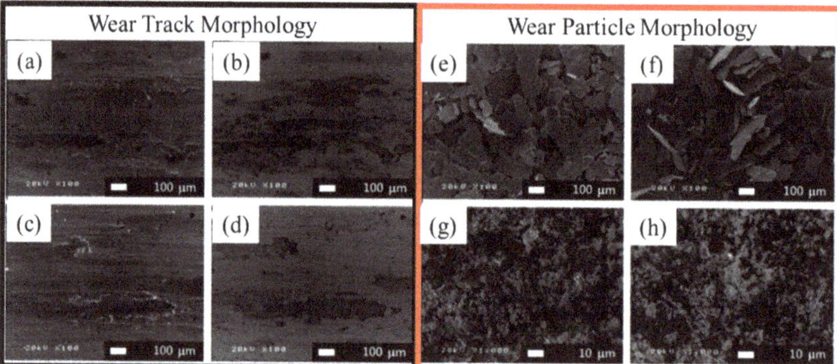

Figure 28. (**a–d**) Wear track morphology and wear particle morphology showing clear (**e**,**f**) delamination wear and (**g**,**h**) partial oxidation wear on the $Al_xCo_{1.5}CrFeNi_{1.5}Ti_y$ alloys [91] (reprinted with permission from Elsevier).

Al and Ti promoted the formation of ordered phases and had a significant effect on the hardness and wear resistance of alloys. Extensive microstructure characterization was done and wear behavior was studied for the $Al_{0.25}CoCrFeNiTi_{0.75}$ alloy [92]. The $Al_{0.25}CoCrFeNiTi_{0.75}$ alloy was seen to have a Cr-Fe-Co rich phase, a Chi (χ) phase with BCC structure, a Ni_2AlTi based L21 ordered phase, and FCC minor phase. A scanning electron microscopy image of the alloy along with the EDS elemental maps is shown in Figure 29. The EDS maps showed that the lighter contrast phase is chi (χ) phase, the darker contrast phase as the FCC based ordered phase, and the greyish phase as disordered FCC. This is distinct from the $Al_{0.2}Co_{1.5}CrFeNi_{1.5}Ti_{1.0}$ alloy that showed blocky complex η-$(Ni,Co)_3Ti$ phase with needle like morphology. The hardness of the alloy was also observed to be around

570 HV for the (L21) phase, while, lighter contrast matrix regions showed a hardness of 1090 HV. Comparatively, these values are lower than the $Al_{0.2}Co_{1.5}CrFeNi_{1.5}Ti_{1.0}$ that showed η phase with Widmanstätten structures having a hardness of 1200 HV, and an overall hardness of 717 HV. The wear behavior of the alloy when tested in sliding reciprocating mode showed long grooves running parallel to the wear track. The wear tracks were shallow, with little to no surface oxidation. The wear behavior of the alloy followed Archard's relation. Between the two phases, the softer dark phase composed of Al-Ti-Ni was observed to wear off preferentially. The size of the wear tracks increased with increasing test load. These features are shown in Figure 30. The $Al_{0.2}Co_{1.5}CrFeNi_{1.5}Ti_{1.0}$ was reported to have improved wear resistance than commercial high hardness wear resistant steels, such as SUJ2 and SKH51 (~65 HR_C), whereas the wear performance of $Al_{0.25}CoCrFeNiTi_{0.75}$ CCA was more comparable to SS 440C (~55 HR_C).

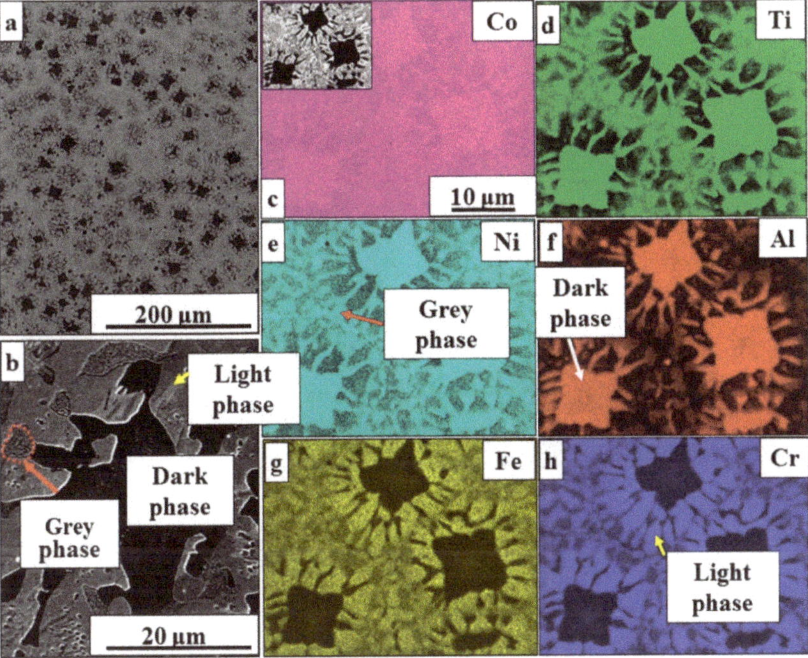

Figure 29. Scanning electron microscope images of $Al_{0.25}CoCrFeNiTi_{0.75}$ alloy in as-cast condition. (**a**) Low magnification scanning electron image of the alloy; (**b**) high magnification image showing the three distinct contrasts - lighter contrast phase is chi (χ) phase, the darker contrast phase as the FCC based ordered phase, and the greyish phase as disordered FCC; EDS elemental maps showing distribution of (**c**) Co (**d**) Ti (**e**) Ni (**f**) Al (**g**) Fe and (**h**) Cr [92] (reprinted with permission from Elsevier).

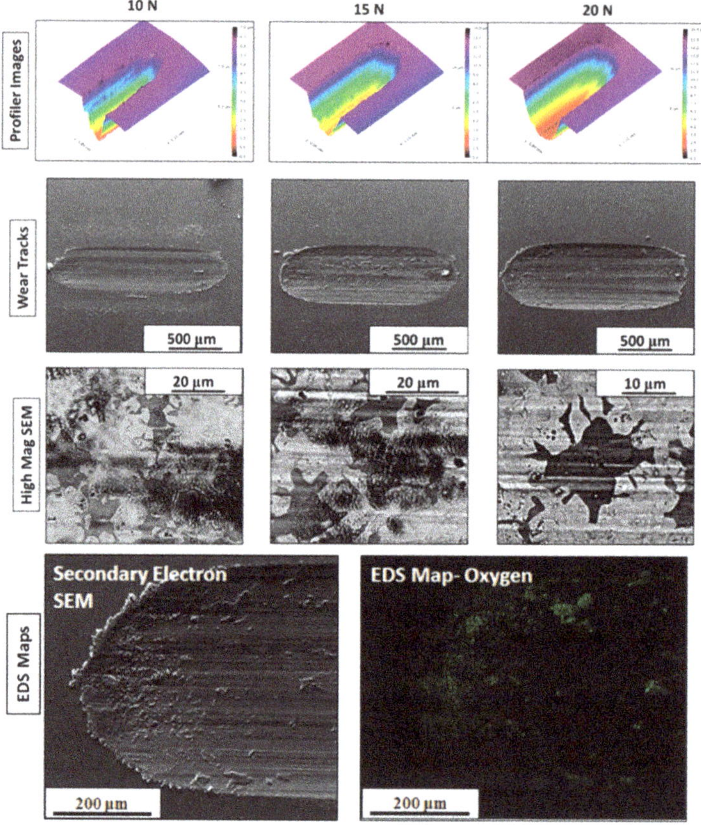

Figure 30. Wear behavior of $Al_{0.2}Co_{1.5}CrFeNi_{1.5}Ti_{1.0}$ alloy. The alloy showed increased wear volume loss with increased load, corresponding well with archard's law. High magnification SEM images, and EDS maps shows oxides on the surface, indicating oxidative wear operating on the sample surface [92] (reprinted with permission from Elsevier).

The wear behavior of high entropy alloys in marine conditions was evaluated for CoCrFeNiMn and $Al_{0.1}$CoCrFeNi alloys. Here, isolating the effects of wear and corrosion that are acting simultaneously on the sample are important. Such problems can be approached by individually assessing the wear in dry condition, corrosion using electrochemial or immersion tests, and comparing the results to marine wear tests. A weighted summation of material loss from each of the wear tests would reveal the predominant mechanism of material loss between the competing mechanisms. In the case of CoCrFeNiMn and $Al_{0.1}$CoCrFeNi alloys, the synergy between wear and corrosion was observed to be negative—implying that corrosion did not aggravate material loss in the alloys. Material loss during marine wear was lower than the summation of dry wear loss and material loss from corrosion [31]. The wear tracks imaged using white light interferometry (Figure 31a,b for dry and Figure 30g,h for wet conditions) show higher wear volume loss on the CoCrFEMnNi alloy in dry and marine condition. High magnification images confirm micro-grooving on CoCrFeNiMn, while spalling and fatigue wear on the $Al_{0.1}$CoCrFeNi alloy. The $Al_{0.1}$CoCrFeNi alloy showed lower wear loss in both dry and wet test conditions (Figure 31c,d for dry and Figure 31i,j for marine condition). Parallel grooves in the wear track indicate two body or three body wear to be operative in the alloys.

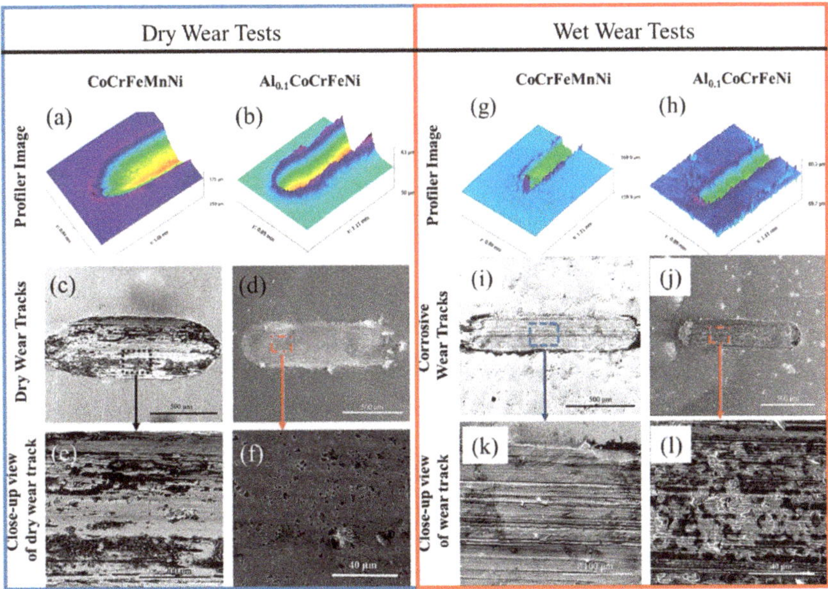

Figure 31. Interferometry images of dry wear track for (**a**) CoCrFeMnNi and (**b**) $Al_{0.1}CoCrFeNi$. Scanning Electron Microscope images of wear track on (**c**) CoCrFeMnNi and (**d**) $Al_{0.1}CoCrFeNi$; Higher magnification images of wear tracks for (**e**) CoCrFeMnNi showing coarse microabrasion/microcutting and (**f**) $Al_{0.1}CoCrFeNi$ showing finer microabrasion, deformation and delamination of oxide layer; The corresponding interferometry images of wear tracks generated during marine wear for (**g**) CoCrFeMnNi and (**h**) $Al_{0.1}CoCrFeNi$. Scanning Electron Microscope images of wear track due to corrosive wear on (**i**) CoCrFeMnNi and (**j**) $Al_{0.1}CoCrFeNi$; Higher magnification images of wear tracks for (**k**) CoCrFeMnNi showing fine microabrasion/microcutting and shallow deformation and (**l**) $Al_{0.1}CoCrFeNi$ showing corrosive wear in terms of break-down of surface passive layers [31] (reprinted with permission from Elsevier).

Wear resistance of the alloys that were surveyed in this study have been plotted against the reported hardness values in Figure 32. The crystal structure of the alloys have also been marked. It can be seen that irrespective of chemistry, FCC alloys typically are softer and have lower wear resistance, followed by two-phase alloys, and highest wear resistance was seen for BCC alloys. This trend changes in case of high wear resistance materials, as shown in Figure 32b. The two-phase alloys showed significantly higher hardness and wear resistance than the BCC alloys. However, there was no significant correlation for the BCC and two-phase alloys. For example, the $Al_{0.5}CoCrCuFeNiV_{1.6}$ alloy has a hardness of 600 HV with a wear resistance of 1.1 m/mm^3, whereas the $AlCoCrFe_{1.5}MoNi$ has a wear resistance of over 1000 m/mm^3 with similar hardness values, although both of the alloys have BCC crystal structure. This implies that a materials response to sliding wear is governed by factors more than just hardness and crystal structure. These critical knowledge gaps need to be addressed in future studies. The effect of lattice distortion from complex compositions on the hardness, wear resistance and friction evolution need further investigations.

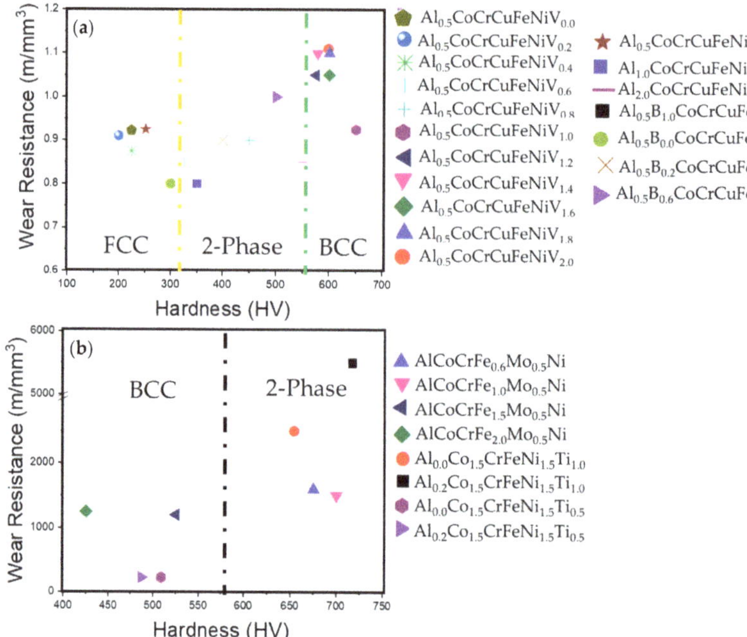

Figure 32. Hardness-wear resistance relationship classified with respect to their magnitude of wear resistance (**a**) less than 2.5 m/mm^3; (**b**) between 500 and 6000 m/mm^3. The wear performance of the alloys do not show a particular dependency on the crystal structure.

In contrast to corrosion behavior, there are limited reports on wear behavior of CCA coatings. Surface cladding was typically employed for wear performance enhancement. AlCoCrNiW and AlCoCrNiSi CCA cladded layers through gas tungsten arc welding (GTAW) showed enhanced wear resistance than AISI 1050 medium carbon steel [93]. The superior wear performance of these coatings was attributed to the strong mechanical interlocking between the dense dendrites and the matrix. During the wear test, dense dendrites can strengthen the structure and prevent plastic flow. The wear performance of AlCoCrNiW layer exceeded that of AlCoCrNiSi due to stronger mechanical interlocking and a more complex microstructure. Mechanical and wear behavior of $Al_{0.5}CoCrFe_2MoNiSi$ CCA coating fabricated by the GTAW method were reported as a function of silicon addition [94]. Superior wear resistance of this cladding layer was attributed to the formation of strong bonds between Si and the other elements in dendritic region and nanoscale precipitation in the inter-dendritic region.

Tungsten inert gas (TIG) was also used to produce CoCrFeMnNbNi CCA coating and showed much lower wear loss when compared with AISI 304 steel due to presence of a FCC Nb-rich Laves phase with nanoscale lamellar spacing [95]. The Laves phase resisted damage during sliding and protected the coating surface from plastic deformation. CuNiSiTiZr CCA coating fabricated by vacuum arc melting showed almost 2.5 times higher hardness than TC11 (typically used in aerospace industry), and superior wear resistance due to various effects, such as solid solution strengthening, precipitation strengthening, and nanocomposite strengthening [96]. A similar phenomenon was observed when AlCrSiTiV CCA coating was deposited on Ti-6Al-4V substrate via the laser cladding technique. CCA coating showed improved wear rate as compared with the Ti-6Al-4V substrate as well as higher hardness values (Figure 33).

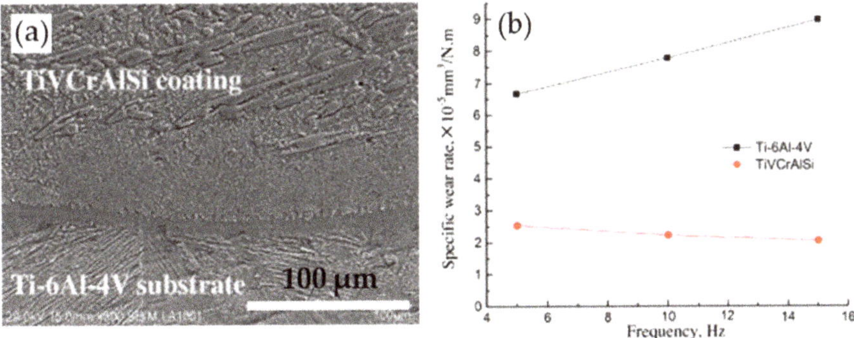

Figure 33. (**a**) SEM micrograph of AlCrSiTiV CCA coating on Ti-6Al-4V substrate and (**b**) specific wear rate of two materials after dry sliding wear test under various frequencies [97] (reprinted with permission from Elsevier).

The high wear resistance of AlCrSiTiV CCA coating was attributed to the formation of hard intermetallic phase in a relatively ductile BCC matrix. The softer matrix limited brittle crack propagation during abrasive and adhesive wear. Plasma-spray has also been used to synthesize several CCA coatings, including $AlCo_{0.6}Cr_yFe_{0.2}Ni_xSiTi_{0.2}$, AlCoCrCuFeNi, AlCoCrFeNi, CoCrFeMnNi, and AlCoCrFeNiTi [98]. The effect of temperature on the wear behavior of AlCoCrFeNiTi CCA coating has also been reported. Adhesive wear with minor abrasion was the main mechanism for AlCoCrFeNiTi coating wear at 25 and 500 °C [98]. More severe adhesive wear was observed at higher temperatures due to the decrease in hardness of the coating. At temperatures higher than 500 °C, both the morphology of wear track and mechanism changed due to oxidation processes and the softening of the coating. The main wear mechanism at 700 °C was tribo-oxidation wear with wide grooves on the wear surface. The presence of micro-cracks and pores in the coating facilitated the diffusion of oxygen at higher temperatures and it caused more accelerated tribo-oxidation wear. AlCoCrFeNiTi CCA coating showed a volume wear loss of about one-ninth of 316 stainless steel at 700 °C [12].

Thermally sprayed $AlCoCrFeMo_{0.5}NiSiTi$ and $AlCrFeMo_{0.5}NiSiTi$ CCA coatings, consisting of a BCC dendrite and a FCC inter-dendritic microstructure, exhibited very good wear resistance [12,99]. Annealed CCA coatings showed even better wear resistance with minimized weight loss. The hardness value for these coatings was lower than bearing SUJ2 and hot-die tool steel SKD61. However, their wear resistance after annealing at 800 °C was much better than these wrought steels. Laser alloying has also been used for synthesizing thicker HEA coatings with strong metallurgical bonding to the substrate [48,97]. AlCoCrCuFe CCA coating fabricated by laser surface alloying showed much better specific wear rate and lower coefficient of friction (COF) than that of Q235 steel substrate, as well as three times higher hardness. It was shown that the addition of certain elements, such as Boron to $AlB_xCoCrFeNi$ coating fabricated by laser cladding, changes the wear mechanism from adhesive to abrasive wear, as the hardness increased. A summary of wear resistance and hardness is shown in Table 3.

Table 3. Sumamry of wear behavior of alloys reported in literature.

System	Alloy	Hardness (HV)	Wear Resistance
$Al_{0.5}CoCrCuFeNiV_x$ [90]	$Al_{0.5}CoCrCuFeNiV_{0.2}$	200	0.910
	$Al_{0.5}CoCrCuFeNiV_{0.4}$	225	0.875
	$Al_{0.5}CoCrCuFeNiV_{0.4}$	325	0.850
	$Al_{0.5}CoCrCuFeNiV_{0.8}$	450	0.900
	$Al_{0.5}CoCrCuFeNiV_{1.0}$	650	0.925
	$Al_{0.5}CoCrCuFeNiV_{1.2}$	575	1.050
	$Al_{0.5}CoCrCuFeNiV_{1.4}$	578	1.100
	$Al_{0.5}CoCrCuFeNiV_{1.6}$	600	1.050
	$Al_{0.5}CoCrCuFeNiV_{1.8}$	600	1.100
	$Al_{0.5}CoCrCuFeNiV_{2.0}$	598	1.110
$Al_xCo_{1.5}CrFeNi_{1.5}Ti_y$ [91]	$Al_0Co_{1.5}CrFeNi_{1.5}Ti_{0.5}$	501	250
	$Al_{0.2}Co_{1.5}CrFeNi_{1.5}Ti_{0.5}$	480	255
	$Al_0Co_{1.5}CrFeNi_{1.5}Ti$	650	2000
	$Al_{0.2}Co_{1.5}CrFeNi_{1.5}Ti$	700	5500
$Al_xCoCrCuFeNi$ [89]	$Al_{0.5}CoCrCuFeNi$	252	0.925
	$Al_{1.5}CoCrCuFeNi$	350	0.825
	$Al_{2.0}CoCrCuFeNi$	550	0.850
$Al_{0.5}B_xCoCrCuFe$ [38]	$Al_{0.5}B_0CoCrCuFe$	300	0.8
	$Al_{0.5}B_{0.2}CoCrCuFe$	400	0.9
	$Al_{0.5}B_{0.6}CoCrCuFe$	500	1.0
	$Al_{0.5}B_{1.0}CoCrCuFe$	725	1.6
$AlCoCrFe_xMo_{0.5}Ni$ [100]	$AlCoCrFe_{0.6}Mo_{0.5}Ni$	675	1600
	$AlCoCrFe_{1.0}Mo_{0.5}Ni$	700	1500
	$AlCoCrFe_{1.5}Mo_{0.5}Ni$	525	1200
	$AlCoCrFe_{2.0}Mo_{0.5}Ni$	425	1250

6. Conclusions

The number of complex concentrated (or high entropy) alloy systems being reported in recent years has exploded because of their tunable microstructures and desirable properties. Improving the surface degradation characteristics of these alloys will make them very attractive in wide ranging commercial applications. Some of the main corrosion, erosion, and wear characteristics in these emerging materials are summarized below:

- Several CCA compositions showed high corrosion resistance in terms of corrosion current density, corrosion potential and pitting resistance. This was primarily attributed to the high wt% of passivating elements, such as Co, Cr, and Ni (cumulatively as high as 40%) in several of the alloys studied.
- Precipitation of secondary phases by either addition of elements or heat treatment deteriorated the corrosion behavior of multi-phase complex concentrated alloys compared to single-phase ones. On the other hand, heat treatment and secondary phase precipitation resulted in surface hardening and improved the wear resistance and erosion characteristics of the alloys.
- Alloying elements that contributed to the precipitation of secondary phases such as B, Cu, Ti, Mo, and Al deteriorated corrosion resistance. The secondary phase precipitates resulted in galvanic corrosion and promoted materials' degradation.
- Erosion and erosion-corrosion resistance of CCAs was superior when compared to stainless steel grades, owing to their strong passivation and relatively higher hardness.
- When compared to conventional alloys, CCAs/HEAs in many cases showed better overall corrosion and erosion resistance in different media. However, there are significant knowledge gaps with respect to surface passivation mechanisms and synergy between the different degradation routes.

- The wear resistance of some CCA compositions was significantly higher than state of the art steels, such as the SJ grades. The wear resistance varied between 0.8–2.0 m/mm^3 as a function of Vanadium, Boron and Aluminum content.
- Two-phase BCC + FCC alloys and single-phase BCC alloys showed orders of magnitude higher wear resistance (~5500 m/mm^3 wear resistance) when compared to single-phase FCC alloys (~1.0 m/mm^3 wear resistance).
- In addition to as-cast and heat treated alloys, thermally sprayed and annealed CCA coatings showed better wear resistance with minimal weight loss when compared to structural steels.
- Certain CCA compositions demonstrated excellent marine corrosion resistance. The wear volume loss was an order of magnitude lower than mild steels.

7. Future Opportunities and Outlook

Complex concentrated alloys present a plethora of opportunities for the development of next generation materials. The scope is not just limited to bulk materials and melt-deposition coatings but also in the form of thin films and powder-metallurgy products. However, critical knowledge gaps in surface degradation mechanisms need to be assessed prior to determining the true application worthiness of these alloys. The effect of processing on surface degradation mechanisms are not well understood for welding and joining, severe-plastic deformation, and hot working. Another area with very limited number of studies includes extreme environments, such as molten/fused salts, heavy ion/neutron irradiation, and high temperatures.

Understanding the nature and chemistry of the surface passivation layer is critically important for corrosion, erosion, and wear applications. This has not been done in a comprehensive way. Surface characterization using X-ray photoelectron spectroscopy (XPS) and ultra-violet photoemission spectroscopy (UPS) could reveal valuable information about the chemistry and electronic structure of the surface passivation layer and help in fundamental understanding of the underlying mechanisms. This knowledge may be utilized to develop specific surface treatments to produce strongly passivating and non-porous oxides that can offer exceptional surface degradation resistance. Slight changes in composition (micro-alloying) or processing conditions have shown large variations in properties, compounding the complexity in analyzing these multi-component systems. In complex precipitation hardened CCAs, composition fluctuations at multiple length-scales (atomic, nano, micro) may lead to "local" effects that nucleate the breakdown of passivation layers. These effects may be captured by phase-specific corrosion and wear tests at the microstructural length-scales, including scanning electrochemical microscopy (SECM) and nano-scratch/wear tests.

There are very limited reports on the lubricity and friction behavior of complex concentrated alloys. This may be of interest in tribology for developing super-lubricity complex composition coatings. Phase-specific friction studies will provide significant insights into surface degradation from multi-body wear in multi-phase CCAs [101]. Ni free complex alloys might be attractive for biomedical applications. Evaluating in-vitro wear behavior and quantifying cytotoxicity of the wear products will significantly help in developing new biomaterials. Recently reported refractory CCAs may be attractive for highly stressed bearing applications, where high temperature wear behavior, evolution of oxide layers, kinetics of spalling, and pesting are of significant fundamental interest.

Advanced additive manufacturing, LASER melt-deposition, and combinatorial development using powder bed and powder feed techniques for CCAs hold tremendous potential towards meeting long-standing challenges like highly corrosion resistant surface materials and thermal barrier coatings. CCA claddings via melt-deposition is yet to be explored. Electrodeposition of CCA/HEA coatings via co-deposition and auto-catalytic reactions could dramatically enhance functional applications of these alloys. CCA thin films could be potentially transformative as diffusion barriers in integrated circuit (IC) manufacturing because of extremely sluggish diffusion and lattice distortion. The future opportunities and outlook for complex concentrated alloys in different areas are summarized in Figure 34.

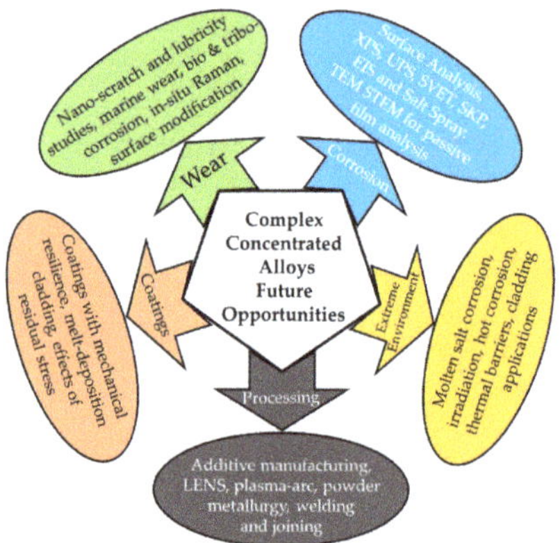

Figure 34. Future opportunities and outlook for complex concentrated alloys.

Author Contributions: A.A. and S.M. conceived and designed the layout of the review paper; A.A. analyzed corrosion and wear sections; V.H. analyzed surface and coatings techniques; H.S.G. and H.A. analyzed erosion, corrosion and erosion behaviors. A.A. and S.M. wrote the manuscript.

Funding: This research received no external funding.

Conflicts of Interest: The authors declare no conflict of interest.

References

1. Murty, B.S.; Yeh, J.; Ranganathan, S. *High-Entropy Alloys*; Butterworth-Heinemann: Oxford, UK, 2014.
2. Yeh, J.; Chen, S.; Lin, S.; Gan, J.; Chin, T.; Shun, T.; Tsau, C.; Chang, S. Nanostructured high entropy alloys with multiple principal elements: Novel alloy design concepts and outcomes. *Adv. Eng. Mater.* **2004**, *6*, 299–303. [CrossRef]
3. Yeh, J.-W. Recent progress in high entropy alloys. *Ann. Chim. Sci. Mater.* **2006**, *31*, 633–648. [CrossRef]
4. Senkov, O.; Wilks, G.; Miracle, D.; Chuang, C.; Liaw, P. Refractory high-entropy alloys. *Intermetallics* **2010**, *18*, 1758–1765. [CrossRef]
5. Yeh, J.W.; Chen, Y.L.; Lin, S.J.; Chen, S.K. High-entropy alloys—A new era of exploitation. In *Materials Science Forum*; Trans Tech Publications: Zürich, Switzerland, 2007; Volume 560, pp. 1–9.
6. Cantor, B. Multicomponent and high entropy alloys. *Entropy* **2014**, *16*, 4749–4768. [CrossRef]
7. Chen, S.Y.; Yang, X.; Dahmen, K.A.; Liaw, P.K.; Zhang, Y. Microstructures and crackling noise of Al_xNbTiMoV high entropy alloys. *Entropy* **2014**, *16*, 870–884. [CrossRef]
8. Gludovatz, B.; Hohenwarter, A.; Catoor, D.; Chang, E.H.; George, E.P.; Ritchie, R.O. A fracture-resistant high-entropy alloy for cryogenic applications. *Science* **2014**, *345*, 1153–1158. [CrossRef] [PubMed]
9. Mishra, R.; Kumar, N.; Komarasamy, M. Lattice strain framework for plastic deformation in complex concentrated alloys including high entropy alloys. *Mater. Sci. Technol.* **2015**, *31*, 1259–1263. [CrossRef]
10. Cantor, B.; Chang, I.; Knight, P.; Vincent, A. Microstructural development in equiatomic multicomponent alloys. *Mater. Sci. Eng. A* **2004**, *375*, 213–218. [CrossRef]
11. Hemphill, M.A.; Yuan, T.; Wang, G.; Yeh, J.; Tsai, C.; Chuang, A.; Liaw, P. Fatigue behavior of Al0. 5CoCrCuFeNi high entropy alloys. *Acta Mater.* **2012**, *60*, 5723–5734. [CrossRef]
12. Huang, P.; Yeh, J.; Shun, T.; Chen, S. Multi-principal-element alloys with improved oxidation and wear resistance for thermal spray coating. *Adv. Eng. Mater.* **2004**, *6*, 74–78. [CrossRef]
13. Grewal, H.S.; Sanjiv, R.M.; Arora, H.S.; Kumar, R.; Ayyagari, A.; Mukherjee, S.; Singh, H. Activation energy and high temperature oxidation behavior of multi-principal element alloy. *Adv. Eng. Mater.* **2017**, 19. [CrossRef]
14. Pickering, E.; Jones, N.G. High-entropy alloys: A critical assessment of their founding principles and future prospects. *Int. Mater. Rev.* **2016**, *61*, 183–202. [CrossRef]

15. Miracle, D.; Senkov, O. A critical review of high entropy alloys and related concepts. *Acta Mater.* **2017**, *122*, 448–511. [CrossRef]
16. Miracle, D.B. High-entropy alloys: A current evaluation of founding ideas and core effects and exploring "nonlinear alloys". *JOM* **2017**, *69*, 2130–2136. [CrossRef]
17. Shi, Y.; Yang, B.; Liaw, P.K. Corrosion-resistant high-entropy alloys: A review. *Metals* **2017**, *7*, 43. [CrossRef]
18. Chen, Y.; Duval, T.; Hung, U.; Yeh, J.; Shih, H. Microstructure and electrochemical properties of high entropy alloys—A comparison with type-304 stainless steel. *Corros. Sci.* **2005**, *47*, 2257–2279. [CrossRef]
19. Levy, A.V. *Solid Particle Erosion and Erosion-Corrosion of Materials*; ASM International: Almere, The Netherlands, 1995.
20. Hutchings, I. *Tribology: Friction and Wear of Engineering Materials*, 1st ed.; Elsevier Butterworth-Heinemann: Oxford, UK, 1992.
21. Stachowiak, G.W.; Batchelor, A.W. Corrosive and oxidative wear. In *Engineering Tribology*, 3rd ed.; Elsevier Butterworth-Heinemann: Amsterdam, The Netherlands, 2005; pp. 649–651.
22. Arndt, R.E. Cavitation in fluid machinery and hydraulic structures. *Ann. Rev. Fluid Mech.* **1981**, *13*, 273–326. [CrossRef]
23. Wood, R.J. Marine wear and tribocorrosion. *Wear* **2017**, *376*, 893–910. [CrossRef]
24. Finnie, I. Some reflections on the past and future of erosion. *Wear* **1995**, *186*, 1–10. [CrossRef]
25. ASTM International. *Standard Guide for Determining Synergism between Wear and Corrosion*; ASTM G119-93; ASTM International: West Conshohocken, PA, USA, 1994; pp. 507–512.
26. Nair, R.; Selvam, K.; Arora, H.; Mukherjee, S.; Singh, H.; Grewal, H. Slurry erosion behavior of high entropy alloys. *Wear* **2017**, *386*, 230–238. [CrossRef]
27. Zhao, J.; Ji, X.; Shan, Y.; Fu, Y.; Yao, Z. On the microstructure and erosion-corrosion resistance of AlCrFeCoNiCu high-entropy alloy via annealing treatment. *Mater. Sci. Technol.* **2016**, *32*, 1271–1275. [CrossRef]
28. Wu, C.; Zhang, S.; Zhang, C.; Zhang, H.; Dong, S. Phase evolution and cavitation erosion-corrosion behavior of FeCoCrAlNiTix high entropy alloy coatings on 304 stainless steel by laser surface alloying. *J. Alloy. Compd.* **2017**, *698*, 761–770. [CrossRef]
29. Hsu, Y.; Chiang, W.; Wu, J. Corrosion behavior of FeCoNiCrCux high-entropy alloys in 3.5% sodium chloride solution. *Mater. Chem. Phys.* **2005**, *92*, 112–117. [CrossRef]
30. Liaw, P.; Egami, T.; Zhang, C.; Zhang, F.; Zhang, Y. Radiation Behavior of High-Entropy Alloys for Advanced Reactors, Technical Report. US Department of Energy, University of Tennessee/Oak Ridge Natonal Laboratory. 2008. Available online: http://www.iaea.org/inis/collection/NCLCollectionStore/_Public/46/119/46119545.pdf (accessed on July 28 2018).
31. Ayyagari, A.; Barthelemy, C.; Gwalani, B.; Banerjee, R.; Scharf, T.W.; Mukherjee, S. Reciprocating sliding wear behavior of high entropy alloys in dry and marine environments. *Mater. Chem. Phys.* **2018**, *210*, 162–169. [CrossRef]
32. Kumar, N.; Fusco, M.; Komarasamy, M.; Mishra, R.; Bourham, M.; Murty, K. Understanding effect of 3.5 wt. % NaCl on the corrosion of $Al_{0.1}$CoCrFeNi high-entropy alloy. *J. Nucl. Mater.* **2017**, *495*, 154–163. [CrossRef]
33. Kao, Y.; Lee, T.; Chen, S.; Chang, Y. Electrochemical passive properties of AlxCoCrFeNi (x = 0, 0.25, 0.50, 1.00) alloys in sulfuric acids. *Corros. Sci.* **2010**, *52*, 1026–1034. [CrossRef]
34. Shi, Y.; Yang, B.; Xie, X.; Brechtl, J.; Dahmen, K.A.; Liaw, P.K. Corrosion of Al_xCoCrFeNi high-entropy alloys: Al-content and potential scan-rate dependent pitting behavior. *Corros. Sci.* **2017**, *119*, 33–45. [CrossRef]
35. Lin, C.; Tsai, H. Evolution of microstructure, hardness, and corrosion properties of high-entropy Al0. 5CoCrFeNi alloy. *Intermetallics* **2011**, *19*, 288–294. [CrossRef]
36. Soare, V.; Mitrica, D.; Constantin, I.; Badilita, V.; Stoiciu, F.; Popescu, A.; Carcea, I. Influence of remelting on microstructure, hardness and corrosion behaviour of AlCoCrFeNiTi high entropy alloy. *Mater. Sci. Technol.* **2015**, *31*, 1194–1200. [CrossRef]
37. Qiu, X. Microstructure and properties of AlCrFeNiCoCu high entropy alloy prepared by powder metallurgy. *J. Alloy. Compd.* **2013**, *555*, 246–249. [CrossRef]
38. Hsu, C.; Yeh, J.; Chen, S.; Shun, T. Wear resistance and high-temperature compression strength of Fcc CuCoNiCr$Al_{0.5}$Fe alloy with boron addition. *Metall. Mater. Trans. A* **2004**, *35*, 1465–1469. [CrossRef]
39. Lee, C.; Chen, Y.; Hsu, C.; Yeh, J.; Shih, H. The effect of boron on the corrosion resistance of the high entropy alloys $Al_{0.5}$CoCrCuFeNiB x. *J. Electrochem. Soc.* **2007**, *154*, C424–C430. [CrossRef]

40. Xiao, D.; Zhou, P.; Wu, W.; Diao, H.; Gao, M.; Song, M.; Liaw, P. Microstructure, mechanical and corrosion behaviors of AlCoCuFeNi-(Cr,Ti) high entropy alloys. *Mater. Des.* **2017**, *116*, 438–447. [CrossRef]
41. Li, B.; Kun, P.; Hu, A.; Zhou, L.; Zhu, J.; Li, D. Structure and properties of $FeCoNiCrCu_{0.5}Al_x$ high-entropy alloy. *Trans. Nonferr. Met. Soc. China* **2013**, *23*, 735–741. [CrossRef]
42. Chou, Y.; Yeh, J.; Shih, H. The effect of molybdenum on the corrosion behaviour of the high-entropy alloys $Co_{1.5}CrFeNi_{1.5}Ti_{0.5}Mo_x$ in aqueous environments. *Corros. Sci.* **2010**, *52*, 2571–2581. [CrossRef]
43. Chou, Y.; Wang, Y.; Yeh, J.; Shih, H. Pitting corrosion of the high-entropy alloy $Co_{1.5}CrFeNi_{1.5}Ti_{0.5}Mo_{0.1}$ in chloride-containing sulphate solutions. *Corros. Sci.* **2010**, *52*, 3481–3491. [CrossRef]
44. Ren, B.; Liu, Z.; Li, D.; Shi, L.; Cai, B.; Wang, M. Corrosion behavior of CuCrFeNiMn high entropy alloy system in 1 M sulfuric acid solution. *Mater. Corros.* **2012**, *63*, 828–834. [CrossRef]
45. Cheng, J.; Liang, X.; Wang, Z.; Xu, B. Formation and mechanical properties of CoNiCuFeCr high-entropy alloys coatings prepared by plasma transferred arc cladding process. *Plasma Chem. Plasma Process.* **2013**, *33*, 979–992. [CrossRef]
46. Liu, L.; Zhu, J.; Hou, C.; Li, J.; Jiang, Q. Dense and smooth amorphous films of multicomponent FeCoNiCuVZrAl high-entropy alloy deposited by direct current magnetron sputtering. *Mater. Des.* **2013**, *46*, 675–679. [CrossRef]
47. Li, X.; Zheng, Z.; Dou, D.; Li, J. Microstructure and properties of coating of FeAlCuCrCoMn high entropy alloy deposited by direct current magnetron sputtering. *Mater. Res.* **2016**, *19*, 802–806. [CrossRef]
48. Zhang, H.; Pan, Y.; He, Y. Synthesis and characterization of FeCoNiCrCu high-entropy alloy coating by laser cladding. *Mater. Des.* **2011**, *32*, 1910–1915. [CrossRef]
49. Qiu, X.; Zhang, Y.; He, L.; Liu, C. Microstructure and corrosion resistance of AlCrFeCuCo high entropy alloy. *J. Alloy. Compd.* **2013**, *549*, 195–199. [CrossRef]
50. Zhang, S.; Wu, C.; Zhang, C.; Guan, M.; Tan, J. Laser surface alloying of FeCoCrAlNi high-entropy alloy on 304 stainless steel to enhance corrosion and cavitation erosion resistance. *Opt. Laser Technol.* **2016**, *84*, 23–31. [CrossRef]
51. Qiu, X.; Zhang, Y.; Liu, C. Effect of Ti content on structure and properties of $Al_2CrFeNiCoCuTi_x$ high-entropy alloy coatings. *J. Alloy. Compd.* **2014**, *585*, 282–286. [CrossRef]
52. Zhang, H.; Pan, Y.; He, Y.; Jiao, H. Microstructure and properties of 6FeNiCoSiCrAlTi high-entropy alloy coating prepared by laser cladding. *Appl. Surf. Sci.* **2011**, *257*, 2259–2263. [CrossRef]
53. Shon, Y.; Joshi, S.S.; Katakam, S.; Rajamure, R.S.; Dahotre, N.B. Laser additive synthesis of high entropy alloy coating on aluminum: Corrosion behavior. *Mater. Lett.* **2015**, *142*, 122–125. [CrossRef]
54. Marcus, P. On some fundamental factors in the effect of alloying elements on passivation of alloys. *Corros. Sci.* **1994**, *36*, 2155–2158. [CrossRef]
55. Cheng, J.; Liang, X.; Xu, B. Effect of Nb addition on the structure and mechanical behaviors of CoCrCuFeNi high-entropy alloy coatings. *Surf. Coat. Technol.* **2014**, *240*, 184–190. [CrossRef]
56. Ye, X.; Ma, M.; Cao, Y.; Liu, W.; Ye, X.; Gu, Y. The property research on high-entropy alloy AlxFeCoNiCuCr coating by laser cladding. *Phys. Procedia* **2011**, *12*, 303–312. [CrossRef]
57. Li, W.; Liu, G.; Guo, J. Microstructure and electrochemical properties of $Al_xFeCoNiCrTi$ high-entropy alloys. *Foundry* **2009**, *58*, 431–435.
58. Qiu, X.; Liu, C. Microstructure and properties of Al2CrFeCoCuTiNix high-entropy alloys prepared by laser cladding. *J. Alloy. Compd.* **2013**, *553*, 216–220. [CrossRef]
59. Qiu, X.; Wu, M.; Liu, C.; Zhang, Y.; Huang, C. Corrosion performance of $Al_2CrFeCo_xCuNiTi$ high-entropy alloy coatings in acid liquids. *J. Alloy. Compd.* **2017**, *708*, 353–357. [CrossRef]
60. Wu, C.; Zhang, S.; Zhang, C.; Chen, J.; Dong, S. Phase evolution characteristics and corrosion behavior of $FeCoCrAlCu\text{-}X_{0.5}$ coatings on cp Cu by laser high-entropy alloying. *Opt. Laser Technol.* **2017**, *94*, 68–71. [CrossRef]
61. Ren, B.; Zhao, R.; Liu, Z.; Guan, S.; Zhang, H. Microstructure and properties of $Al_{0.3}CrFe_{1.5}MnNi_{0.5}Ti_x$ and $Al_{0.3}CrFe_{1.5}MnNi_{0.5}Si_x$ high-entropy alloys. *Rare Met.* **2014**, *33*, 149–154. [CrossRef]
62. Li, Q.; Yue, T.; Guo, Z.; Lin, X. Microstructure and corrosion properties of AlCoCrFeNi high entropy alloy coatings deposited on AISI 1045 steel by the electrospark process. *Metall. Mater. Trans. A* **2013**, *44*, 1767–1778. [CrossRef]

63. Soare, V.; Mitrica, D.; Constantin, I.; Popescu, G.; Csaki, I.; Tarcolea, M.; Carcea, I. The mechanical and corrosion behaviors of as-cast and re-melted AlCrCuFeMnNi multi-component high-entropy alloy. *Metall. Mater. Trans. A* **2015**, *46*, 1468–1473. [CrossRef]
64. Argade, G.R.; Joshi, S.S.; Ayyagari, A.V.; Mukherjee, S.; Mishra, R.S.; Dahotre, N.B. Tribocorrosion Performance of Laser Additively Processed High Entropy Alloy Coatings on Aluminum. *Appl. Surf. Sci.*. under review.
65. Yuan, Y.; Faqin, X.; Tiebang, Z.; Hongchao, K.; Rui, H.; Jinshan, L. Microstructure control and corrosion properties of $AlCoCrFeNiTi_{0.5}$ high-entropy alloy. *Rare Met. Mater. Eng.* **2012**, *5*, 025.
66. Lee, C.; Chang, C.; Chen, Y.; Yeh, J.; Shih, H. Effect of the aluminium content of $Al_xCrFe_{1.5}MnNi_{0.5}$ high-entropy alloys on the corrosion behaviour in aqueous environments. *Corros. Sci.* **2008**, *50*, 2053–2060. [CrossRef]
67. Zhang, C.; Chen, G.; Dai, P. Evolution of the microstructure and properties of laser-clad $FeCrNiCoB_x$ high-entropy alloy coatings. *Mater. Sci. Technol.* **2016**, *32*, 1666–1672. [CrossRef]
68. Lin, C.; Tsai, H.; Bor, H. Effect of aging treatment on microstructure and properties of high-entropy $Cu_{0.5}CoCrFeNi$ alloy. *Intermetallics* **2010**, *18*, 1244–1250. [CrossRef]
69. Zheng, Z.; Li, X.; Zhang, C.; Li, J. Microstructure and corrosion behaviour of $FeCoNiCuSn_x$ high entropy alloys. *Mater. Sci. Technol.* **2015**, *31*, 1148–1152. [CrossRef]
70. Li, J.; Yang, X.; Zhu, R.; Zhang, Y. Corrosion and serration behaviors of $TiZr_{0.5}NbCr_{0.5}V_xMo_y$ high entropy alloys in aqueous environments. *Metals* **2014**, *4*, 597–608. [CrossRef]
71. Ji, X.; Duan, H.; Zhang, H.; Ma, J. Slurry erosion resistance of laser clad $NiCoCrFeAl_3$ high-entropy alloy coatings. *Tribol. Trans.* **2015**, *58*, 1119–1123. [CrossRef]
72. Zhang, L.S.; Ma, G.L.; Fu, L.C.; Tian, J.Y. Recent progress in high-entropy alloys. *Eur. J. Control* **2013**, *631*, 227–232. [CrossRef]
73. Komarasamy, M.; Kumar, N.; Tang, Z.; Mishra, R.; Liaw, P. Effect of microstructure on the deformation mechanism of friction stir-processed $Al_{0.1}CoCrFeNi$ high entropy alloy. *Mater. Res. Lett.* **2015**, *3*, 30–34. [CrossRef]
74. Liu, J.; Chen, C.; Xu, Y.; Wu, S.; Wang, G.; Wang, H.; Fang, Y.; Meng, L. Deformation twinning behaviors of the low stacking fault energy high-entropy alloy: An in-situ TEM study. *Scr. Mater.* **2017**, *137*, 9–12. [CrossRef]
75. Nair, R.B.; Arora, H.S.; Ayyagari, A.; Mukherjee, S.; Grewal, H.S. High entropy alloys: Prospective materials for tribo-corrosion applications. *Adv. Eng. Mater.* **2018**, 1700946. [CrossRef]
76. Nair, R.; Arora, H.; Mukherjee, S.; Singh, S.; Singh, H.; Grewal, H. Exceptionally high cavitation erosion and corrosion resistance of a high entropy alloy. *Ultrason. Sonochem.* **2018**, *41*, 252–260. [CrossRef] [PubMed]
77. Grewal, H.; Agrawal, A.; Singh, H. Slurry erosion mechanism of hydroturbine steel: Effect of operating parameters. *Tribol. Lett.* **2013**, *52*, 287–303. [CrossRef]
78. Ang, A.S.M.; Berndt, C.C.; Sesso, M.L.; Anupam, A.; Praveen, S.; Kottada, R.S.; Murty, B. Plasma-sprayed high entropy alloys: Microstructure and properties of AlCoCrFeNi and MnCoCrFeNi. *Metall. Mater. Trans. A* **2015**, *46*, 791–800. [CrossRef]
79. Liao, W.; Lan, S.; Gao, L.; Zhang, H.; Xu, S.; Song, J.; Wang, X.; Lu, Y. Nanocrystalline high-entropy alloy ($CoCrFeNiAl_{0.3}$) thin-film coating by magnetron sputtering. *Thin Solid Films* **2017**, *638*, 383–388. [CrossRef]
80. Ye, Q.; Feng, K.; Li, Z.; Lu, F.; Li, R.; Huang, J.; Wu, Y. Microstructure and corrosion properties of CrMnFeCoNi high entropy alloy coating. *Appl. Surf. Sci.* **2017**, *396*, 1420–1426. [CrossRef]
81. Zhang, M.; Zhou, X.; Yu, X.; Li, J. Synthesis and characterization of refractory TiZrNbWMo high-entropy alloy coating by laser cladding. *Surf. Coat. Technol.* **2017**, *311*, 321–329. [CrossRef]
82. Nair, R.B.; Arora, H.S.; Mandal, P.; Das, S.; Grewal, H.S. High-performance microwave-derived multi-principal element alloy coatings for tribological application. *Adv. Eng. Mater.* **2018**, 1800163. [CrossRef]
83. Taillon, G.; Pougoum, F.; Lavigne, S.; Ton-That, L.; Schulz, R.; Bousser, E.; Savoie, S.; Martinu, L.; Klemberg-Sapieha, J. Cavitation erosion mechanisms in stainless steels and in composite metal–ceramic HVOF coatings. *Wear* **2016**, *364*, 201–210. [CrossRef]
84. Hong, S.; Wu, Y.; Zhang, J.; Zheng, Y.; Zheng, Y.; Lin, J. Synergistic effect of ultrasonic cavitation erosion and corrosion of WC–CoCr and FeCrSiBMn coatings prepared by HVOF spraying. *Ultrason. Sonochem.* **2016**, *31*, 563–569. [CrossRef] [PubMed]

85. Cuppari, M.D.V.; Souza, R.; Sinatora, A. Effect of hard second phase on cavitation erosion of Fe–Cr–Ni–C alloys. *Wear* **2005**, *258*, 596–603. [CrossRef]
86. Jiang, G.; Zheng, Y.; Yang, Y.; Fang, H. Caviation erosion of bainitic steel. *Wear* **1998**, *215*, 46–53. [CrossRef]
87. Sakamoto, A.; Yamasaki, T.; Matsumura, M. Erosion-corrosion tests on copper alloys for water tap use. *Wear* **1995**, *186*, 548–554. [CrossRef]
88. Man, H.; Kwok, C.; Yue, T. Cavitation erosion and corrosion behaviour of laser surface alloyed MMC of SiC and Si_3N_4 on Al alloy AA6061. *Surf. Coat. Technol.* **2000**, *132*, 11–20. [CrossRef]
89. Wu, J.; Lin, S.; Yeh, J.; Chen, S.; Huang, Y.; Chen, H. Adhesive wear behavior of $Al_xCoCrCuFeNi$ high-entropy alloys as a function of aluminum content. *Wear* **2006**, *261*, 513–519. [CrossRef]
90. Chen, M.; Lin, S.; Yeh, J.; Chuang, M.; Chen, S.; Huang, Y. Effect of vanadium addition on the microstructure, hardness, and wear resistance of $Al_{0.5}CoCrCuFeNi$ high-entropy alloy. *Metall. Mater. Trans. A* **2006**, *37*, 1363–1369. [CrossRef]
91. Chuang, M.; Tsai, M.; Wang, W.; Lin, S.; Yeh, J. Microstructure and wear behavior of $Al_xCo_{1.5}CrFeNi_{1.5}Ti_y$ high-entropy alloys. *Acta Mater.* **2011**, *59*, 6308–6317. [CrossRef]
92. Gwalani, B.; Ayyagari, A.; Choudhuri, D.; Scharf, T.; Mukherjee, S.; Gibson, M.; Banerjee, R. Microstructure and wear resistance of an intermetallic-based $Al_{0.25}Ti_{0.75}CoCrFeNi$ high entropy alloy. *Mater. Chem. Phys.* **2018**, *210*, 197–206. [CrossRef]
93. Lin, Y.; Cho, Y. Elucidating the microstructure and wear behavior for multicomponent alloy clad layers by in situ synthesis. *Surf. Coat. Technol.* **2008**, *202*, 4666–4672. [CrossRef]
94. Chen, J.; Chen, P.; Lin, C.; Chang, C.; Chang, Y.; Wu, W. Microstructure and wear properties of multicomponent alloy cladding formed by gas tungsten arc welding (GTAW). *Surf. Coat. Technol.* **2009**, *203*, 3231–3234. [CrossRef]
95. Huo, W.; Shi, H.; Ren, X.; Zhang, J. Microstructure and wear behavior of CoCrFeMnNbNi high-entropy alloy coating by TIG cladding. *Adv. Mater. Sci. Eng.* **2015**, *2015*. [CrossRef]
96. Wang, X.; Wang, Z.; He, P.; Lin, T.; Shi, Y. Microstructure and wear properties of CuNiSiTiZr high-entropy alloy coatings on TC11 titanium alloy produced by electrospark—Computer numerical control deposition process. *Surf. Coat. Technol.* **2015**, *283*, 156–161. [CrossRef]
97. Huang, C.; Zhang, Y.; Vilar, R.; Shen, J. Dry sliding wear behavior of laser clad TiVCrAlSi high entropy alloy coatings on Ti–6Al–4V substrate. *Mater. Des.* **2012**, *41*, 338–343. [CrossRef]
98. Tian, L.; Xiong, W.; Liu, C.; Lu, S.; Fu, M. Microstructure and wear behavior of atmospheric plasma-sprayed AlCoCrFeNiTi high-entropy alloy coating. *J. Mater. Eng. Perform.* **2016**, *25*, 5513–5521. [CrossRef]
99. Wang, L.; Chen, C.; Yeh, J.; Ke, S. The microstructure and strengthening mechanism of thermal spray coating $Ni_xCo_{0.6}Fe_{0.2}Cr_ySi_zAlTi_{0.2}$ high-entropy alloys. *Mater. Chem. Phys.* **2011**, *126*, 880–885. [CrossRef]
100. Hsu, C.; Sheu, T.; Yeh, J.; Chen, S. Effect of iron content on wear behavior of $AlCoCrFe_xMo_{0.5}Ni$ high-entropy alloys. *Wear* **2010**, *268*, 653–659. [CrossRef]
101. Ayyagari, A.; Hasannaeimi, V.; Arora, H.; Mukherjee, S. Electrochemical and friction characteristics of metallic glass composites at the microstructural length-scales. *Sci. Rep.* **2018**, *8*, 906. [CrossRef] [PubMed]

Article

Corrosion Behavior of Selectively Laser Melted CoCrFeMnNi High Entropy Alloy

Jie Ren [1,†], Chaitanya Mahajan [2,†], Liang Liu [1], David Follette [3], Wen Chen [1,*] and Sundeep Mukherjee [2,*]

1 Department of Mechanical and Industrial Engineering, University of Massachusetts, Amherst, MA 01003, USA; jren@umass.edu (J.R.); liang@umass.edu (L.L.)
2 Department of Materials Science and Engineering, University of North Texas, Denton, TX 76203, USA; ChaitanyaMahajan@my.unt.edu
3 Advanced Digital Design and Fabrication Lab (ADDFab), University of Massachusetts, Amherst, MA 01003, USA; follette@umass.edu
* Correspondence: wenchen@umass.edu (W.C.); Sundeep.Mukherjee@unt.edu (S.M.)
† These authors contribute equally to this work.

Received: 21 August 2019; Accepted: 19 September 2019; Published: 23 September 2019

Abstract: CoCrFeMnNi high entropy alloys (HEAs) were additively manufactured (AM) by laser powder bed fusion and their corrosion resistance in 3.5 wt% NaCl solution was studied by potentiodynamic polarization and electrochemical impedance spectroscopy tests. A systematic study of AM CoCrFeMnNi HEAs' porosity under a wide range of laser processing parameters was conducted and a processing map was constructed to identify the optimal laser processing window for CoCrFeMnNi HEAs. The near fully dense AM CoCrFeMnNi HEAs exhibit a unique non-equilibrium microstructure consisting of tortuous grain boundaries, sub-grain cellular structures, columnar dendrites, associated with some processing defects such as micro-pores. Compared with conventional as-cast counterpart, the AM CoCrFeMnNi HEAs showed higher pitting resistance (ΔE) and greater polarization resistance (R_p). The superior corrosion resistance of AM CoCrFeMnNi HEAs may be attributed to the homogeneous elemental distribution and lower density of micro-pores. Our study widens the toolbox to manufacture HEAs with exceptional corrosion resistance by additive manufacturing.

Keywords: CoCrFeMnNi high entropy alloys; additive manufacturing; corrosion behavior; non-equilibrium microstructure; micro-pores

---

1. Introduction

In recent years, high entropy alloys (HEAs) have received remarkable attention from both academia and industry due to a portfolio of unusual properties including high specific strength [1–6], high fracture resistance [7,8], and excellent corrosion and oxidation resistance [9,10]. The superior corrosion behavior has been attributed to the locally disordered chemical environment obtained from random arrangement of multi-principal elements in solid solution [9,11]. Because of the formation of protective passive films on the surface, the corrosion behavior of Cr-, Ni-, and Mo-based HEAs have been widely investigated [12–16]. Despite the technical potential, processing of HEAs is generally a challenge. Most HEAs suffer from inferior flowability or formability in the liquid state to be shaped into useful components by conventional manufacturing routes such as casting [17]. Recently, the rapid development of additive manufacturing (AM) has enabled precise and versatile manufacturing of geometrically complex components necessary for practical applications [18]. Here, we apply a laser powder bed fusion based AM technique, namely selective laser meting (SLM), to produce CoCrFeMnNi HEAs. Unlike the conventional manufacturing technique, SLM prints materials and components directly from a computer-aided design file and offers unique advantages of design freedom for three-dimensional complex geometry. The highly localized melting, strong temperature gradient, and high cooling rate (~10^6 K/s) [19] during laser melting often give rise to unique ultra-fine and non-equilibrium microstructures that deliver superior mechanical properties such as high strength and high ductility that are not accessible via conventional methods [20,21]. For example, AM CoCrFeNiMn HEA demonstrated a yield strength of 510 ± 10 MPa with a uniform elongation of 32.4%, in contrast to 205 ± 5 MPa and 50.2% for the as-cast counterpart [21].

Corrosion resistance is often an additional core performance factor for structural metals. Fundamental understanding of the microstructure-electrochemical property is therefore critical to the AM technology. The corrosion resistance of AM 316L stainless steels [22–26] and Ti-6Al-4V alloys [27,28] have so far been widely studied. Contradictory results have been reported for the corrosion resistance of AM 316L stainless steels [22–26]. For example, Sun et al. [22] and Ziętala et al. [23] showed that the corrosion behavior of AM 316L stainless steels were similar to that of conventionally manufactured counterparts and the AM samples were more susceptible to pitting corrosion. Trelewicz et al. [24] and Geenen et al. [25] reported reduced corrosion resistance of AM 316L stainless steels. In contrary, Kazemipour et al. [26] reported superior pitting resistance and reduced metastable pitting rate in SLM-fabricated samples compared to wrought alloy. While for Ti-6Al-4V alloy, the constituent phase is the dominant factor for corrosion resistance, followed by grain size and morphology. Therefore, the SLM alloy with more acicular α' and less β–Ti phase exhibited worse corrosion resistance than Grade 5 alloy [27,28]. However, for CoCrFeMnNi, one of the most notable HEAs till date, corrosion studies have been mainly focused on as-cast bulk samples [29,30] and laser cladded coatings [31], for which elemental segregation makes Cr-depleted inter-dendrites especially vulnerable to pitting corrosion. Therefore, there is limited understanding of the effect of highly non-equilibrium microstructure obtained in AM on the corrosion resistance of this alloy.

In this study, the corrosion behavior of AM CoCrFeMnNi HEA was compared with the as-cast counterpart in 3.5 wt% NaCl solution by potentiodynamic polarization and electrochemical impedance spectroscopy measurements. The microstructures and surface morphologies of both alloys were analyzed to explain the difference in corrosion resistance. Our study reveals that the AM CoCrFeMnNi HEA with unique non-equilibrium microstructure, homogeneous elemental distribution, and smaller defect (micro-pore) density exhibit superior corrosion resistance to the as-cast counterpart.

2. Materials and Methods

The SLM as-printed samples were manufactured by M290 (EOS GmbH, Munich, Germany), which is equipped with a Yb-fiber laser with a maximum power of 400 W and a focus diameter of 100 μm. Gas atomized equiatomic CoCrFeMnNi HEA powders with the particle size ranging from 15 to 53 μm (D_{10} = 22.8 μm, D_{50} = 36.8 μm, D_{90} = 58.4 μm) were used. As shown in Figure 1, 36 cubes of 8 × 8 × 6 mm^3 were printed with different processing parameters by bi-directional and chessboard scan strategies at a constant layer thickness of 0.04 mm and a hatching distance of 0.08 mm with varying laser powers (250–370 W) and scan speeds (500–2500 m/s). To reduce the anisotropy of mechanical properties, laser scan direction was rotated alternately 90° for successive layers. The as-cast counterpart with the same composition was prepared by arc melting constituent elements with 99.99% purity under argon atmosphere. The ingots were re-melted at least five times to assure chemical homogeneity and subsequently sucked into a water-cooled copper mold.

The relative density of the as-printed samples was measured by gas pycnometer (AccuPyc II 1340, Micromeritics, Norcross, GA, USA). Optical microscope (OM, BX53M, Olympus, Tokyo, Japan) and scanning electron microscopy (SEM, Magellan 400 XHR, FEI, Hillsboro, OR, USA) were used to examine the microstructures of the samples. The metallurgical samples were polished with SiC abrasive papers with different grits of 400, 800, and 1200 respectively, and finally with 1 μm diamond suspension. To reveal the melt pool boundaries and sub-microstructures, the samples were etched in a mixture of 50% aqua regia and 50% ethanol (vol.%).

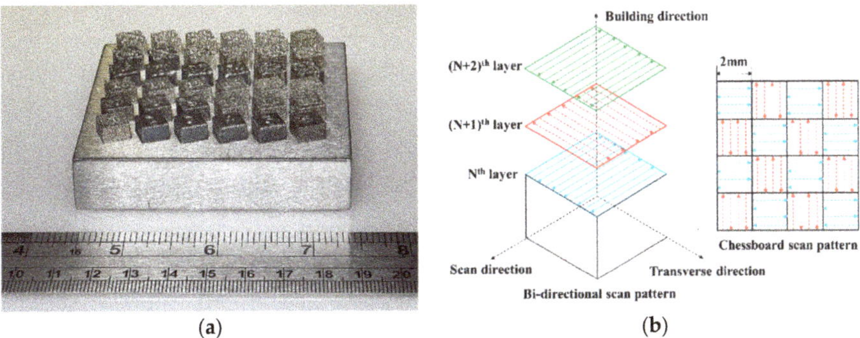

(a) (b)

Figure 1. (**a**) As-printed samples with different processing parameters; (**b**) Schematics of scan strategies in selective laser melting process.

Prior to the corrosion experiments, samples were cleaned with deionized water, acetone, and then dried in air. The specimen surface was masked with a tape exposing 0.283 cm^2 area to 3.5 wt.% NaCl electrolyte solution. Electrochemical studies were performed with a potentiostat (Reference 3000, Gamry, Warminster, PA, USA) at room temperature. A graphite rod with an area of 4.9 cm^2 was used as the counter electrode and saturated calomel electrode (SCE) was used as the reference electrode. Electrochemical impedance spectroscopy (EIS) was performed after the open circuit potential (OCP) stabilized after ~6000 s. EIS was started with an AC voltage amplitude of 10 mV and frequency was swept from 10 mHz to 100 kHz. The impedance data was interpreted using Gamry software and equivalent circuits followed by cyclic polarization tests which were performed starting from −0.25 mV versus OCP to the upper threshold limit of 10 $mA \cdot cm^{-2}$ for the reverse scan. The scan rate for both forward and reverse scan was 0.25 $mV \cdot s^{-1}$. The corrosion rate (*CR*) was calculated by [30,32]:

$$CR = \frac{I_{corr} \times K \times EW}{\rho} \tag{1}$$

where, I_{corr} is the corrosion current density, K is a constant for use in Faraday's penetration rate equation which equals 3.27×10^{-3} mm·g/(μA·cm·year), ρ is the density, and EW is the equivalent weight which can be regarded as the mass of alloy in grams that will be oxidized by the passage of one Faraday (96489 ± 2 C) of electric charge:

$$EW = \frac{1}{\sum \frac{n_i f_i}{W_i}}, \tag{2}$$

where f_i is the mass fraction of the ith element in the alloy, W_i is the atomic weight of the ith element in the alloy, and n_i is the valence of the ith element of the alloy.

3. Results and Discussion

3.1. Porosity and Relative Density

With the bulk density of 8.05 g/cm^3, the relationship of relative density of AM CoCrFeMnNi samples versus volume energy density (VED) is shown in Figure 2. When VED is in the range of 62.5–115.6 J/mm^3, the relative densities of AM CoCrFeMnNi samples were all above 99.5%. When the VED was further increased, the relative density decreased dramatically due to keyhole effect [33] and partial evaporation of some constituent elements [34], particularly Mn which has higher vapor pressure and lower melting point compared to the other elements [35]. Chemical evaporation and over-penetration of laser beam in the high VED regime (>115.6 J/mm^3) can cause entrapped gas and formation of micro-pores inside the melt pool. In contrast, at low VED (<62.5 J/mm^3), inadequate penetration of the melt pool into the previously deposited layers may leave voids in the sample [36]. In this study, the samples with relative density of 99.5% (VED of 77.08 J/mm^3, laser power of 370 W, scan speed of 1500 mm/s, and bi-directional scan strategy) were selected for microstructural characterization and electrochemical corrosion tests.

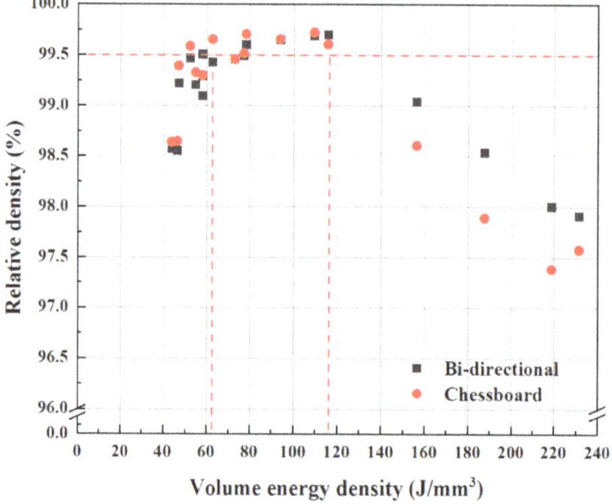

Figure 2. Variation of relative density of additively manufactured (AM) CoCrFeMnNi samples with different volume energy densities.

3.2. Microstructural Characterization

As shown in Figure 3, the gas-atomized CoCrFeMnNi powders are composed of irregular needle-like dendrites after etching. Some internal micro-pores were also observed, which is one root of the micro-pores (1–18 μm) in the eventual printed samples (Figure 4a,b). After etching, more pores were observed especially at the melt pool boundary (Figure 4c,d), which indicates the microstructure near the melt pool boundary has weaker chemical resistance. The purple arrows in Figure 4c display the epitaxial growth of the elongated grain with tortuous grain boundaries. Melt pools with the width and depth of about 165 μm and 124 μm, respectively, were generated during the rapid solidification.

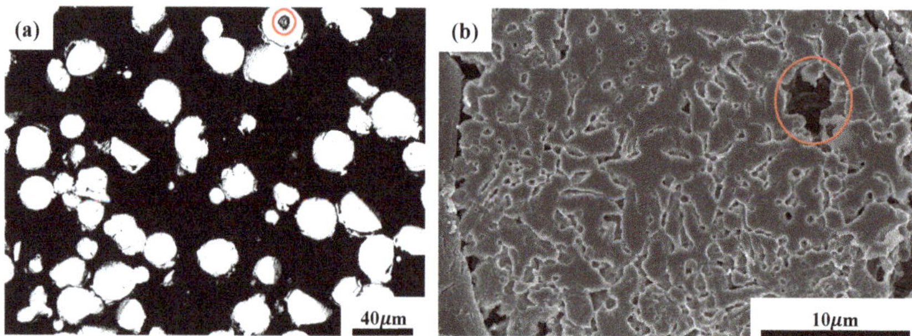

Figure 3. (**a**) Optical microscopic (OM) cross-section image of gas-atomized CoCrFeMnNi powders with pores inside; (**b**) SEM cross-section image of a single gas-atomized CoCrFeMnNi powder with needle-like dendrites (after etching) and internal pores. Micro-pores are highlighted by red circles.

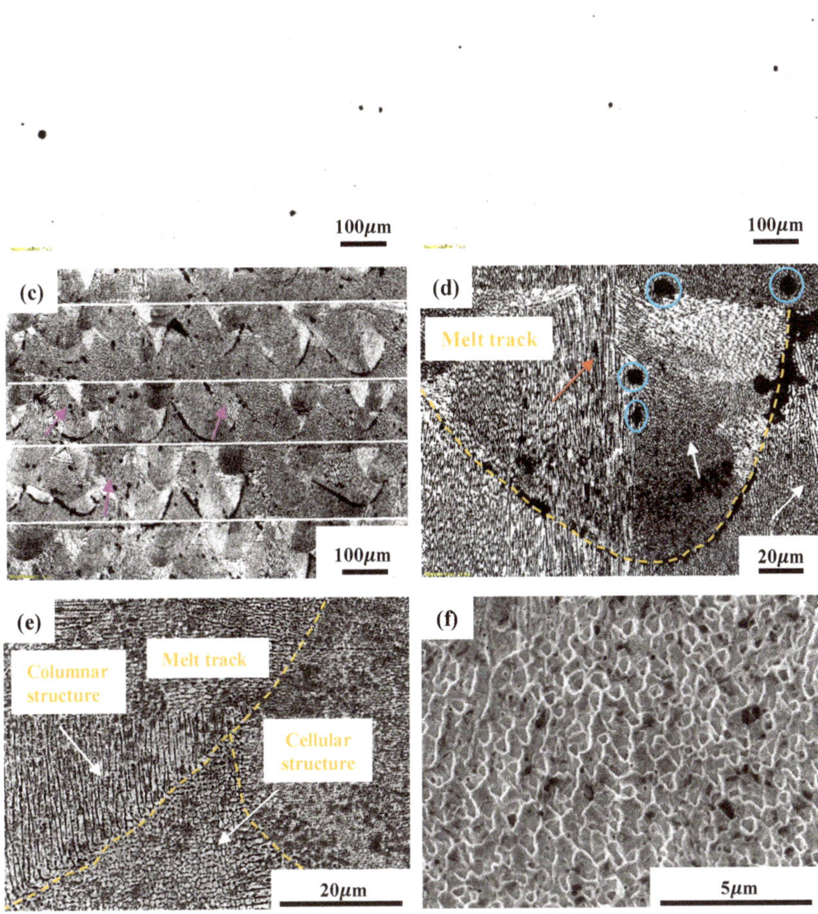

Figure 4. Non-equilibrium microstructure of AM CoCrFeMnNi as-cast sample: (**a**,**b**) OM images showing the porosity profile of the side and top surfaces of the AM sample; (**c**) OM image showing the microstructures of the side surface of the AM sample. The laser melting layers and elongated grains are depicted by white lines and purple arrows, respectively; (**d**) high magnification OM image showing the heterogeneous microstructures of the side surface of the AM sample. The melt pool boundaries, micro-pores, columnar and equiaxed cellular sub-structures are represented by orange dash lines, blue circles, red and white arrows, respectively; (**e**) Secondary electrons (SE) mode SEM image showing the melt pool boundaries as well as columnar and equiaxed cellular sub-structures; (**f**) high magnification SEM image of equiaxed cellular structures.

As displayed in Figure 4d,e, the AM sample is primarily composed of columnar dendrites and equiaxed cellular sub-structures. The columnar sub-grain structures in Figure 4d display epitaxial growth across the melt pool centerlines and the boundary. The average size of the equiaxed cellular structure is in the range of 0.69–0.91 μm. Compared to the as-cast sample, the refinement of the microstructure was attributed to the high cooling rate (~10^6 K/s) [19] and the re-melting of the previous deposited layer. Based on solidification theory [37], the metallurgical morphology and cooling rate are dependent on the temperature gradient (G) and solid/liquid interface growth rate (R). As shown in

Figure 4d,e, the cellular sub-grain structures mainly formed near the melt pool boundary where the temperature gradient (G) is higher and the solid/liquid interface growth rate (R) is relative lower. While at the melt pool center where the G/R ratio is lower than the boundary, the columnar sub-structures were observed.

3.3. Electrochemical Corrosion Behavior

The open circuit potential (OCP) for the two alloys are shown in Figure 5a. For both the alloys, the potential increased to positive (more noble) values indicating the formation of passive film in 3.5 wt.% NaCl solution [38]. Both the alloys showed multiple small spikes in open circuit potential, suggesting breakdown and re-passivation prior to the stabilization of the OCP after ~6000 s. The OCP value was very similar for the alloys prepared by the two different routes, indicating overall identical chemical composition. The passivation range and the corrosion current density were extracted from cyclic polarization tests (Figure 5b). The corrosion potential (E_{corr}) for AM and as-cast HEAs were −189 mV and −179 mV, respectively. The values of corrosion current density (I_{corr}) were similar for both alloys. In both samples, several spikes in current density were observed indicating metastable pitting followed by re-passivation. Note that the AM HEA showed a wider passive range ($\Delta E_{resistance} = E_{pit} - E_{corr}$) of ~386 mV compared to the as-cast HEA (~200 mV). Figure 5b shows that both samples did not re-passivate, which suggests that no protective passive film was formed at the large active pits.

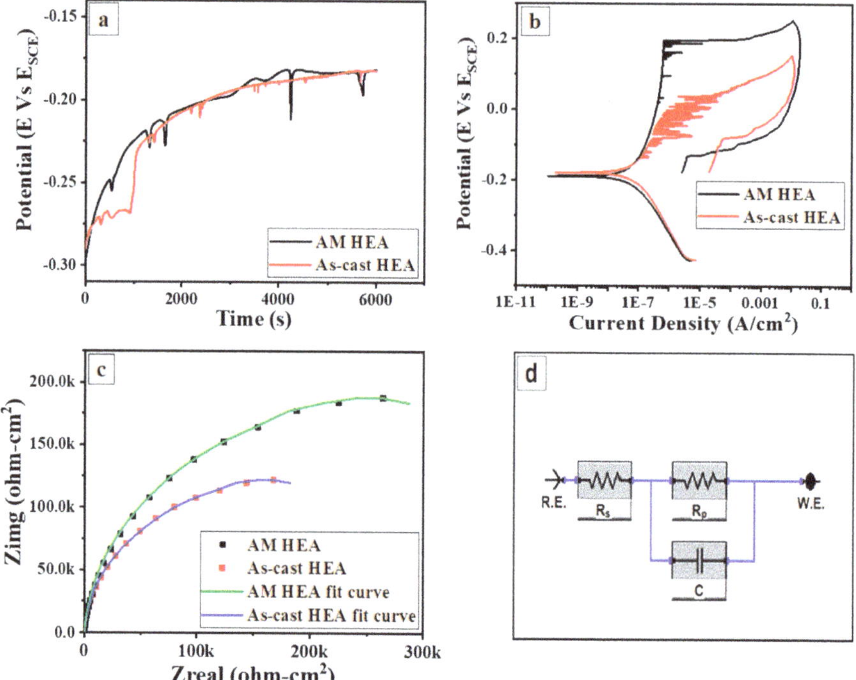

Figure 5. (**a**) Open-circuit potential (E_{OCP}) with respect to saturated calomel electrode (E_{SCE}) as a function of time; (**b**) cyclic polarization plots at a scan rate of 0.25mV/s; (**c**) Nyquist plot; and (**d**) the equivalent circuit used for fitting is shown, where R.E. is the reference electrode, R_s is the solution resistance, R_p is the polarization resistance, C is the constant phase element and W.E. is the working electrode.

The Nyquist plots depicted in Figure 5c was obtained from electrochemical impedance spectroscopy (EIS) in steady-state condition in which the capacitive arc trend corresponds to the double layer and passive film formation on the surface. The radius of the arc is directly proportional to the polarization resistance of the material. A modified Randles circuit (Figure 5d) was fit to experimental data to calculate the electrochemical parameters which includes solution resistance (R_s), a double layer capacitor (C) to represent the double layer charge capacitance absorbed onto the sample surface, and polarization resistance (R_p). All the parameters are summarized in Table 1. The simulation showed that the polarization resistance of AM HEA alloy was higher than the as-cast alloy, while the solution resistance and double layer capacitance values were approximately similar.

Table 1. Electrochemical and equivalent circuit parameters for AM and as-cast CoCrFeMnNi alloys in 3.5 wt% NaCl solution.

Corrosion Parameter	AM CoCrFeMnNi	As-cast CoCrFeMnNi
E_{corr} (mV)	−189	−179
I_{corr} (μA/cm^2)	0.09	0.11
E_{pit} (mV)	197	37
$\Delta E_{resistance}$ (mV)	386	216
Corrosion Rate (μm/yr)	0.7304	0.909
Polarization resistance, R_p (kilo ohm/cm^2)	278	186
Solution resistance, R_s (ohm/cm^2)	5.4	5.2
Double Layer Capacitance, C (μF/cm^2)	16.8	26

3.4. Morphology of Corroded Surfaces

SEM micrographs of the alloys before and after corrosion tests are shown in Figure 6. The AM alloy exhibited tortuous grain boundaries and micro-pores (Figure 6a). The as-cast alloy showed large grains of the order of 50–100 μm in diameter and high density of micro-pores randomly distributed in the microstructure (Figure 6c). Figure 6b,d show corrosion attack by pitting mechanism in both HEAs which may be correlated with the differences in porosity in microstructure and the elemental distribution between the AM and the as-cast alloy. The AM alloy demonstrated less and smaller current fluctuations in the metastable pitting region due to the lower density of large pores which acted as active anodic sites and prevented nucleation and growth of new pits. As the potential reaches the breaking point of the passive film, the stable pitting of the surface likely initiates at the existing large pores. The as-cast alloy on the other hand showed extensive metastable pitting due to the presence of high density of micro-pores. In another study, CoCrFeMnNi was 72% cold rolled and annealed at 900 °C for 20 h resulting in a much more homogenized microstructure with equiaxed grains. This microstructure showed similar corrosion current density but much larger pitting resistance (ΔE ~420 mV) [39] compared to the AM CoCrFeMnNi in the present study.

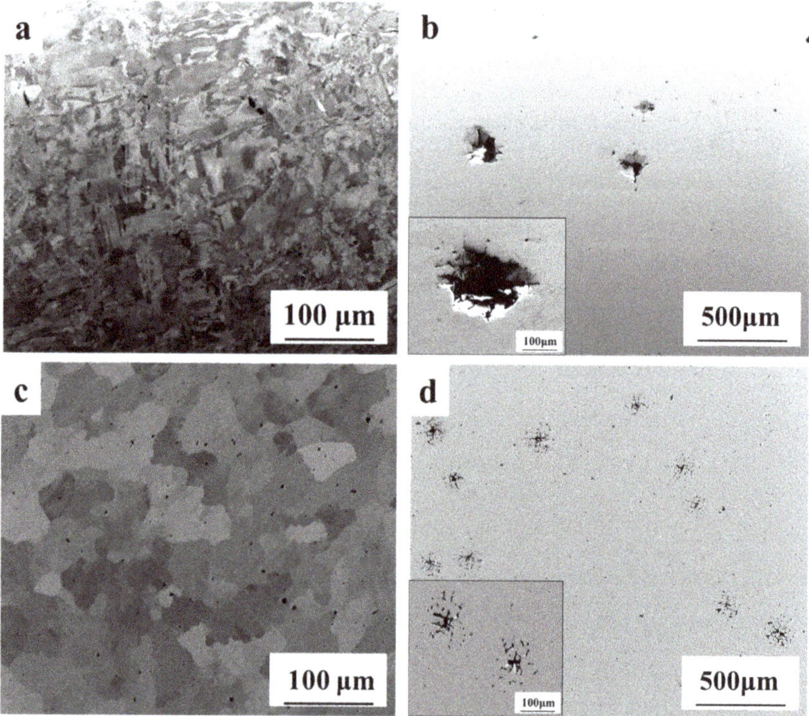

Figure 6. Back scattered electrons(BSE) mode SEM images of CoCrFeMnNi surface before and after cyclic polarization test in 3.5 wt.% NaCl solution: (**a**) AM HEA (top-surface) before corrosion; (**b**) AM HEA (top-surface) after corrosion; (**c**) As-cast HEA before corrosion; (**d**) As-cast HEA after corrosion. The inset in (**b**) and (**d**) are high magnification images.

Elemental mapping was conducted by energy-dispersive spectroscopy (EDS) for both AM and as-cast HEAs. Table 2 shows the EDS quantitative analysis of the both alloys, where impurities such as C, N, and O were less than 0.1 at.%. The AM HEA showed uniform distribution of the elements throughout the surface (Figure 7) which may be due to better homogenization from repeated melting or thermal annealing of previous layers during the multi-layer deposition process. The higher pitting resistance of the AM HEA can be attributed to the lower density of large pores and the uniform distribution of Cr, which leads to the formation of a more stable passive film over the surface. In contrast, the as-cast HEA exhibited non-uniform elemental distribution with significant chemical micro-segregation (Figure 8). There were two distinct segregation regions, one with higher Cr, Co, and Fe and the other with higher Mn and Ni. Similar segregation behavior was reported for as-cast $CoCrFeMn_{0.5}Ni$ [40], which is typical for multi-principal element alloys that often have a wide freezing range [41]. The elemental map for the as-cast HEA revealed Cr depleted regions acting as the active sites for corrosion initiation and leading to the formation of non-uniform and less stable passive film. The EDS mapping also confirmed the presence of less Cr content at the micro-pores, which is potentially the main reason for the severe corrosion attack by pitting mechanism on as-cast HEA.

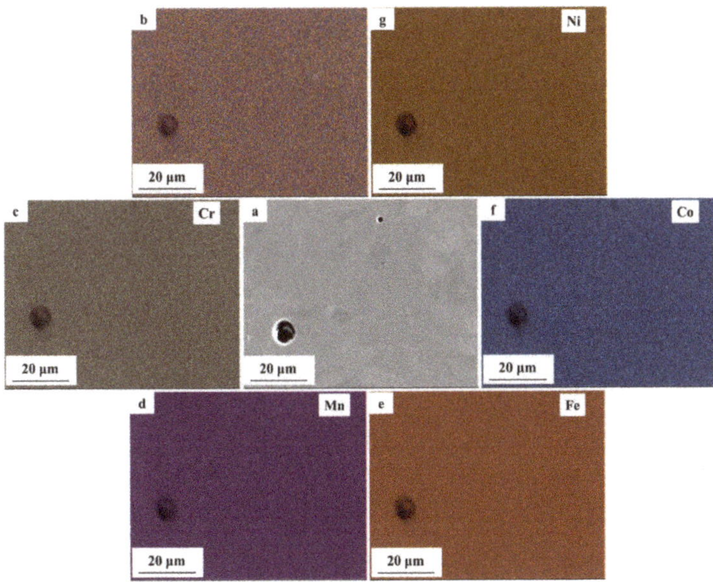

Figure 7. Energy-dispersive spectroscopy (EDS) map of additive manufactured CoCrFeMnNi HEA surface.

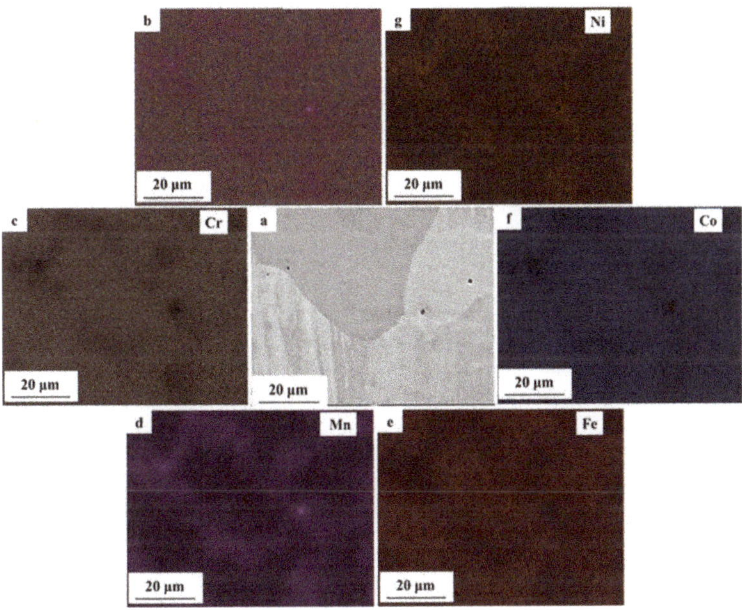

Figure 8. EDS map of as-cast CoCrFeMnNi HEA surface.

Table 2. EDS quantitative analysis of AM and as-cast CoCrFeMnNi alloy.

Composition (at. %)	Cr	Mn	Fe	Co	Ni
AM top surface	20.61 ± 2.61	18.84 ± 2.58	20.25 ± 2.8	20.52 ± 2.89	19.78 ± 3.12
AM side surface	20.26 ± 2.63	18.35 ± 2.59	20.41 ± 2.76	20.93 ± 2.88	20.06 ± 3.13
As-cast	19.32 ± 2.2	19.89 ± 2.03	20.74 ± 2.2	20.22 ± 2.29	19.83 ± 2.47

4. Conclusions

In this study, CoCrFeMnNi HEA was additively manufactured by SLM and its corrosion resistance in 3.5 wt% NaCl solution was evaluated by potentiodynamic polarization and electrochemical impedance spectroscopy measurements and compared with the as-cast counterpart. The main conclusions are as follows:

(1) By varying laser power and scan speed, near fully dense CoCrFeMnNi HEA was manufactured by SLM when the VED was in the range of 62.5–115.6 J/mm^3. It exhibited a unique non-equilibrium microstructure consisting of tortuous grain boundaries, sub-grain cellular structures, and some processing defects such as micro-pores.

(2) With wider passive region (ΔE), higher polarization resistance (R_p) and pitting potential (E_{pit}), the AM CoCrFeMnNi HEA showed superior corrosion resistance over the as-cast counterpart, which was attributed to the homogeneous elemental distribution at the investigated scale and lower density of micro-pores. The large micro-pores in AM CoCrFeMnNi HEA acted as active anodic sites and prevented nucleation and growth of new pits. The high density of micro-pores in the as-cast alloy resulted in extensive metastable pitting and the micro-segregation regions acted as active sites for corrosion initiation.

Author Contributions: Conceptualization, W.C. and S.M.; Methodology, W.C. and S.M.; Formal analysis, J.R. and C.M.; Investigation, J.R., C.M., L.L. and D.F.; Writing—original draft preparation, J.R. and C.M.; Writing—review and editing, L.L., W.C. and S.M.; Supervision, L.L., W.C. and S.M.; Project administration, W.C. and S.M.

Funding: This research received no external funding.

Acknowledgments: W.C. acknowledges the support by the NASA Marshall Space Flight Center Cooperative Agreement Notice (80MSFC19M0030) and UMass Amherst faculty startup.

Conflicts of Interest: The authors declare no conflict of interest.

References

1. Li, Z.M.; Pradeep, K.G.; Deng, Y.; Raabe, D.; Tasan, C.C. Metastable high-entropy dual-phase alloys overcome the strength–ductility trade-off. *Nature* **2016**, *534*, 227–230. [CrossRef] [PubMed]
2. Liang, Y.J.; Wang, L.J.; Wen, Y.R.; Cheng, B.Y.; Wu, Q.L.; Cao, T.Q.; Xiao, Q.; Xue, Y.F.; Sha, G.; Wang, Y.D.; et al. High-content ductile coherent nanoprecipitates achieve ultrastrong high-entropy alloys. *Nat. Commun.* **2018**, *9*, 4063. [CrossRef] [PubMed]
3. Shi, P.J.; Ren, W.L.; Zheng, T.X.; Ren, Z.M.; Hou, X.L.; Peng, J.C.; Hu, P.F.; Gao, Y.F.; Zhong, Y.B.; Liaw, P.K. Enhanced strength–ductility synergy in ultrafine-grained eutectic high-entropy alloys by inheriting microstructural lamellae. *Nat. Commun.* **2019**, *10*, 489. [CrossRef] [PubMed]
4. He, J.Y.; Wang, H.; Huang, H.L.; Xu, X.D.; Chen, M.W.; Wu, Y.; Liu, X.J.; Nieh, T.G.; An, K.; Lu, Z.P. A precipitation-hardened high-entropy alloy with outstanding tensile properties. *Acta Mater.* **2016**, *102*, 187–196. [CrossRef]
5. Ye, Y.F.; Wang, Q.; Lu, J.; Liu, C.T.; Yang, Y. High-entropy alloy: Challenges and prospects. *Mater. Today* **2016**, *19*, 349–362. [CrossRef]
6. Kilmametov, A.; Kulagin, R.; Mazilkin, A.; Seils, S.; Boll, T.; Heilmaier, M.; Hahn, H. High-pressure torsion driven mechanical alloying of CoCrFeMnNi high entropy alloy. *Scr. Mater.* **2019**, *158*, 29–33. [CrossRef]
7. Gludovatz, B.; Hohenwarter, A.; Catoor, D.; Chang, E.H.; George, E.P.; Ritchie, R.O. A fracture-resistant high-entropy alloy for cryogenic applications. *Science* **2014**, *345*, 1153–1158. [CrossRef]

8. Zhang, Z.J.; Mao, M.M.; Wang, J.W.; Gludovatz, B.; Zhang, Z.; Mao, S.X.; George, E.P.; Yu, Q.; Ritchie, R.O. Nanoscale origins of the damage tolerance of the high-entropy alloy CrMnFeCoNi. *Nat. Commun.* **2015**, *6*, 10143. [CrossRef]
9. Shi, Y.Z.; Yang, B.; Liaw, P.K. Corrosion-resistant high-entropy alloys: A review. *Metals* **2017**, *7*, 43. [CrossRef]
10. Tsai, M.H.; Yeh, J.W. High-entropy alloys: A critical review. *Mater. Res. Lett.* **2014**, *2*, 107–123. [CrossRef]
11. Diao, H.Y.; Santodonato, L.J.; Tang, Z.; Egami, T.; Liaw, P.K. Local structures of high-entropy alloys (HEAs) on atomic scales: An overview. *JOM* **2015**, *67*, 2321–2325. [CrossRef]
12. Qiu, Y.; Thomas, S.; Gibson, M.A.; Fraser, H.L.; Birbilis, N. Corrosion of high entropy alloys. *NPJ Mater. Degrad.* **2017**, *1*, 15. [CrossRef]
13. Chou, Y.L.; Yeh, J.W.; Shih, H.C. The effect of molybdenum on the corrosion behaviour of the high-entropy alloys $Co_{1.5}CrFeNi_{1.5}Ti_{0.5}Mo_x$ in aqueous environments. *Corros. Sci.* **2010**, *52*, 2571–2581. [CrossRef]
14. Kao, Y.F.; Lee, T.D.; Chen, S.K.; Chang, Y.S. Electrochemical passive properties of $Al_xCoCrFeNi$ (x = 0, 0.25, 0.50, 1.00) alloys in sulfuric acids. *Corros. Sci.* **2010**, *52*, 1026–1034. [CrossRef]
15. Qiu, X.W.; Liu, C.G. Microstructure and properties of $Al_2CrFeCoCuTiNi_x$ high-entropy alloys prepared by laser cladding. *J. Alloy. Compd.* **2013**, *553*, 216–220. [CrossRef]
16. Hsu, Y.J.; Chiang, W.C.; Wu, J.K. Corrosion behavior of FeCoNiCrCux high-entropy alloys in 3.5% sodium chloride solution. *Mater. Chem. Phys.* **2005**, *92*, 112–117. [CrossRef]
17. Lu, Y.P.; Dong, Y.; Guo, S.; Jiang, L.; Kang, H.J.; Wang, T.M.; Wen, B.; Wang, Z.J.; Jie, J.C.; Cao, Z.Q.; et al. A promising new class of high-temperature alloys: Eutectic high-entropy alloys. *Sci. Rep.* **2014**, *4*, 6200. [CrossRef]
18. Estrin, Y.; Beygelzimer, Y.; Kulagin, R. Design of Architectured Materials Based on Mechanically Driven Structural and Compositional Patterning. *Adv. Eng. Mater.* **2019**, 1900487. [CrossRef]
19. Thijs, L.; Kempen, K.; Kruth, J.P.; Humbeeck, J.V. Fine-structured aluminium products with controllable texture by selective laser melting of pre-alloyed AlSi10Mg powder. *Acta Mater.* **2013**, *61*, 1809–1819. [CrossRef]
20. Wang, Y.M.; Voisin, T.; McKeown, J.T.; Ye, J.C.; Calta, N.P.; Li, Z.; Zeng, Z.; Zhang, Y.; Chen, W.; Roehling, T.T.; et al. Additively manufactured hierarchical stainless steels with high strength and ductility. *Nat. Mater.* **2018**, *17*, 63–71. [CrossRef]
21. Zhu, Z.G.; Nguyen, Q.B.; Ng, F.L.; An, X.H.; Liao, X.Z.; Liaw, P.K.; Nai, S.M.L.; Wei, J. Hierarchical microstructure and strengthening mechanisms of a CoCrFeNiMn high entropy alloy additively manufactured by selective laser melting. *Scr. Mater.* **2018**, *154*, 20–24. [CrossRef]
22. Sun, Y.; Moroz, A.; Alrbaey, K. Sliding wear characteristics and corrosion behaviour of selective laser melted 316L stainless steel. *J. Mater. Eng. Perform.* **2014**, *23*, 518–526. [CrossRef]
23. Ziętala, M.; Durejko, T.; Polański, M.; Kunce, I.; Płociński, T.; Zieliński, W.; Łazińska, M.; Stępniowski, W.; Czujko, T.; Kurzydłowski, K.J.; et al. The microstructure, mechanical properties and corrosion resistance of 316 L stainless steel fabricated using laser engineered net shaping. *Mater. Sci. Eng. A* **2016**, *677*, 1–10. [CrossRef]
24. Trelewicz, J.R.; Halada, G.P.; Donaldson, O.K.; Manogharan, G. Microstructure and corrosion resistance of laser additively manufactured 316L stainless steel. *JOM* **2016**, *68*, 850–859. [CrossRef]
25. Geenen, K.; Röttger, A.; Theisen, W. Corrosion behavior of 316L austenitic steel processed by selective laser melting, hot-isostatic pressing, and casting. *Mater. Corros.* **2017**, *68*, 764–775. [CrossRef]
26. Kazemipour, M.; Mohammadi, M.; Mfoumou, E.; Nasiri, A.M. Microstructure and corrosion characteristics of selective-laser melted 316L stainless steel: The impact of process-induced porosities. *JOM* **2019**, *71*, 3230–3240. [CrossRef]
27. Dai, N.W.; Zhang, L.C.; Zhang, J.X.; Chen, Q.M.; Wu, M.L. Corrosion behavior of selective laser melted Ti-6Al-4 V alloy in NaCl solution. *Corros. Sci.* **2016**, *102*, 484–489. [CrossRef]
28. Yang, J.J.; Yang, H.H.; Yu, H.C.; Wang, Z.M.; Zeng, X.Y. Corrosion behavior of additive manufactured Ti-6Al-4V alloy in NaCl solution. *Metall. Mater. Trans. A* **2017**, *48*, 3583–3593. [CrossRef]
29. Luo, H.; Li, Z.M.; Mingers, A.M.; Raabe, D. Corrosion behavior of an equiatomic CoCrFeMnNi high-entropy alloy compared with 304 stainless steel in sulfuric acid solution. *Corros. Sci.* **2018**, *134*, 131–139. [CrossRef]
30. Ayyagari, A.; Hasannaeimi, V.; Grewal, H.S.; Arora, H.; Mukherjee, S. Corrosion, erosion and wear behavior of complex concentrated alloys: A review. *Metals* **2018**, *8*, 603. [CrossRef]

31. Ye, Q.F.; Feng, K.; Li, Z.G.; Lu, F.G.; Li, R.F.; Huang, J.; Wu, Y.X. Microstructure and corrosion properties of CrMnFeCoNi high entropy alloy coating. *Appl. Surf. Sci.* **2017**, *396*, 1420–1426. [CrossRef]
32. ASTM International. *G102-89 (2015)e1 Standard Practice for Calculation of Corrosion Rates and Related Information from Electrochemical Measurements*; American Society for Testing and Materials (ASTM): West Conshohocken, PA, USA, 2015.
33. Cunningham, R.; Zhao, C.; Parab, N.; Kantzos, C.; Pauza, J.; Fezzaa, K.; Sun, T.; Rollett, A.D. Keyhole threshold and morphology in laser melting revealed by ultrahigh-speed x-ray imaging. *Science* **2019**, *363*, 849–852. [CrossRef] [PubMed]
34. Khairallah, S.A.; Anderson, A.T.; Rubenchik, A.; King, W.E. Laser powder-bed fusion additive manufacturing: Physics of complex melt flow and formation mechanisms of pores, spatter, and denudation zones. *Acta Mater.* **2016**, *108*, 36–45. [CrossRef]
35. Li, R.D.; Niu, P.D.; Yuan, T.C.; Cao, P.; Chen, C.; Zhou, K.C. Selective laser melting of an equiatomic CoCrFeMnNi high-entropy alloy: Processability, non-equilibrium microstructure and mechanical property. *J. Alloy. Compd.* **2018**, *746*, 125–134. [CrossRef]
36. Mukherjee, T.; Zuback, J.S.; De, A.; DebRoy, T. Printability of alloys for additive manufacturing. *Sci. Rep.* **2016**, *6*, 19717. [CrossRef] [PubMed]
37. DuPont, J.N. Fundamentals of weld solidification. In *ASM Handbook, Welding Fundamentals and Processes*; Lienert, T., Siewert, T., Babu, S., Acoff, V., Eds.; ASM International: Materials Park, OH, USA, 2011; Volume 6A, pp. 96–114.
38. Shang, X.L.; Wang, Z.J.; He, F.; Wang, J.C.; Li, J.J.; Yu, J.K. The intrinsic mechanism of corrosion resistance for FCC high entropy alloys. *Sci. Chin. Technol. Sci.* **2018**, *61*, 189–196. [CrossRef]
39. Ayyagari, A.; Barthelemy, C.; Gwalani, B.; Banerjee, R.; Scharf, T.W.; Mukherjee, S. Reciprocating sliding wear behavior of high entropy alloys in dry and marine environments. *Mater. Chem. Phys.* **2018**, *210*, 162–169. [CrossRef]
40. Christofidou, K.A.; Pickering, E.J.; Orsatti, P.; Mignanelli, P.M.; Slater, T.J.A.; Stone, H.J.; Jones, N.G. On the influence of Mn on the phase stability of the CrMnxFeCoNi high entropy alloys. *Intermetallics* **2018**, *92*, 84–92. [CrossRef]
41. Flemings, M.C. Solidification processing. *Metall. Mater. Trans. B* **1974**, *5*, 2121–2134. [CrossRef]

Perspective

High Entropy Alloys: Ready to Set Sail?

Indranil Basu [1,2,*] and Jeff Th. M. De Hosson [1,*]

1 Department of Applied Physics, Zernike Institute for Advanced Materials, University of Groningen, 9747AG Groningen, The Netherlands
2 Laboratory of Metal Physics and Technology, Department of Materials, ETH Zurich, 8093 Zurich, Switzerland
* Correspondence: ibasu@ethz.ch (I.B.); j.t.m.de.hosson@rug.nl (J.T.M.D.H.)

Received: 24 December 2019; Accepted: 27 January 2020; Published: 29 January 2020

Abstract: Over the past decade, high entropy alloys (HEAs) have transcended the frontiers of material development in terms of their unprecedented structural and functional properties compared to their counterpart conventional alloys. The possibility to explore a vast compositional space further renders this area of research extremely promising in the near future for discovering society-changing materials. The introduction of HEAs has also brought forth a paradigm shift in the existing knowledge about material design and development. It is in this regard that a fundamental understanding of the metal physics of these alloys is critical in propelling mechanism-based HEA design. The current paper highlights some of the critical viewpoints that need greater attention in the future with respect to designing mechanically and functionally advanced materials. In particular, the interplay of large compositional gradients and defect topologies in these alloys and their corresponding impact on overall mechanical response are highlighted. From the point of view of functional response, such chemistry vis-à-vis topology correlations are extended to novel class of nano-porous HEAs that beat thermal coarsening effects despite a high surface to volume ratio owing to retarded diffusion kinetics. Recommendations on material design with regards to their potential use in diverse applications such as energy storage, actuators, and as piezoelectrics are additionally considered.

Keywords: serrated flow; thermal coarsening; actuators; phase transformation; nanoporous metals and alloys

---

1. Introduction

The classical rationale behind complex concentrated alloys (CCAs), more popularly referred to as high entropy alloys, comprises the addition of four or five elements in equiatomic and near-equiatomic proportions that eventually generate a single-phase solid solution [1,2]. The theoretical feasibility of such a counterintuitive alloy design stems from the concept of entropic stabilization, wherein a large number of elements would raise the configurational entropy and simultaneously overcome the enthalpies associated with the formation of intermetallic compounds. Owing to the inherent complexity of entropic stabilization, such multicomponent alloys have been observed to display peculiar characteristics that were summarized by Yeh and co-workers [1] as the four core effects: (i) The maximization of configurational entropy, owing to which the term high entropy alloys (HEAs) was coined; (ii) lattice strain due to large variation in atomic sizes of constituent elements; (iii) sluggish diffusion kinetics due to frustrated crystal structures; and iv) unusual properties displayed due to diverse interatomic interactions, also referred to as the "cocktail" effect [3].

However, the universality of occurrence of the abovementioned core effects in HEAs is debatable [4–9]. For instance, the very theory describing entropic stabilization has found little experimental validation, indicated by the fact that majority of HEAs that have fabricated till now either exist as multiphase alloys or decompose to more than one phase at thermodynamic equilibrium [6–8,10–12]. This is primarily because of the fact that the postulation by Yeh et al. on the maximization of configurational entropy remains valid at melting temperatures [13]. On the other hand, most experimental alloys are characterized at room temperatures, which renders significant microstructural and phase reordering driven by local stress/strain distributions, local compositional fluctuations, and interatomic interactions among constituent elements under diverse thermomechanical processing schemes [13,14]. Nevertheless, due to the widespread popularity of the postulated core effects within the community, they are still being actively associated with HEAs.

While the search for stable single-phase HEAs via combinatorial approaches continues [9,15], a large part of the research in the field delves into exploiting multiphase HEAs as means to design materials with significantly enhanced structural and functional properties [10,13]. Of special note in this regard are the less stringent alloying routes that provide access to a much larger compositional space in comparison with conventional alloys. This facilitates a mechanism driven compositional tuning of alloys, wherein the high solid solution content of individual elements can be tailored in order to generate unique features, such as spinodal structures [16], nano-scale coherent precipitates [17], and the modification of stacking fault energies [18,19], that give rise to interesting metal physics at intrinsic length scales. The advantages of this are clearly visible from the wide-spread future applications proposed for HEAs with unprecedented combinations of structural and functional properties [13,20–23].

In light of the aforementioned aspects of HEAs, it has become imperative to understand the independent and interdependent effects of the spatial distribution of linear and planar defects (i.e., defect topology) and compositional fluctuations existing in these alloys on the local mechanical and functional responses.

2. Crystallographic Defects versus Compositional Variation: A Tale of Two Effects

A critical aspect that influences mechanical responses in materials is the dependence of defect topologies and compositional gradients upon length scale dynamics. When considering crystal defects, the particular game players with respect to local plasticity are line defects and surface/interface defects, as well as their mutual interactions. Adding to this, the effect of local compositional fluctuations that could be considerable in HEAs further adds another layer of complexity in terms of local stress evolution and corresponding strain accommodation. Owing to such compositional diversity, metal physics in HEAs therefore offers an interesting playground wherein a superposition of multiple strengthening mechanisms that can interact and augment each other's contribution takes place. It must be understood that even though the observed metal physics in HEAs are derived from those that are seen for conventional alloys and superalloys, the compositional flexibility and simultaneous activation of different strengthening modes in the former potentially places them as more suitable candidates for mechanism-based alloy design.

The influence of local chemistry on meso-scale microstructures serves as a key design aspect in HEAs in regard to tailoring alloys with significantly improved structural responses. Owing to the absence of a primary solvent, local elemental partitioning in these alloys often results in the formation of phase interfaces. Depending upon whether such interfaces are crystallographically similar or dissimilar, the corresponding impact on the plasticity and strengthening behavior is distinct. For instance, one of the commonly studied systems in this regard is the $Al_xCoCrFeNi$, wherein increasing the Al content drives a crystallographic transition from an Face-centered cubic (FCC) solid solution to a Body-centered cubic (BCC) phase [1]. Moreover, the BCC phases that exist in this alloy are further known to spinodally decompose into ordered B2 (enriched with Al and Ni) and disordered BCC structures [16,23–25]. It must be highlighted that the underlying strength contribution and metal physics are entirely different when considering such different interface types, and these become critical to appraise when engineering HEAs with strengthening across different length scales.

2.1. Defect Generation and Strengthening Behavior across Crystallographically Similar Interfaces

In a recent study [16], the nano-indentation response of BCC grains in $Al_{0.7}CoCrFeNi$ alloys indicated series of random displacement bursts. The observed pop-ins showed varying amplitudes and seemed to be more obvious at lower indentation loads and smaller penetration depths (Figure 1c). Serrated flow characteristics are the fingerprint of jerky dislocation kinetics, as these arise from intermittent intervals of obstructed dislocation motion. In the present case, it was shown that the spinodally decomposed ordered B2 and disordered A2 phases generated interfaces that gave rise to simultaneous spinodal hardening and order-hardening effects (Figure 1a,b). These effects typically manifest at deformation length scales that are comparable to the mean size of the A2 phase, that is ~100–200 nm (Figure 1d,e). Considering the fact that spinodal strengthening in BCC crystal structures can be significantly larger than FCC spinodal alloys due to the sizable contribution of both elastic coherency strains and hardening from the modulus differential in the former, it has been shown that the strengthening potential in spinodal BCC HEAs can be as high as 0.5 GPa. In another study [17], it was shown that the addition of Ti and Al to single phase FCC CoCrFeNi HEAs leads to precipitation hardening effects due to presence of ordered FCC precipitates in a random FCC matrix, and these contribute to a strengthening increment between 0.3 and 0.4 GPa, which is significantly larger than counterpart contributions from strain hardening, grain boundary hardening, and solid solution strengthening. Lately, the concept of utilizing such spinodally-induced strengthening and order hardening effects in HEAs has given rise to a new generation of modulated, nano-phase structured, BCC-refractory HEAs that mimic super alloy type microstructures [26,27].

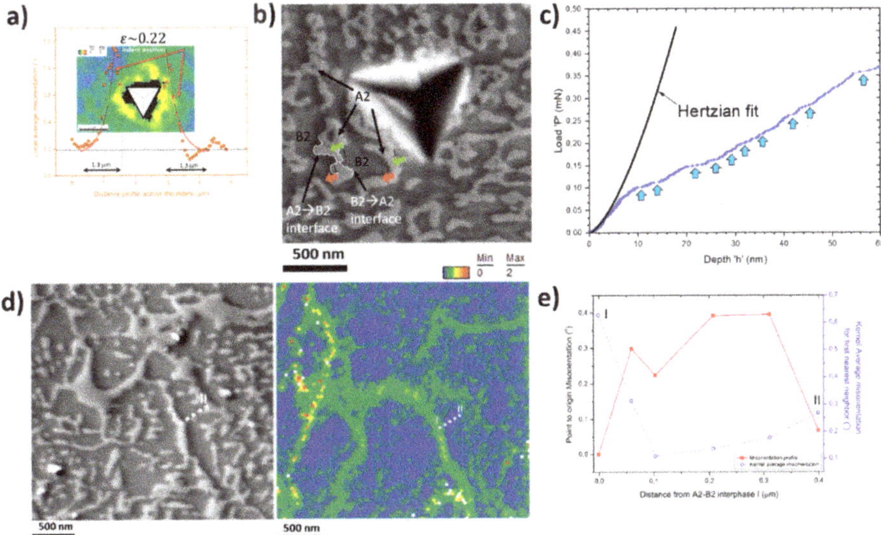

Figure 1. (**a**) Indent strain profile and (**b**) corresponding microstructure showing the dislocation motion across an A2/B2 interface. Dislocations experience strengthening when moving from the A2 to the ordered B2 phase, owing to spinodal and order hardening effects. The corresponding jerky dislocation motion is highlighted in nano-indentation load-displacement curves in (**c**). (**d**) shows the strain distribution with bright A2 phases being more plastic than the elastically stiffer dark B2 phase. (**e**) Corresponding misorientation profile and kernel average misorientation gradient on moving away from the interphase into the B2 phase. Figure 1d,e was adapted from [16] with permission from Elsevier, 2018.

2.2. Strengthening Mechanisms across Crystallographically Dissimilar Phase Boundaries

Unlike grain boundaries, the mechanics of strengthening across heterophase interfaces involve diverse contributing factors. Classical grain boundaries govern strain transmission, primarily on the basis of grain boundary geometry and the alignment of active slip systems in pile-up and emission grains [28–31]. On the other hand, interphase boundaries involve interface-dependent strengthening mechanisms [32] that add to the overall extent of dislocation pile-up and internal stress configurations.

In theory, heterophase interphases derive strength from three main contributing mechanisms, apart from the geometric slip transmission criterion.

2.2.1. Image forces or Koehler forces:

The mismatch in shear moduli between neighboring phases gives rise to a Koehler force barrier between the dislocation and the interface [33,34]. The underlying effect that is responsible for this is the variation of strain energy per unit length of dislocation with changing modulus, such that a dislocation that moves from a stiffer grain to a softer grain would experience an attractive force, and the opposite scenario would result in repulsion between the incoming dislocation and the interface (Figure 2c). Mathematically, Koehler forces ($\tau_{Koehler}$) at an interphase boundary between phase A and phase B can be expressed as

$$\tau_{Koehler} = \frac{G_A(G_B - G_A)}{4\pi(G_B + G_A)} \cdot \frac{b}{h} \tag{1}$$

where G_A and G_B are the shear moduli values of incident and emission grains, respectively; b is the magnitude of the Burgers vector of active slip system in the incident grain; and h is the normal distance between the dislocation and the interface. The exerted force is hypothetically similar to the stress field that is exerted by a negative image dislocation that is positioned at the other side of the interface, hence the term of image forces. The presence of Koehler forces significantly impacts the dislocation pile-up characteristics at the phase boundary that subsequently influence the strengthening that is imparted from the interfaces, as shown in Figure 2a,b. For instance, it has been shown in a BCC/FCC dual-phase $Al_xCoCrFeNi$ alloy that the elastic modulus difference in FCC and ordered BCC phases (E_{BCC} = 275 GPa vs. E_{FCC} = 252 GPa) results in an attractive image force on incoming BCC dislocations and repulsive image forces on incoming FCC dislocations [35].

2.2.2. Misfit Stresses

Crystallographically dissimilar phase boundaries also result in interfacial stresses that arise from lattice parameter mismatch (Δa) between adjacent phases (Figure 2d). The size misfit is compensated by a grid of van der Merwe dislocations that give rise to coherency strain hardening effects at the interface [36]. Coherency stresses typically dampen as a function of $1/\lambda$ upon moving away from the interface, and they are mathematically given as,

$$\tau_{misfit} = 0.5G^* \sqrt{\frac{2b(\delta - \varepsilon)}{\lambda}} \tag{2}$$

where $\delta = \frac{\Delta a}{\bar{a}}$; $\bar{a}$ is the mean lattice parameter $(a_{phase\_A} + a_{phase\_B})/2$, $\varepsilon = 0.76\delta$ is the residual elastic strain that was determined to agree for most heterophase interface types, G^* is the average shear modulus for the two phases, and λ is the grain dimension over which misfit stresses are determined, i.e., the distance between a dislocation and the interface. Coherency stresses exert a Peach–Koehler force on incoming glide dislocations, which can be either attractive or repulsive depending on the sense of applied stress with respect to the dislocation slip system. Apart from affecting the dislocation glide stresses, the coherency stresses aid in strengthening by additionally influencing the non-glide stress components of the dislocation stress field, whereby they can locally modify the dislocation core energy that directly influences the ease of a dislocation in overcoming an obstacle.

2.2.3. Chemical Mismatch Effect

Another aspect that contributes to interfacial strengthening is a mismatch in chemical energy or gamma surfaces, as this mismatch directly determines the stacking fault energies in adjacent phases [37]. When the leading partial in a stacking fault moves across an interface, the dislocation configuration experiences an abrupt change in stacking fault energy. This manifests as an effective stress that is exerted upon the leading dislocation in the pile-up (Figure 2e). The resultant stacking fault strengthening stresses is described as

$$\tau_{chemical} = \frac{\Delta\gamma}{b} \tag{3}$$

where $\Delta\gamma$ is the stacking fault energy differential between neighboring phases. Overall, the change in energy of dislocation as it moves across the phase boundary involves an elastic energy contribution (that is a combination of $\tau_{Koehler}$ and τ_{misfit}) and a chemical contribution in the form of $\tau_{chemical}$.

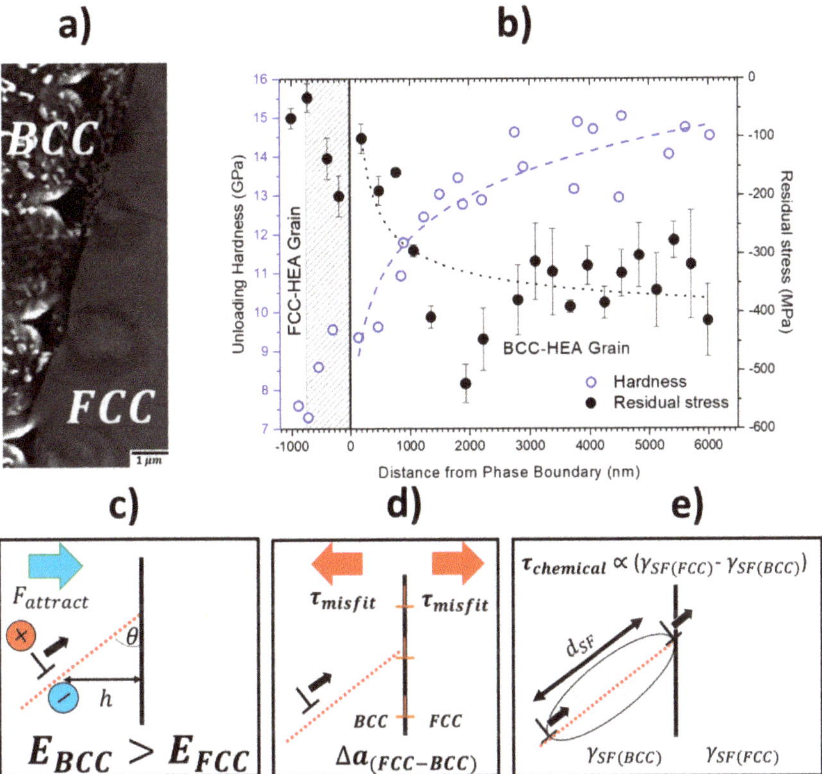

Figure 2. (**a**) BCC–FCC interface in an $Al_xCoCrFeNi$ high entropy alloy, with indent profile. (**b**) Variation of hardness and residual stress as a function of distance from phase boundary. The role of image forces is highlighted by local maximum in stress values on the BCC side (attractive image forces) and a local minimum on the FCC side (repulsive image forces). (**c–e**) illustrate different interfacial-dependent strengthening mechanisms. Figure 2b was adapted from [2] with permission from Elsevier, 2018.

Owing to the strong compositional fluctuations and propensity of single-phase decomposition in HEAs, the contribution of interface-dependent strengthening could be significant in terms of augmenting overall material strength at both local and global scales. Recent studies have now made effective use of such hardening mechanisms to tailor microstructural designs that give rise to substantially stronger HEAs in comparison with their single-phase counterparts or with respect to conventional alloys. It has been shown that the BCC–FCC interfaces in HEAs could give rise to strengths of the order of 4 GPa that are nearly four times the measured values of conventional

BCC–FCC interfaces [35]. The underlying contribution for the augmented strengthening in HEAs has primarily been attributed to the enhanced interfacial-dependent strengthening that is caused by the strong compositional gradients in these alloys. Investigations on HEAs that comprised of BCC/FCC multilayers has indicated yield strengths of the order of 3.3. GPa, with more than two-thirds of this strength coming from interfacial strengthening effects and the remaining coming from solid solution strengthening [38]. In another study [39], it was shown that BCC/FCC interfacial strengthening mechanisms could be further enhanced by tuning the multilayer thickness in HEAs, whereby strengths of the order of ~13 GPa can be reached.

3. An Outlook to HEAs: Structural Properties

Compositionally, HEAs can be described as a concoction of multiple elements, an arrangement that often results in a frustrated crystal structure. Moreover, chemical gradients further trigger local rearrangements and the shuffling of elements, thus influencing the stability of the existing phases.

In short, HEAs are considerably more prone to phase transformation under applied temperature or stress, which could be a potent mechanism to trigger interesting plasticity mechanisms as well as to accommodate larger strains. For instance, in a seminal work by Li et al. on non-equiatomic compositions [10] based on the FCC single phase cantor alloy, it was shown that under plastic deformation, the dynamic transformation from an FCC to an HCP crystal structure is achieved that simultaneously enhances strength and ductility. In a more recent study [16], dynamic indentation-induced phase transition from BCC to FCC was observed in BCC $Al_{0.7}CoCrFeNi$ HEAs (Figure 3a,b). The underlying reason behind the transformation was attributed to the spinodal decomposition of the BCC phase in Al, Ni rich-ordered B2 phases and random A2 phases (Figure 3a). Under applied stress, the A2 phases that are locally depleted in Al content could displacively transform and revert back to the more stable and ductile FCC phase (Figure 3c).

The results once again provide an opportunity to exploit the compositional fluctuations in tandem with thermomechanical treatment to dynamically trigger strength and ductility enhancing mechanisms. Displacive phase transformation effects or TRIP effects in HEAs could be exciting focal points in novel advances of HEAs in structural properties and applications.

Another mechanistic design criterion that employs compositional fluctuations is through the intrinsic modification of stacking fault energies. It was shown by Ritchie and co-workers [40,41] that by tuning local chemical ordering in HEAs, considerable variation in the intrinsic and extrinsic stacking fault energy values can be realized. Such design pathways are critical in terms of triggering new deformation mechanisms such as deformation twinning, whereby additional strain accommodation mechanisms are dynamically activated. Twinning not only contributes to plasticity but can also promote dynamic Hall–Petch-driven strengthening behavior, owing to grain fragmentation caused by twin boundary formation. For instance, it was shown by Deng et al. that for non-equiatomic $Fe_{40}Mn_{40}Co_{10}Cr_{10}$ HEA deformation, twinning is triggered as an additional mechanism for higher strains, thereby contributing to the overall strength–ductility increment [18].

Finally, the possibility to exploit simultaneous TRIP and TWIP effects along with multiscale strengthening effects that encompass interfacial strengthening and solid solution hardening effects can be envisioned in upcoming HEAs. In this regard, it has been shown that for the non-equiatomic FeMnCoCr alloy, dilute additions of C (~0.6 at. %) already trigger simultaneous twinning and phase transformation along with an interstitial hardening response [19]. Similar alloying strategies have been implemented on non-equiatomic BCC HfNbTaTiZr, wherein simultaneous plasticity-induced displacive transformation from the BCC phase to the HCP phase was observed, along with twinning in the HCP phase [42].

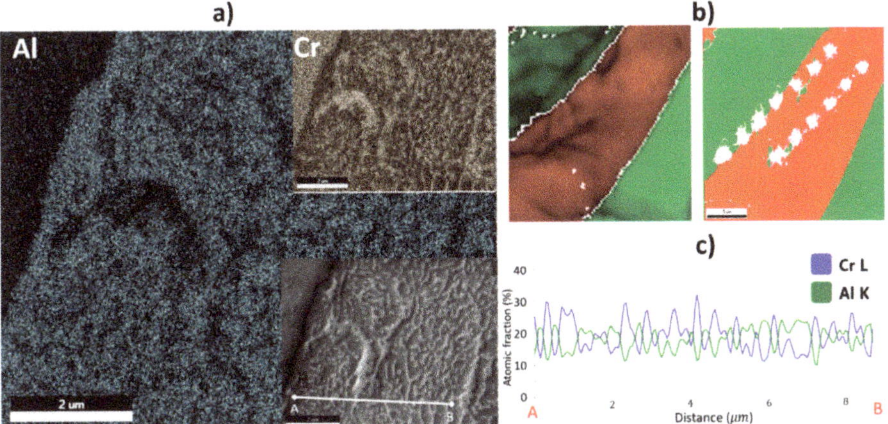

Figure 3. (**a**) EDS maps showing selective partitioning of Al and Cr into B2 and A2 phases, respectively. (**b**) Nano-indentation-induced phase transformation of A2 BCC phase (shown in red) to FCC phase (shown in green). Grain boundaries shown in white in the left image, and the non-indexed areas that are shown in white in the right image correspond to the experimentally made indents; local compositional fluctuations of Al and Cr indicated in (**c**), highlighting the instability of A2 BCC phases in regions of depleted Al, whereby such dynamic phase transformation to FCC at room temperature deformation is facilitated. Experimental data for the figures were derived from [16]. Figure 3b was adapted from [16] with permission from Elsevier, 2018.

The abovementioned impact of compositional fluctuations and strengthening modes related to phase formation also needs to be incorporated into current solid solution strengthening models in HEAs. There are now sufficient studies that have revealed that the strengthening of dislocation motion in HEAs is strongly dependent upon its susceptibility to display either short or long range ordering effects rather than simple lattice friction-induced hardening responses [43,44]. This was validated by a recent study by Robert Maaß and collaborators, wherein the peak dislocation velocities in FCC $Al_{0.3}CoCrFeNi$ and pure Au did not show much difference, indicating that dislocation motion was not significantly sluggish in single phase solid solution HEAs (Rizzardi et al. [45]). Moreover, the contributions of interfacial-dependent strengthening and solute strengthening modes need to be appraised, as these could be critical in driving application-based future multiphase HEA alloy design. In this regard, greater efforts are needed in understanding the influence of alloying chemistry on engineering interphase boundaries in HEAs rather than focusing upon solid-solution strengthening as the primary strength contributor in these alloys. Indeed, the outcomes look promising and may open a new paradigm of structurally advanced HEAs, as was recently shown in study [46] where a compositionally graded $Al_xCoCrFeNi$ bar was additively manufactured with increasing Al contents from $x = 0.3$ to $x = 0.7$ along the longitudinal direction, such that one end of the material was a single phase FCC and the other end formed a dual phase B2–FCC microstructure. From the point of view of mechanical response, the dual phase microstructure clearly highlighted the positive role of interfaces with significantly larger strengthening potentials compared to the single-phase FCC solid solutions.

4. An Outlook to HEAs: Functional Properties

Thus far, the emphasis in our feature paper has lied upon mechanical performance, plasticity and damage control. Attractive and rather unexplored frontiers of HEAs concern applications of functional properties, e.g., magnetic and electrical including thermoelectricity (Seebeck effect) [40], in the field of microelectronics and bio-medicine [23,47].

Several ideas are currently under investigation, and, in particular, we like to refer to possibilities of HEAs as cellular/porous materials that can be used in 'energy materials' (H-storage) but also explored as radiation resistant sensors and actuators. Actuation refers to mechanical displacement due to an electric signal. The opposite is also possible when an electrical current is generated by mechanical deformation, which creates what is known as piezoelectric materials. In general, these piezoelectrics need high voltages, with an order of 100 V, and, at present, methods are being developed in medicine and biology to manufacture high-precision actuators that work at lower voltages on cell manipulation [48]. Structural stability over a large range of temperatures is essential.

In particular, highly porous metallic systems can mimic the properties of muscles upon an outside stimulus, and they have been coined 'artificial muscles' in analogy to human skeletal muscles, which are ideal actuators with a high energy efficiency, fast strain-rate response, and high durability. The common use of existing materials as actuators like piezoceramics and electroactive polymers are limited by several factors, including low energy efficiency, low strain amplitudes, fatigue limits, and the high actuation voltages needed. In our recent work, we have shown that nanoporous organometallic materials can operate as actuators, thereby offering a unique combination of relatively large strain amplitudes, high stiffness and strength, and importantly (see above) low operating voltages—say, at a few volts. However, a serious concern in this field of applications is (thermal) stability and the effects of coarsening.

Because of their high porosity and surface areas, the stability of cellular/porous systems is a major issue. In fact, depending on temperature (low versus high) and environment, the stability might be questionable, and, as a consequence, the functional properties might be not very stable and may deteriorate over time. Because of the suppression of the diffusional processes of defects at the surface of a (nano) ligament, we believe that HEAs might provide a very interesting and effective remedy to address these essential problems for applications of unique properties of functional HEA materials.

4.1. Porous/Cellular Systems

To illustrate the problem in a bit more detail, we discuss here the characteristics of cellular and porous media that possess a lower density and a higher surface area-to-volume ratio. The terminology is in macrofoams a bit different from that of nanofoams. In the former, besides pores, the material is made up of struts, and, in the latter, we call the struts ligaments. The topology of nodes and struts/ligaments can be anisotropic as well as isotropic. Besides these structural differences, nanoporous foams have been applied in nanofiltration systems, drug delivery platforms, catalysis, sensing and actuation [49–54]. In contrast, macro foams have been explored in macroscopic applications of the transport, automotive and aerospace industries.

A popular way of making metal nanofoams is based on dealloying through leaching. Preferably, the base material is a solid solution of a noble and a less noble element. Unfortunately, many alloys form intermetallics and many metals do not easily form solid solutions, which limits the dealloying methodology. Recently, nanoporous HEAs were produced through a rather novel method [55] by using liquid metal dealloying (LMD), a technique to fabricate non-noble porous materials by suppressing oxidation in a metallic melt [56–62]. It turned out that the structure of nanoporous TiVNbMoTa HEAs [55] can be described as nanoscale ligaments of a solid-solution phase, the stability of which is due to suppressed surface diffusion.

4.2. Suppression of Coarsening

It should be realized that the actuation mechanism based on nanoporous metals with high surface-to-volume ratios is different to piezoceramics. For details, one can look to [63]. The physical principles in the case of metals are based on the lower coordination of the surface atoms. Therefore, to gain 'density,' the atoms move inwards, and a positive displacement is necessary to bring them back to the equilibrium interatomic distances of the situation in the bulk. Therefore, a positive, i.e., tensile stress state at the surface, is generated that is compensated for by a compressive stress state inside the strut/ligament. Clearly, a positive charge injection can equilibrate the existing excess negative surface charge. As a consequence, a positive charge lowers the positive tensile stress and, importantly, relaxes the negative compressive stress in the ligament, i.e., the negative compressive stress becomes a bit more 'positive', i.e., a positive charge will generate a positive displacement that can be detected by optical means by using a small laser. It is noteworthy that these rather small displacements as a result of stress relaxations are not detectable in macrofoams, but nanofoams are completely different because a much large surface area to volume ratio exists. To give a rough estimate: If the size of the ligaments is of the order of 5–10 nm, a substantial fraction of the total number of atoms, say 20%–10%, is on the surface, meaning that a 100 nm thick ligament already reflects the bulk situation, and the actuating properties are hardly detectable. It is important to realize that the electronic charge distribution at a nanoporous metal interface can effectively be controlled during cycling voltammetry experiments. Only small electrical voltages of the order of 1 V are needed to bring positive or negative charge carriers (ions) from the electrolyte to the nanoporous metal [64,65]. Of course, working with liquid electrolytes can be cumbersome, and [65] we recently demonstrated that metallic muscles can operate in a dry environment, even at high strain rates, i.e., much higher than allowed for in electrochemical artificial muscles [65].

Nonetheless, electrochemical processes may lead to severe coarsening (undesired growth) of the ligaments [66–68], which is a major concern because actuation is hampered or completely lost, as is shown in Figure 4, where strain amplitudes are plotted as a function of the average ligament size. As expected, there has been a strong size effect and the strain amplitude recorded on nanoporous systems with different ligament sizes, and these may decrease upon increasing the ligament size (Figure 4). Even as a sensor (i.e., without external applied electric fields), the response is not stable due to coarsening effects (Figure 5), and, at higher temperatures, these effects are amplified tremendously.

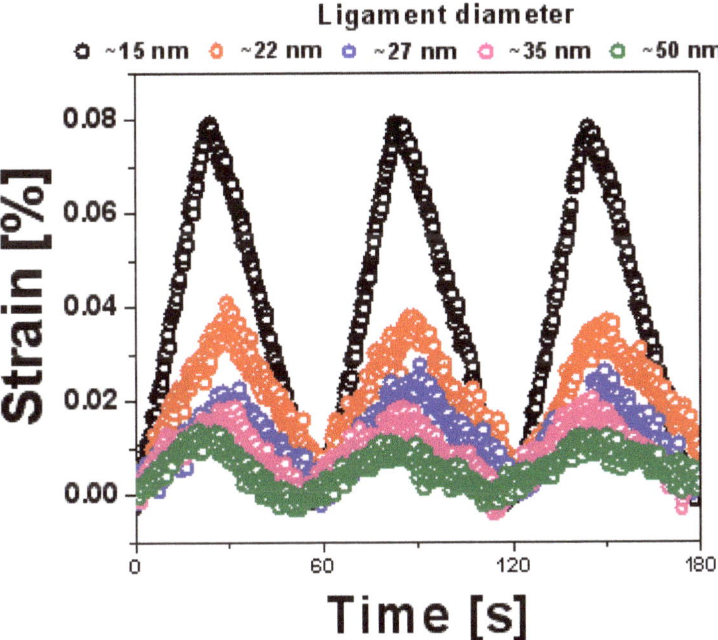

Figure 4. Ligament size-dependence of the charge-induced strain in nanoporous metals (Au). The strain amplitude that was recorded on five specimens with different ligament sizes decreased with increasing ligament size. This shows that ligament growth during electrochemical actuation is undesirable [67,68].

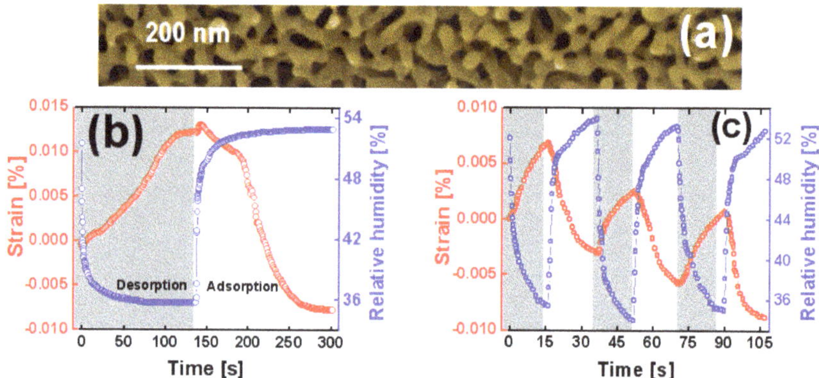

Figure 5. (**a**) Typical scanning electron micrograph of a nanoporous material (Au) that was synthesized by the dealloying process. The diameter of the ligaments was about 20 nm. (**b**) Display of the changes in relative humidity versus time for alternations of humid and dry air (blue curve refers to right ordinate) and corresponding strain versus time (red curve refers to left ordinate). (**c**) Responses of relative humidity versus time for long alternations of dry and humid air (blue curve refers to right ordinate) and corresponding strain versus time (red curve refers to left ordinate [67]).

The solution to these instabilities is basically to reduce surface diffusion, and, for that reason, HEAs may offer a suitable solution. Recently, Soo-Hyun Joo and collaborators found an exceptional stability against coarsening of an MoNbTaTiV nanoporous HEA at elevated temperatures [55]. The ligament size and distribution of the HEA versus dealloying time at various temperatures are displayed in Figure 6.

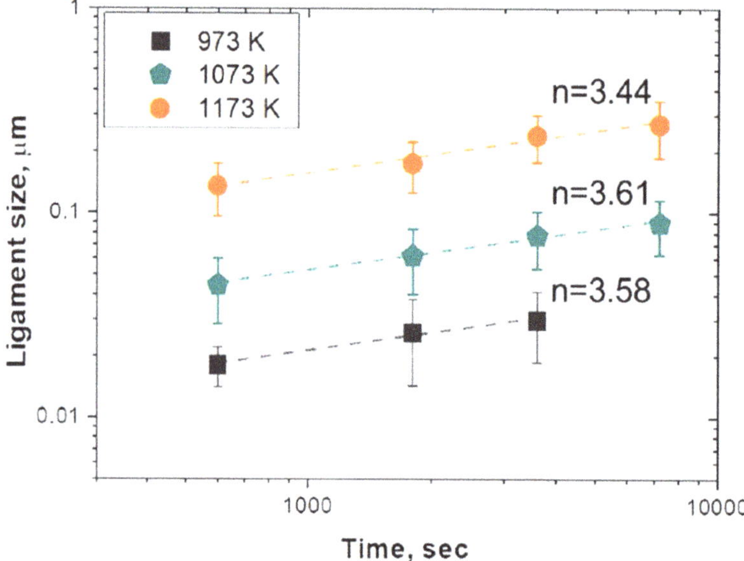

Figure 6. Ligament size in a TiVNbMoTa nanoporous high entropy alloy (HEA) versus dealloying time at various temperatures. The dealloying time, t, and ligament size, d, are correlated through a power function $d^n = ktD$, where n is the coarsening component, k is a constant, and D is the surface diffusivity. By plotting the $ln[d(t)]$ vs. ln t curve, the coarsening exponent, n, can be obtained. Error bars denote the distribution of ligament sizes. (Figure reprinted from [55] with permission from Wiley, 2019.)

To make the coarsening behavior a bit more quantitative: Dealloying time, t, and ligament size, d, are written as a power function: $d^n = ktD$, where n is the coarsening component, k is a constant, and D is the surface diffusivity. By plotting the $ln[d(t)]$ vs. $ln\ t$ curve, the coarsening exponent, n, can be measured. Figure 6 displays the measured specific surface areas of 55.7 (<10 nm), 38.8 (14 nm), and 3.6 (155 nm) m^2/g, depending on the ligament size as obtained in [55] based on an analytical model we designed in the past [68]. The adsorption/desorption isotherm curve (inset in Figure 7) corresponds to a ligament size of about 10 nm. The hysteresis loop in the isotherm is associated with capillary condensation in nanopores (<50 nm).

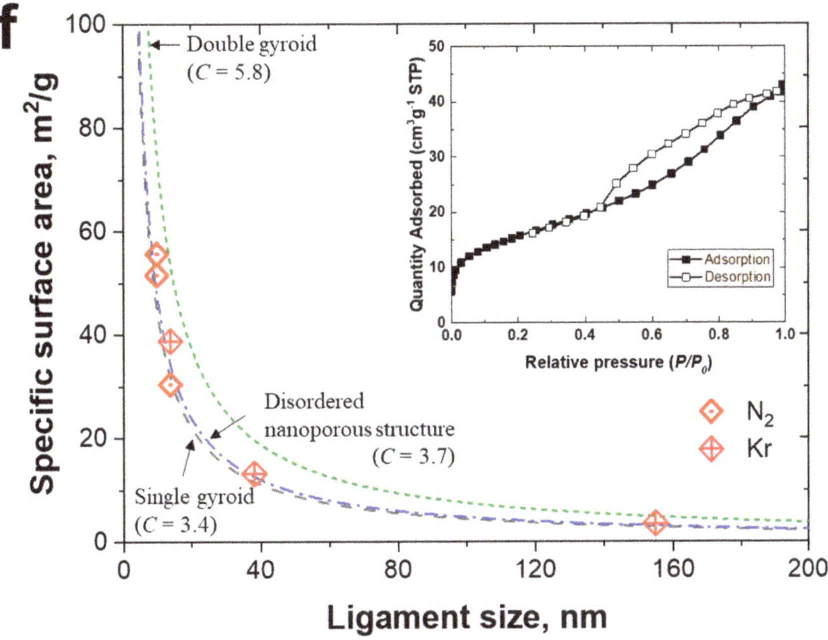

Figure 7. Specific surface areas measured with the Brunauer–Emmett–Teller (BET) method as a function of the ligament size in a TiVNbMoTa nanoporous HEA. The lines (dashed and dashed-dot) correspond to the predictions based on the analytic model $S = C/\rho d$, as designed and tested in [68]. The inset shows the nitrogen adsorption/desorption curves with an average ligament size of ~10 nm. The solid bulk density ρ was assumed to be 7.7 g·cm^{-3} from the average atomic weight and the radius of the constituent elements (Figure reprinted from [55] with permission from Wiley, 2019.)

As far as the physical explanations of these phenomena are concerned, a first and easy conclusion would be: The suppression of coarsening is due to what has been called sluggish diffusion in HEA. However, this might be a too simple and naïve reason. There is no literature on the details of diffusional processes in nano-porous HEA, but many papers have been published on bulk HEA [69], e.g., CoCrFeMnNi [70]. Dezso Beke [71] analyzed the diffusion coefficients of elements in CoCrFeMnNi HEA. Interestingly, sluggish diffusion could be explained as not being based on high activation energies but on correlation effects. Indeed [72], it has been reported that the diffusion of Ni in both HEAs follow an Arrhenius behavior. In fact, the tracer diffusion in HEAs does not become sluggish at absolute temperature; it only becomes so if it is considered at a homologous temperature. As a sequence, it can be concluded that diffusion in HEAs is not sluggish, as such, but other factors such as frequency factors can explain a slower diffusion rate in HEAs. From a scientific viewpoint, more in-depth analyses regarding the diffusional processes are necessary, not only for bulk HEA but also for surface diffusional processes in nanoporous HEAs.

A stimulating and exciting field of research would be to devise specific rules for the design of highly porous HEAs that can be used over a large range of temperatures in the field of sensors and actuators, in 'energy materials' (H-storage), radiation resistant materials, and many other scenarios. It is important to note that in these cellular and highly porous HEAs materials, it is not just the specific materials properties of strength and surface diffusion matter—the local topology and connectivity of struts and nano-ligaments also does [73].

As a consequence, topology provides an additional design parameter, in addition to the extrinsic and intrinsic material size-dependent properties [73,74]. This approach fits the term 'architected material,' which was, to the best of our knowledge, proposed for the first time by Mike Ashby and Yves Bréchet and which bridges the structural engineering of topology and good practice in architecture [75]. More recently, a special issue of excellent contributions to the Materials Research Society (MRS) Bulletin, edited by Julia R. Greer and Vikram S. Deshpande [76], was published on the design, fabrication and mechanical performance of three-dimensional architected materials and structures. It would be exciting to explore HEAs among these lines of 'architected materials' for the optimization of the 'integral' of both structural and functional properties together for novel applications.

Funding: This research received no external funding.

Acknowledgments: The work was supported by the Applied Physics-Materials Science group of the Zernike Institute for Advanced Materials of the University of Groningen, the Netherlands. Discussions and contributions by Eric Detsi (now at UPENN, Philadelphia, USA) and Václav Ocelík (Groningen, the Netherlands) are gratefully acknowledged.

Conflicts of Interest: The authors declare no conflict of interest.

References

1. Tsai, M.-H.; Yeh, J.-W. High-Entropy Alloys: A Critical Review. *Mater. Res. Lett.* **2014**, *2*, 107–123. [CrossRef]
2. Yeh, J.-W.; Chen, S.-K.; Lin, S.-J.; Gan, J.-Y.; Chin, T.-S.; Shun, T.-T.; Tsau, C.-H.; Chang, S.-Y. Nanostructured High-Entropy Alloys with Multiple Principal Elements: Novel Alloy Design Concepts and Outcomes. *Adv. Eng. Mater.* **2004**, *6*, 299–303. [CrossRef]
3. Ranganathan, S. Alloyed pleasures: Multimetallic cocktails. *Curr. Sci.* **2003**, *85*, 1404–1406.
4. Christofidou, K.A.; Pickering, E.J.; Orsatti, P.; Mignanelli, P.M.; Slater, T.J.A.; Stone, H.J.; Jones, N.G. On the influence of Mn on the phase stability of the CrMnxFeCoNi high entropy alloys. *Intermetallics* **2018**, *92*, 84–92. [CrossRef]
5. Pickering, E.J.; Jones, N.G. High-entropy alloys: A critical assessment of their founding principles and future prospects. *Int. Mater. Rev.* **2016**, *61*, 183–202. [CrossRef]
6. Pickering, E.J.; Muñoz-Moreno, R.; Stone, H.J.; Jones, N.G. Precipitation in the equiatomic high-entropy alloy CrMnFeCoNi. *Scr. Mater.* **2016**, *113*, 106–109. [CrossRef]
7. Jones, N.G.; Frezza, A.; Stone, H.J. Phase equilibria of an Al0.5CrFeCoNiCu High Entropy Alloy. *Mater. Sci. Eng. A* **2014**, *615*, 214–221. [CrossRef]
8. Jones, N.G.; Izzo, R.; Mignanelli, P.M.; Christofidou, K.A.; Stone, H.J. Phase evolution in an Al0.5CrFeCoNiCu High Entropy Alloy. *Intermetallics* **2016**, *71*, 43–50. [CrossRef]
9. Senkov, O.N.; Miller, J.D.; Miracle, D.B.; Woodward, C. Accelerated exploration of multi-principal element alloys with solid solution phases. *Nat. Commun.* **2015**, *6*, 6529. [CrossRef]
10. Li, Z.; Pradeep, K.G.; Deng, Y.; Raabe, D.; Tasan, C.C. Metastable high-entropy dual-phase alloys overcome the strength–ductility trade-off. *Nature* **2016**, *534*, 227–230. [CrossRef]
11. Otto, F.; Dlouhý, A.; Pradeep, K.G.; Kuběnová, M.; Raabe, D.; Eggeler, G.; George, E.P. Decomposition of the single-phase high-entropy alloy CrMnFeCoNi after prolonged anneals at intermediate temperatures. *Acta Mater.* **2016**, *112*, 40–52. [CrossRef]
12. Otto, F.; Yang, Y.; Bei, H.; George, E.P. Relative effects of enthalpy and entropy on the phase stability of equiatomic high-entropy alloys. *Acta Mater.* **2013**, *61*, 2628–2638. [CrossRef]
13. George, E.P.; Raabe, D.; Ritchie, R.O. High-entropy alloys. *Nat. Rev. Mater.* **2019**, *4*, 515–534. [CrossRef]
14. Basu, I.; Ocelík, V.; De Hosson, J.T.M. Size effects on plasticity in high-entropy alloys. *J. Mater. Res.* **2018**, *33*, 3055–3076. [CrossRef]

15. Miracle, D.B. High entropy alloys as a bold step forward in alloy development. *Nat. Commun.* **2019**, *10*, 1805. [CrossRef] [PubMed]
16. Basu, I.; Ocelík, V.; De Hosson, J.T.M. Size dependent plasticity and damage response in multiphase body centered cubic high entropy alloys. *Acta Mater.* **2018**, *150*, 104–116. [CrossRef]
17. He, J.Y.; Wang, H.; Huang, H.L.; Xu, X.D.; Chen, M.W.; Wu, Y.; Liu, X.J.; Nieh, T.G.; An, K.; Lu, Z.P. A precipitation-hardened high-entropy alloy with outstanding tensile properties. *Acta Mater.* **2016**, *102*, 187–196. [CrossRef]
18. Deng, Y.; Tasan, C.C.; Pradeep, K.G.; Springer, H.; Kostka, A.; Raabe, D. Design of a twinning-induced plasticity high entropy alloy. *Acta Mater.* **2015**, *94*, 124–133. [CrossRef]
19. Li, Z.; Tasan, C.C.; Springer, H.; Gault, B.; Raabe, D. Interstitial atoms enable joint twinning and transformation induced plasticity in strong and ductile high-entropy alloys. *Sci. Rep.* **2017**, *7*, 40704. [CrossRef]
20. Gludovatz, B.; Hohenwarter, A.; Catoor, D.; Chang, E.H.; George, E.P.; Ritchie, R.O. A fracture-resistant high-entropy alloy for cryogenic applications. *Science* **2014**, *345*, 1153–1158. [CrossRef]
21. Li, Z.; Zhao, S.; Ritchie, R.O.; Meyers, M.A. Mechanical properties of high-entropy alloys with emphasis on face-centered cubic alloys. *Prog. Mater. Sci.* **2019**, *102*, 296–345. [CrossRef]
22. Zhang, Z.; Mao, M.M.; Wang, J.; Gludovatz, B.; Zhang, Z.; Mao, S.X.; George, E.P.; Yu, Q.; Ritchie, R.O. Nanoscale origins of the damage tolerance of the high-entropy alloy CrMnFeCoNi. *Nat. Commun.* **2015**, *6*, 10143. [CrossRef] [PubMed]
23. Zhang, Y.; Zuo, T.T.; Tang, Z.; Gao, M.C.; Dahmen, K.A.; Liaw, P.K.; Lu, Z.P. Microstructures and properties of high-entropy alloys. *Prog. Mater. Sci.* **2014**, *61*, 1–93. [CrossRef]
24. Rao, J.C.; Ocelík, V.; Vainchtein, D.; Tang, Z.; Liaw, P.K.; De Hosson, J.T.M. The fcc-bcc crystallographic orientation relationship in AlxCoCrFeNi high-entropy alloys. *Mater. Lett.* **2016**, *176*, 29–32. [CrossRef]
25. Wang, W.-R.; Wang, W.-L.; Wang, S.-C.; Tsai, Y.-C.; Lai, C.-H.; Yeh, J.-W. Effects of Al addition on the microstructure and mechanical property of AlxCoCrFeNi high-entropy alloys. *Intermetallics* **2012**, *26*, 44–51. [CrossRef]
26. Soni, V.; Senkov, O.N.; Gwalani, B.; Miracle, D.B.; Banerjee, R. Microstructural Design for Improving Ductility of An Initially Brittle Refractory High Entropy Alloy. *Sci. Rep.* **2018**, *8*, 8816. [CrossRef]
27. Senkov, O.N.; Isheim, D.; Seidman, D.N.; Pilchak, A.L. Development of a Refractory High Entropy Superalloy. *Entropy* **2016**, *18*, 102. [CrossRef]
28. Clark, W.A.T.; Wagoner, R.H.; Shen, Z.Y.; Lee, T.C.; Robertson, I.M.; Birnbaum, H.K. On the criteria for slip transmission across interfaces in polycrystals. *Scr. Metall. Mater.* **1992**, *26*, 203–206. [CrossRef]
29. Shen, Z.; Wagoner, R.H.; Clark, W.A.T. Dislocation and grain boundary interactions in metals. *Acta Metall.* **1988**, *36*, 3231–3242. [CrossRef]
30. Kacher, J.; Eftink, B.P.; Cui, B.; Robertson, I.M. Dislocation interactions with grain boundaries. *Curr. Opin. Solid State Mater. Sci.* **2014**, *18*, 227–243. [CrossRef]
31. Basu, I.; Ocelík, V.; De Hosson, J.T.M. Measurement of spatial stress gradients near grain boundaries. *Scr. Mater.* **2017**, *136*, 11–14. [CrossRef]
32. Rao, S.I.; Hazzledine, P.M. Atomistic simulations of dislocation–interface interactions in the Cu-Ni multilayer system. *Philos. Mag. A* **2000**, *80*, 2011–2040. [CrossRef]
33. Head, A.K.X. The Interaction of Dislocations and Boundaries. *London Edinburgh Dublin Philos. Mag. J. Sci.* **1953**, *44*, 92–94. [CrossRef]
34. Koehler, J.S. Attempt to Design a Strong Solid. *Phys. Rev. B* **1970**, *2*, 547–551. [CrossRef]
35. Basu, I.; Ocelík, V.; De Hosson, J.T. BCC-FCC interfacial effects on plasticity and strengthening mechanisms in high entropy alloys. *Acta Mater.* **2018**, *157*, 83–95. [CrossRef]
36. Frank, F.C.; Van der Merwe, J. One-dimensional dislocations. II. Misfitting monolayers and oriented overgrowth. *Proc. R. Soc. London Ser. A Math. Phys. Sci.* **1949**, *198*, 216–225.
37. Vitek, V. Theory of the core structures of dislocations in BCC metals. *Cryst. Lattice Defects* **1974**, *5*, 1–34.
38. Cai, Y.P.; Wang, G.J.; Ma, Y.J.; Cao, Z.H.; Meng, X.K. High hardness dual-phase high entropy alloy thin films produced by interface alloying. *Scripta Mater.* **2019**, *162*, 281–285. [CrossRef]
39. Cao, Z.H.; Ma, Y.J.; Cai, Y.P.; Wang, G.J.; Meng, X.K. High strength dual-phase high entropy alloys with a tunable nanolayer thickness. *Scripta Materialia* **2019**, *173*, 149–153. [CrossRef]
40. Ding, J.; Yu, Q.; Asta, M.; Ritchie, R.O. Tunable stacking fault energies by tailoring local chemical order in CrCoNi medium-entropy alloys. *Proc. Natl. Acad. Sci. USA* **2018**, *115*, 8919–8924. [CrossRef]

41. Ding, Q.; Zhang, Y.; Chen, X.; Fu, X.; Chen, D.; Chen, S.; Gu, L.; Wei, F.; Bei, H.; Gao, Y.; et al. Tuning element distribution, structure and properties by composition in high-entropy alloys. *Nature* **2019**, *574*, 223–227. [CrossRef] [PubMed]
42. Lilensten, L.; Couzinié, J.-P.; Bourgon, J.; Perrière, L.; Dirras, G.; Prima, F.; Guillot, I. Design and tensile properties of a bcc Ti-rich high-entropy alloy with transformation-induced plasticity. *Mater. Res. Lett.* **2017**, *5*, 110–116. [CrossRef]
43. Bracq, G.; Laurent-Brocq, M.; Varvenne, C.; Perrière, L.; Curtin, W.A.; Joubert, J.M.; Guillot, I. Combining experiments and modeling to explore the solid solution strengthening of high and medium entropy alloys. *Acta Mater.* **2019**, *177*, 266–279. [CrossRef]
44. Osetsky, Y.N.; Pharr, G.M.; Morris, J.R. Two modes of screw dislocation glide in fcc single-phase concentrated alloys. *Acta Mater.* **2019**, *164*, 741–748. [CrossRef]
45. Rizzardi, Q.; Sparks, G.; Maaß, R. Fast Slip Velocity in a High-Entropy Alloy. *JOM* **2018**, *70*, 1088–1093. [CrossRef]
46. Gwalani, B.; Gangireddy, S.; Shukla, S.; Yannetta, C.J.; Valentin, S.G.; Mishra, R.S.; Banerjee, R. Compositionally graded high entropy alloy with a strong front and ductile back. Mater. *Today Commun.* **2019**, *20*, 100602.
47. Miracle, D.B. Critical Assessment 14: High entropy alloys and their development as structural materials. *Mater. Sci. Technol.* **2015**, *31*, 1142–1147. [CrossRef]
48. Kudoh, K. Development of piezo micromanipulator for cell micromanipulation. *J. Mamm. Ova Res.* **1990**, *7*, 7–12.
49. Wittstock, A.; Zielasek, V.; Biener, J.; Friend, C.M.; Bäumer, M. Nanoporous Gold Catalysts for Selective Gas-Phase Oxidative Coupling of Methanol at Low Temperature. *Science* **2010**, *327*, 319. [CrossRef]
50. Nagle, L.C.; Rohan, J.F. Nanoporous gold anode catalyst for direct borohydride fuel cell. *Int. J. Hydrogen Energy* **2011**, *36*, 10319–10326. [CrossRef]
51. Lang, X.Y.; Yuan, H.T.; Iwasa, Y.; Chen, M.W. Three-dimensional nanoporous gold for electrochemical supercapacitors. *Scripta Mater.* **2011**, *64*, 923–926. [CrossRef]
52. Detsi, E.; Chen, Z.G.; Vellinga, W.P.; Onck, P.R.; Hosson, J.T.M.D. Actuating and Sensing Properties of Nanoporous Gold. *J. Nanosci. Nanotechnol.* **2012**, *12*, 4951–4955. [CrossRef] [PubMed]
53. Yavuz, M.S.; Cheng, Y.; Chen, J.; Cobley, C.M.; Zhang, Q.; Rycenga, M.; Xie, J.; Kim, C.; Song, K.H.; Schwartz, A.G.; et al. Gold nanocages covered by smart polymers for controlled release with near-infrared light. *Nat. Mater.* **2019**, *8*, 935–939. [CrossRef]
54. Au, L.; Zheng, D.; Zhou, F.; Li, Z.-Y.; Li, X.; Xia, Y. A Quantitative Study on the Photothermal Effect of Immuno Gold Nanocages Targeted to Breast Cancer Cells. *ACS Nano* **2008**, *2*, 1645–1652. [CrossRef] [PubMed]
55. Joo, S.-H.; Bae, J.W.; Park, W.-Y.; Shimada, Y.; Wada, T.; Kim, H.S.; Takeuchi, A.; Konno, T.J.; Kato, H.; Okulov, I.V. Beating Thermal Coarsening in Nanoporous Materials via High-Entropy Design. *Adv. Mater.* **2019**, 1906160. [CrossRef]
56. Wada, T.; Yubuta, K.; Inoue, A.; Kato, H. Dealloying by metallic melt. *Mater. Lett.* **2011**, *65*, 1076–1078. [CrossRef]
57. Wada, T.; Ichitsubo, T.; Yubuta, K.; Segawa, H.; Yoshida, H.; Kato, H. Bulk-Nanoporous-Silicon Negative Electrode with Extremely High Cyclability for Lithium-Ion Batteries Prepared Using a Top-Down Process. *Nano Lett.* **2014**, *14*, 4505–4510. [CrossRef]
58. Geslin, P.-A.; McCue, I.; Gaskey, B.; Erlebacher, J.; Karma, A. Topology-generating interfacial pattern formation during liquid metal dealloying. *Nat. Commun.* **2015**, *6*, 8887. [CrossRef]
59. Okulov, I.V.; Weissmüller, J.; Markmann, J. Dealloying-based interpenetrating-phase nanocomposites matching the elastic behavior of human bone. *Sci. Rep.* **2017**, *7*, 20. [CrossRef]
60. Okulov, I.V.; Okulov, A.V.; Soldatov, I.V.; Luthringer, B.; Willumeit-Römer, R.; Wada, T.; Kato, H.; Weissmüller, J.; Markmann, J. Open porous dealloying-based biomaterials as a novel biomaterial platform. *Mater. Sci. Eng. C* **2018**, *88*, 95–103. [CrossRef]
61. Joo, S.-H.; Wada, T.; Kato, H. Development of porous FeCo by liquid metal dealloying: Evolution of porous morphology and effect of interaction between ligaments and melt. *Mater. Des.* **2019**, *180*, 107908.
62. Joo, S.-H.; Yubuta, K.; Kato, H. Ordering kinetics of nanoporous FeCo during liquid metal dealloying and the development of nanofacets. *Scr. Mater.* **2020**, *177*, 38–43. [CrossRef]
63. Jayachandran, K.P.; Guedes, J.M.; Rodrigues, H.C. Ferroelectric materials for piezoelectric actuators by optimal design. *Acta Mater.* **2011**, *59*, 3770–3778. [CrossRef]

64. Saane, S.S.R.; Mangipudi, K.R.; Loos, K.U.; De Hosson, J.T.M.; Onck, P.R. Multiscale modeling of charge-induced deformation of nanoporous gold structures. *J. Mech. Phys. Solids* **2014**, *66*, 1–15. [CrossRef]
65. Detsi, E.; Punzhin, S.; Rao, J.; Onck, P.R.; De Hosson, J.T.M. Enhanced Strain in Functional Nanoporous Gold with a Dual Microscopic Length Scale Structure. *ACS Nano* **2012**, *6*, 3734–3744. [CrossRef] [PubMed]
66. Zhang, J.; Liu, P.; Ma, H.; Ding, Y. Nanostructured Porous Gold for Methanol Electro-Oxidation. *J. Phys. Chem. C* **2007**, *111*, 10382–10388. [CrossRef]
67. Detsi, E.; Onck, P.; De Hosson, J.T.M. Metallic Muscles at Work: High Rate Actuation in Nanoporous Gold/Polyaniline Composites. *ACS Nano* **2013**, *7*, 4299–4306. [CrossRef]
68. Detsi, E.; De Jong, E.; Zinchenko, A.; Vuković, Z.; Vuković, I.; Punzhin, S.; Loos, K.; ten Brinke, G.; De Raedt, H.A.; Onck, P.R.; et al. On the specific surface area of nanoporous materials. *Acta Mater.* **2011**, *59*, 7488–7497. [CrossRef]
69. Praveen, S.; Kim, H.S. High-Entropy Alloys: Potential Candidates for High-Temperature Applications—An Overview. *Adv. Eng. Mater.* **2018**, *20*, 1700645. [CrossRef]
70. Tsai, K.Y.; Tsai, M.H.; Yeh, J.W. Sluggish diffusion in Co–Cr–Fe–Mn–Ni high-entropy alloys. *Acta Mater.* **2013**, *61*, 4887–4897.
71. Beke, D.L.; Erdélyi, G. On the diffusion in high-entropy alloys. *Mater. Lett.* **2016**, *164*, 111–113. [CrossRef]
72. Vaidya, M.; Trubel, S.; Murty, B.S.; Wilde, G.; Divinski, S.V. Ni tracer diffusion in CoCrFeNi and CoCrFeMnNi high entropy alloys. *J. Alloys Compd.* **2016**, *688*, 994–1001. [CrossRef]
73. Michelsen, K.; De Raedt, H.; De Hosson, J.T.M. Aspects of mathematical morphology. In *Advances in Imaging and Electron Physics*; Hawkes, P.W., Kazan, B., Mulvey, T., Eds.; Elsevier: Amsterdam, Netherlands, 2003; pp. 119–194.
74. Yang, C.; Xu, P.; Xie, S.; Yao, S. Mechanical performances of four lattice materials guided by topology optimization. *Scr. Mater.* **2020**, *178*, 339–345. [CrossRef]
75. Ashby, M.F.; Bréchet, Y.J.M. Designing hybrid materials. *Acta Mater.* **2003**, *51*, 5801–5821. [CrossRef]
76. Greer, J.R.; Deshpande, V.S. Three-dimensional architected materials and structures: Design, fabrication, and mechanical behavior. *MRS Bull.* **2019**, *44*, 750–757. [CrossRef]

MDPI
St. Alban-Anlage 66
4052 Basel
Switzerland
Tel. +41 61 683 77 34
Fax +41 61 302 89 18
www.mdpi.com

Metals Editorial Office
E-mail: metals@mdpi.com
www.mdpi.com/journal/metals

www.ingramcontent.com/pod-product-compliance
Lightning Source LLC
LaVergne TN
LVHW070152170726
843515LV00007B/1903

* 9 7 8 3 0 3 9 4 3 4 7 4 9 *